Lecture Notes in Computer Science 13831

Founding Editors

Gerhard Goos
Juris Hartmanis

Editorial Board Members

The series Lecture Notes in Computer Science (LNCS), including its subseries Lecture Notes in Artificial Intelligence (LNAI) and Lecture Notes in Bioinformatics (LNBI), has established itself as a medium for the publication of new developments in computer science and information technology research, teaching, and education.

LNCS enjoys close cooperation with the computer science R & D community, the series counts many renowned academics among its volume editors and paper authors, and collaborates with prestigious societies. Its mission is to serve this international community by providing an invaluable service, mainly focused on the publication of conference and workshop proceedings and postproceedings. LNCS commenced publication in 1973.

Thang N. Dinh · Minming Li

Editors

Computational Data and Social Networks

11th International Conference, CSoNet 2022
Virtual Event, December 5–7, 2022
Proceedings

Editors
Thang N. Dinh ⓘ
Virginia Commonwealth University
Richmond, VA, USA

Minming Li ⓘ
Yeung Kin Man Academic Building
City University Hong Kong
Kowloon Tong, Hong Kong

ISSN 0302-9743   ISSN 1611-3349 (electronic)
Lecture Notes in Computer Science
ISBN 978-3-031-26302-6   ISBN 978-3-031-26303-3 (eBook)
https://doi.org/10.1007/978-3-031-26303-3

This Springer imprint is published by the registered company Springer Nature Switzerland AG
The registered company address is: Gewerbestrasse 11, 6330 Cham, Switzerland

Preface

The 11th International Conference on Computational Data and Social Networks (CSoNet 2022), held online, was a premier interdisciplinary forum that brought together researchers and practitioners from all fields of big data and social networks, such as billion-scale network computing, social network/media analysis, mining, security and privacy, and deep learning. CSoNet 2022 aimed to address emerging, yet important computational problems with a focus on the fundamental background, theoretical technology development, and real-world applications associated with big data network analysis, modelling, and deep learning. CSoNet 2022 welcomed both the presentation of original research results, the exchange and dissemination of truly innovative theoretical advancements, as well as outcomes of practical deployments and real-world applications in the broad area of information networks.

The core research topics include: theories of network organization; influence modeling, propagation, and maximization; adversarial attacks of network; NLP and affective computing; computational methods for social good; and security, trust, and privacy, among others. We selected 18 regular full papers, 7 short papers, along with 1 two-page extended abstracts for presentation and publication. One invited paper from active researchers in the related fields is also included. A number of selected best papers were invited for publication in the Journal of Combinatorial Optimization and IEEE Transactions on Network Science and Engineering.

This conference would not have been possible without the support of a large number of individuals. First, we sincerely thank all authors for submitting their high-quality work to the conference, especially as the Covid-19 pandemic continued to ravage communities and countries around the world. We fully understand the unique challenges facing authors during the pandemic. Our thanks also go to all Technical Program Committee members and sub-reviewers for their willingness to provide timely and detailed reviews of all submissions. Their hard work during the pandemic made the success of the conference possible. We also offer our special thanks to the Publicity and Publication Chairs for their dedication in disseminating the call and encouraging participation in such challenging times, in addition to the preparation of the proceedings. Special thanks are also due to the Publicity Co-chairs, and the Web Chair. Lastly, we acknowledge the support and patience of Springer staff members throughout the process.

December 2022

Thang N. Dinh
Minming Li

Preface

Organization

General Chairs

Trung Q. Duong   Queen's University Belfast, UK
Xiaofeng Gao   Shanghai JiaoTong University, China

Program Committee Chairs

Thang N. Dinh   Virginia Commonwealth University, USA
Minming Li   City University of Hong Kong, China

Publicity Chairs

Yong Zhang   Shenzhen Institute of Advanced Technology, China
Qin Hu   Indiana University–Purdue University Indianapolis, USA

Web Chair

Phuc D. Thai   Virginia Commonwealth University, USA

Steering Committee

My T. Thai (Chair)   University of Florida, USA
Kim-Kwang Raymond Choo   University of Texas at San Antonio, USA
Zhi-Li Zhang   University of Minnesota, USA
Weili Wu   University of Texas, Dallas, USA

Program Committee

Mohammed Abuhamad   Loyola University Chicago, USA
Gianni D'Angelo   University of Salerno, Italy

Impact of Geographic Factors on Friendship Link Patterns and Influence Cascade Propagation in Social Media Networks: A Case of VK.com (Extended Abstract)

Alexander Semenov[1], Alexander Nikolaev[2], Eduardo L. Pasiliao[3],
and Vladimir Boginski[4]

[1] University of Florida, Gainesville, FL, USA
asemenov@ufl.edu
[2] University at Buffalo, Buffalo, NY, USA
anikolae@buffalo.edu
[3] Air Force Research Laboratory, Eglin AFB, FL, USA
eduardo.pasiliao@us.af.mil
[4] University of Central Florida, Orlando, FL USA
vladimir.boginski@ucf.edu

Extended Abstract

Recent studies suggest the existence of dependencies between geographic proximity of social actors and their online connections in social media networks. We conduct a large-scale study of the anonymized dataset corresponding to the entire social network of VK.com (data collected in December 2016 and originally analyzed in [1]): the largest social media platform is Eastern Europe, where a significant percentage of users self-report their geographic location. In our study we address the following broad research questions:

- Are "global" friendship link patterns in the online social media network (e.g., degree distributions) affected by geographic factors, or are these patterns "geographically invariant"?
- From a "local" perspective, does physical (geographic) proximity of users affect the volume of friendship links between those users?
- How do patterns of influence cascade propagation over the social media network depend on geographic locations of seed nodes?

Background. VK.com (referred to simply as VK) is the largest European social media portal with over 400M users. VK users are located mainly in Eastern Europe, and it is the most popular social media portal in the so-called "post-Soviet space". VK is formerly known as "VKontakte", which means "in touch" in English. The site was founded in 2006, and is often referred to as "Russian Facebook". Our recent study analyzing VK network characteristics is presented in [2]. In this work, we build on our previous results and further study the impact of geographic factors on friendship link patterns and simulated cascade propagation.

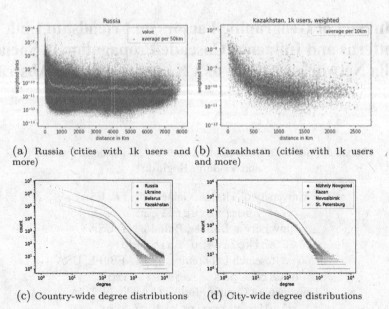

(a) Russia (cities with 1k users and more)

(b) Kazakhstan (cities with 1k users and more)

(c) Country-wide degree distributions

(d) City-wide degree distributions

Fig. 1: (a), (b): Dependence of volume of friendship links vs. geographical distance between cities: Russia and Kazakhstan; (c), (d): "Geographically scale free" nature of degree distributions in the VK network: examples of individual countries and major cities.

Work Summary. Figures 1a and 1b show that volume of links drops with increase of distance between the cities: this pattern is observed both for the entire VK and for each country with substantial number of VK users. Interestingly, the drop effect is slightly more pronounced in countries with larger populations (e.g., Russia); however, it is still observed in all considered geographic regions, and the drop "saturates" at around 1,000 km distance. These results suggest that the geographic proximity effect on social media connections manifests itself only on relatively short physical distances (under 1,000 km). Figures 1c and 1d show the *"geographically scale free"* nature of degree distributions in the VK network: the "global" pattern of the entire VK[1] is also observed at smaller scale in individual countries and cities. In addition, we investigate characteristics of cascading propagation of influence over the VK network. We conduct within-city cascade simulations, and country-wide cascade simulations. Our simulation approach allows for studying hundreds of thousands cascades at once, through graph partitioning. The approach is applicable for both Linear Threshold and Independent Cascade models.

Reference

1. Semenov, A., et al.: Exploring social media network landscape of post-soviet space. IEEE Access **7**, 411–426 (2019). https://doi.org/10.1109/ACCESS.2018.2885479

Contents

Security and Blockchain

Fact-checking, Fake News, and Hate Speech

Network Analysis

Optimization

Machine Learning and Prediction

Incorporating Neighborhood Information and Sentence Embedding Similarity into a Repost Prediction Model in Social Media Networks

Zhecheng Qiang[1]($\boxtimes$) (iD), Eduardo L. Pasiliao[2] (iD), Alexander Semenov[3] (iD), and Qipeng P. Zheng[1] (iD)

[1] Department of Industrial Engineering and Management Systems, University of Central Florida, Orlando, FL, USA
jenniferqiangs@gmail.com
[2] Munitions Directorate, Air Force Research Laboratory, Shalimar, FL, USA
[3] Department of Industrial and Systems Engineering, University of Florida, Gainesville, FL, USA

Abstract. Predicting repost behaviors within social media networks plays an important role in human activities analysis and influence maximization decision making. Traditional methods for repost prediction can be categorized into stochastic diffusion based models and user profile or content features based machine learning models. In this paper, we propose a new framework combining user profile, content similarity and the neighborhood information around each target link as input features to make the prediction. Here neighborhood information can be interpreted as the combination of neighbors' user profile. Two different kinds of graph based combination models are introduced in the article. After collecting the input features, we implement the state-of-the-art machine learning methods, e.g., Logistic Regression, K-nearest Neighbors, Gaussian Naive Bayes, Deep Neural Network, Random Forest, XGBoosting and Stacking Model to predict repost probability. We evaluate our model on real dataset Weibo to compare the performance with different features and machine learning methods.

Keywords: Repost probability prediction · Machine learning · Graph theory · Social media networks

1 Introduction

Mass adoption of mobile devices forever changed the way people communicate and interact with each other. The planet suddenly became unexpectedly smaller for its seven billion residents, especially in the sense of communication. One irreversible result is that people have adopted social media platforms as the major venue for sharing their opinions and news, and for spreading their influence. This poses a series of challenges and opportunities. For example, understanding how the information propagates within social media platforms is essential

T. N. Dinh and M. Li (Eds.): CSoNet 2022, LNCS 13831, pp. 3–14, 2023.
https://doi.org/10.1007/978-3-031-26303-3_1

for successful implementation of viral marketing [6] and cyber security [3] in social media networks. To this end, researchers have defined various diffusion tasks to tackle the real world problems. Researchers have defined various tasks to tackle the opportunities and challenges in social media. The tasks include buzz prediction [4], volume prediction [23], infection prediction [2], spread prediction, source prediction [21], cascade prediction, link detection [17], influence maximization [5,12,25] and firefighter problem [1]. One of the most essential and fundamental tasks is to predict the information diffusion between each pair of users. On one hand, it has applications in the analysis of human activities and the exploration of the factors that influence information diffusion, which helps business and organizations in crafting more attractive advertisements to boost the business. On the other hand, learning diffusion probability is the first step in tackling the influence maximization problem. Influence maximization problem involves finding a limited number of nodes that have the largest influence for the spread of information. Influence maximization is widely used in the areas of viral marketing and misinformation detection.

Although social media platforms differ slightly from one another in information diffusion and cascade, the predominant dynamic for information diffusion is through the reposting or retweeting of messages. Unfortunately, the reposting probability for each user within a network is not readily available from the massive historical dataset of social media networks. To fill this gap, several research articles have studied the problem of learning repost probability in online social media networks.

Most of the literature are based on the two following stochastic diffusion models: Independent Cascade (IC) [12] and Linear Threshold Model (LT) [9]. The Independent Cascade Model assumes that every node has a single chance to activate another node; the influence is modeled as an impulse function. The Linear Threshold Model assumes that a node is activated when the total weights of its neighbors' influence are at least θ_v; the influence is modeled as a step function. [8] adopted the General Threshold Diffusion Model and proposed three probabilistic models (Bernoulli distribution, Jaccard Index and Partial Credits) to predict the repost probability. [15] adopted the Threshold Propagation Model and built two mathematical learning models (Linear Threshold Learning Model and Random Walk Learning Model) to predict the repost probability. [19] studied the problem of learning repost probability based on the Independent Cascade Model. They used EM algorithm to maximize the likelihood estimate. Besides, there are some variants of Independent Cascade Model based research considering continuous time delays of infection as well [17,18]. The other direction of the literature is based on machine learning models, which learned the repost probability through content and user profile features [10,11,13,20,24,28].

Contributions. For large scale social media networks, the relationship between important factors in a user's repost decision and repost probability is complex and nonlinear. Therefore, we implement several machine learning approaches to estimate the repost probability. We take all the essential factors that might affect the diffusion process into consideration, including user profile, neighborhood information, and content similarity. The contributions of this paper are summarized as follows:

1. We take user neighborhood information into consideration. Each user's friend circle is a good indicator of that user's repost decision. Users who have very active friends tend to be more active in reposting messages. To collect neighborhood information, we introduce two different combination models of neighbors' user profile from a graph theoretic perspective. We analyze both combination models and compare their performance in learning repost probability.
2. Different state-of-the-art machine learning models are implemented to estimate the repost probability. In addition, we analyze the importance of different features based on tree-based ensemble prediction models as well.
3. We use multil-lingual sentence embedding, instead of the more often used LDA in research studies to extract the information embedding, which works better in generating dense vector representations for short sentences.

The rest of the paper is organized as follows: we introduce the repost prediction model including the feature extraction and machine learning models in Sect. 2. In Sect. 3, we introduce the experimental datasets and explain the experimental set-up. Section 4 presents the performance results of the repost prediction model. Finally, Sect. 5 concludes the paper.

2   Repost Prediction Model

In social media, information propagates primarily through the reposting or retweeting of messages. Understanding repost probabilities contributes to better decision making in social media. We model the social media network as a directed graph $G = (V, E)$, where $V = \{v_1, \ldots, v_n\}$ is the set of users of the network and $E \subseteq V \times V$ is the set of edges representing the friend relationships between users. Each user has a list of attributes $A = \{a_1, \ldots, a_m\}$, which is referred to as the user profile in social media. And we define the content of each message or post using M.

Researchers have demonstrated that post content and user profile are major factors in determining a users' repost decisions [22].

Additionally, we think the neighborhood user profile also contributes to the prediction of the repost decisions as well. Active neighbors are expected to have a positive influence in the users, making the users more active in reposting when they are exposed to more posts from the neighbors. Thus we extract the features of user profile, neighborhood information and message embedding similarity from the original data user profile, graph topology and messages. Therefore, we aim to model the reposting or retweeting probability $p(i, j, m)$ between source user

i and target user j, $i, j \in V$ for the message m as a function of user profile(A), neighborhood information(N) and content embedding similarity(S).

$$p(i, j, m) = f(A, N, S, \theta) \tag{1}$$

Considering the complexity of the function, we implement state-of-the-art machine learning methods to approximate the function and get the estimation of parameter θ.

2.1 Feature Extraction

The user profile(A), network topology(G) and messages(M) are original data we could obtain from the dataset directly. Based on the original data, we extract three different kinds of features: user profile, neighborhood information and content embedding similarity as input features to make the prediction.

User Profile: User profile features are generally good predictors of the repost probability. We consider user profile of both source user and target user. The widely used user profile include the total number of followers, the total number of followees, the total number of posts or reposts, the total number of reciprocal relationships, gender and the created time.

Neighborhood Information: Information from a user's circle of friends is a good indicator of repost decision. Commonly, a user with active friends tend to be more active in reposting. It is natural to expect that friends (reciprocal relationships) share similarity in behaviors with each other. Here the neighborhood information ($N \subseteq G \times A$) contains both the topology of neighborhood and the user profile of the neighbors.

Subgraph Extraction: To get the neighborhood for each pair of source user i and target user j, we extract an enclosing subgraph around them as the circle of friends of the pair. The subgraph is extracted from the network by the union of all user i and j's friends nodes within 1 hop. The set of nodes in the 1-hop subgraph is defined as:

$$V_{i,j}^1 = \{v | d(v, i) \leqslant 1 \quad \text{or} \quad d(v, j) \leqslant 1\} \tag{2}$$

We use the adjacency matrix $\tilde{A}$ to represent the structure of the subgraph, where $\tilde{A}_{i,j} = 1$ if $(i, j) \subset E$ and $\tilde{A}_{i,j} = 0$ otherwise. Considering the memory limitation and computation speed, we only consider the first two rows that contain the friendship relationship of source user i and target user j as matrix $\tilde{A}'$. Suppose the 1-hop subgraph contains n users in total, then the extracted adjacency matrix is $\tilde{A}' \subseteq \{1, 0\}^{2 \times n}$. The enclosing subgraph contains rich information about the pair's circle of friends. Figure 1 shows an illustration of subgraph extraction.

Combination Models: After extracting the neighborhood of the pair, we combine user profile features of each user in the neighborhood. The neighborhood information features can be incorporated as a combination of circle of friends'

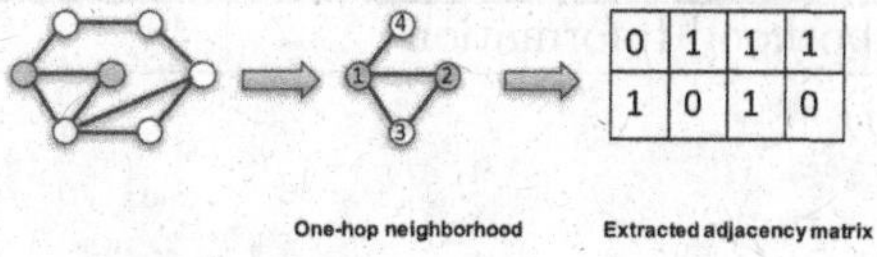

One-hop neighborhood   Extracted adjacency matrix

Fig. 1. An illustration of subgraph extraction

user profile features. Here we introduce different combination models to get the neighborhood information. *Combination Model 1:* The first combination model combines user profile of neighbors using extracted adjacency matrix with added self loop.Then the neighborhood information could be represented as:

$$N = (\tilde{A}' + I')X_G \qquad (3)$$

where $\tilde{A}'$ represents the extracted adjacency matrix of subgraph, I' represents the first two rows of the identity matrix and $X_G \subseteq R^{n \times c}$ is the selected user profile of neighbors of the pair. From another prospective, the neighborhood information could also be interpreted as the sum of the users and their neighbors' user profile features.

Combination Model 2: The second combination model is inspired by graph Laplacian. The formula of graph Laplacian is shown below:

$$L = D - \tilde{A}, \qquad (4)$$

where D is the degree matrix and $\tilde{A}$ is the adjacency matrix of the subgraph. Then we normalize the graph Laplacian. The random walk normalized Laplacian is:

$$L = D^{-1}(D - \tilde{A}) = I - D^{-1}\tilde{A} \qquad (5)$$

To reduce the computation complexity, we introduce the following renormalization trick: $I - D^{-1}\tilde{A} \rightarrow D^{-1}(\tilde{A} + I)$ [27]. Then the neighborhood information can be represented as:

$$N = D'^{-1}(\tilde{A}' + I')X_G \qquad (6)$$

where D'^{-1} represents the inverse of first two rows of degree matrix, I' represents the first two rows of the identity matrix and $X_G \subseteq R^{n \times c}$ is the selected user profile of users of the neighborhood. From another angle, the neighborhood information could also be interpreted as the average of users and their neighbors' user profile features. Algorithm 1 shows the process of getting graph neighborhood information from the second combination model.

Content Embedding Similarity. [7] demonstrated that repost probability increased with an increase in the similarity between the post content and the user interests. User interests could be inferred from the user's historical posts. We define the post content user interest similarity as:

$$S_{p,u} = \frac{C_p I_u}{\| C_p \| \| I_u \|} \qquad (7)$$

Algorithm 1. Neighborhood Information

1: **procedure** $N(i, j)$
2:     $V_{i,j}^1 = \{i, j\}$
3:     friends= $\Gamma(i) \cup \Gamma(j)$
4:     $V_{i,j}^1 = V_{i,j}^1 \cup$ friends
5:     **for** $x = 1, 2$ **do**   ▷ x represents node i or j
6:         **for** $y = 1, 2, \ldots, n$ **do**   ▷ n is the total number of nodes in $V_{i,j}^1$
7:             **if** node i and node j are friends **then:**
8:                 $\tilde{A}'[x][y] = 1$
9:             **if** y=1 or y=2 **then:**
10:                 $I'[x][y] = 1$
11:                 $D'[x][y] = |\Gamma(i)| \text{ or } |\Gamma(j)|$
12:     **for** $y = 1, 2, \ldots, n$ **do**
13:         $X_G[y]$=UserFeatures
14:     **return** $D'^{-1}(\tilde{A}' + I')X_G$

Here, C_p represents the content embedding of post and I_u represents the content embedding of user interests, i.e., content embedding of user's historical posts.

[14,22] have shown that content propagation across links occurs more frequently between the users sharing common interests. User interests could be inferred from their posts. We define the user and user interest similarity as:

$$S_{u1,u2} = \frac{I_{u1} I_{u2}}{\| I_{u1} \| \| I_{u2} \|} \tag{8}$$

Here, I_{u1} represents the content embedding of source user interests, i.e., content embedding of historical posts from source user $u1$ and I_{u2} represents the content embedding of target user interests, i.e., content embedding of historical posts from target user $u2$.

We implement a multi-lingual sentence embedding model [16] using knowledge distillation to extract the content embedding. In the multi-lingual sentence embedding model, the sentences with similar meanings are close in vector space. The multi-lingual model uses an English Sentence Bert Embedding (SBERT) model as a teacher model and uses XLM-RoBERTa (XLM-R) as a multilingual student model. Here the content embedding of post just refers to the sentence embedding of the post. In addition, we extract sentence embedding of the latest 50 posts of each user and get the average of the vector representations as the content embedding of user interests.

2.2 Prediction Models

To predict the repost probability, here we implement different machine learning methods, e.g., Logistic Regression, Deep Neural Network, Random Forest, XGBoosting and Stacking to make the prediction.

3   Experiments

We evaluate our proposed repost prediction model using the real world data set:
Sina Weibo. Weibo is a large scale publicly available dataset released by [26].

3.1   Datasets

Sina Weibo (http://www.weibo.com) is a Chinese microblogging website allowing users to follow other users and retweet the messages from the followed users
similar to Twitter. Specifically, for experimental purpose, we randomly sample
2000 messages. The total number of involved reposts is 290250 by 228250 different users. The 290250 reposts are taken as positive observations for model
training and testing. We also randomly sample the same amount of negative
observations based on the absence of response from source user's followers for
each post to train the model. The total negative observations we collect is 248770
because some of the posts don't have enough negative observations. In total, we
collect 539020 observations for our experiment.

3.2   Experimental Set-Up

User profile includes both poster and reposter features. The features of the Poster
user profile include Followers of poster, Followees of poster, Gender, Total messages of poster, Number of the reciprocal relationships and created Time. The
features of the Re-Poster user profile include Followers of poster, Followees of
poster, Gender, Total messages of re-poster, Number of the reciprocal relationships and created Time.

Messages propagate from users to users on Sina Weibo Dataset. We believe
the neighborhood information of the pair is a significant indicator of the repost
prediction. We obtain the neighborhood information from the combination of
neighbors' user profile including number of followers, number of followees, total
number of messages and total number of the reciprocal relationships.

Regarding to the feature of similarity, we get the content embedding of both
the post and user interests through using the multi-lingual sentence embedding
model [16]. Then we calculate the cosine similarity of content embedding as the
similarity features.

For all kinds of features including user profile, neighborhood information and
content similarity, we standardize all the features before training to give them
equal importance. The normalization is shown below:

$$\hat{X} = \frac{X - \bar{X}}{\sigma(X)} \tag{9}$$

where X represents the feature of a sample, $\bar{X}$ represents the average value of
the feature, $\sigma(X)$ is the standard deviation of the feature.

For training the deep neural network, we split the dataset into training set,
validation set and testing set with 80% of training set and validation set, 20% of

testing set. The validation set is 50000 for both datasets. ADAM optimization algorithm is implemented to update the parameters and get the function. We train the network in 250 epochs with the batch size of 50. In practice, we make use of Keras for a CPU-based implementation of the neural network. We used 6 i7 cores for training.

4 Results

In our experiments, we analyze the computational performance when we introduce the neighborhood information features. Besides, we also compare the performance and computational time of each prediction model as well. Lastly, we analyze the importance of each input feature by implementing tree-based ensemble prediction methods.

4.1 Combination Model Comparison

As we assume the neighborhood information collecting from the combination of user profile of neighbors is a good indicator of user's repost decision, thus we integrate neighborhood information into repost prediction. Then we come up with two different combination models to extract the neighborhood information. Table 1 shows the performance of different prediction models without neighborhood information and with neighborhood information using different combination models.

Table 1. Combination models comparison

Combination model	LR	KNN	GB	NN	RF	XGB	Stacking
NA	68.44%	74.30%	64.67%	76.51%	82.41%	83.88%	84.56%
$(A^{'} + I^{'})X$	68.62%	74.74%	61.04%	76.72%	83.79%	85.17%	86.06%
$D^{'-1}(A^{'} + I^{'})X$	68.65%	73.97%	65.21%	77.04%	83.98%	85.40%	86.33%

We could conclude that models including the neighborhood information have much better performance especially for random forest, XGB and Stacking models. In addition, the second combination model of $D^{'-1}(A^{'} + I^{'})X$ has better performance for all the models except K nearest neighbor. Therefore, we take the neighborhood information getting from the second combination model as an input feature for future experiments.

4.2 Prediction Models Comparison

In this subsection, we compare the prediction performance of different prediction models, e.g., Logistic Regression, K Nearest Neighbor, Gaussian Bayesian, Deep Neural Network, Random Forest, XGBoosting and Stacking. Measurement used

Table 2. Prediction model comparison

Model	Accuracy	Precision	Recall	F1	Time
LR	68.65%	77.00%	68.60%	72.56%	1.08
KNN	73.97%	73.74%	76.95%	75.31%	925.51
GB	65.21%	71.19%	66.54%	68.78%	0.77
NN	77.04%	77.24%	79.58%	78.06%	672.66
RF	83.98%	84.98%	85.23%	85.10%	152.73
XGB	85.40%	86.08%	86.70%	86.39%	103.27
Stacking	86.33%	87.30%	87.30%	87.30%	1391.81

for evaluating the robustness of binary classifiers are Accuracy, Precision, Recall, F1-score, and AUC, which are described in details in Appendix A.

The prediction performance of different models is shown in Table 2. Overall, the results of different prediction models indicate that the tree-based ensemble models provide better prediction performance than the other models in terms of Accuracy, Precision, Recall and F1-score. XGBoosting provides slightly better prediction performance than the Random Forest Method with much less time. The Stacking method provides the best performance among all the prediction modes. However, it's more time consuming because it trains all the models as the base models.

4.3 Performance of Models Using Different Features

In Fig. 2, we demonstrate the importance of each input feature by implementing tree-based ensemble prediction methods, i.e., Random Forest and XGBoosting. Tree-based methods measure the importance of each feature by collecting how on average each feature decreases the impurity. The importance is calculated for each tree by the amount that each feature split point decreases the impurity. The feature importance are then averaged all of the decision trees within the model.

We could conclude that followees number, created time (how long the user has been in social media) and post user similarity are the most importance features in repost prediction for both Random Forest and XGBoosting. However, source followers number seems to be an inconsistent element for the tree-based learning models, i.e., XGBoosting learning method thinks it's a more important feature than Random Forest learning method. Regarding to the neighborhood information, target user neighborhood information features (i.e., N5, N6,N7 and N8) are more important than source user neighborhood information features (i.e., N1,N2,N3 and N4) for Random Forest learning model. In contrast, source user neighborhood information features are more important than target user neighborhood information features for XGBoosting learning model.

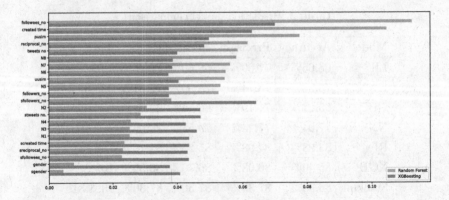

Fig. 2. Feature importance

5   Conclusion

In this work, we integrate neighborhood information and embedding similarity into repost prediction in social media. We come up with different machine learning prediction models and stacking model outperforms the other models in Sina Weibo dataset. In addition, XGBoosting model is also a good choice with comparable performance but much less computational time. In addition, we also prove that neighborhood information is a good indicator for repost prediction. Especially, extracting the neighborhood information using the second combination model inspired by the graph Laplacian has a better performance in repost prediction.

Appendix

A   Prediction Performance Measures

Measure	Definition	Formula
Accuracy	The ratio of correctly predicted observations to the total observations	$\frac{(TP+TN)}{(TP+FP+FN+TN)}$
Precision	The ratio of correctly predicted positive observations to the total predicted positive observations	$\frac{TP}{(TP+FP)}$
Recall	The ratio of correctly predicted positive observations to all the observations in actual class	$\frac{TP}{(TP+FN)}$
F1 Score	The weighted average of Precision and Recall	$\frac{2 \cdot Precision \cdot Recall}{(Precision+Recall)}$
ROCAUC	Compute area under the receiver operating characteristic curve which is True Positive Rate against False Positive Rate curve from prediction scores	–

References

1. Anshelevich, E., Chakrabarty, D., Hate, A., Swamy, C.: Approximation algorithms for the firefighter problem: cuts over time and submodularity. In: Dong, Y., Du, D.-Z., Ibarra, O. (eds.) ISAAC 2009. LNCS, vol. 5878, pp. 974–983. Springer, Heidelberg (2009). https://doi.org/10.1007/978-3-642-10631-6_98
2. Bourigault, S., Lamprier, S., Gallinari, P.: Representation learning for information diffusion through social networks: an embedded cascade model. In: Proceedings of the Ninth ACM International Conference on Web Search and Data Mining, WSDM 2016, pp. 573–582. ACM, New York (2016). https://doi.org/10.1145/2835776.2835817, http://doi.acm.org/10.1145/2835776.2835817
3. Budak, C., Agrawal, D., El Abbadi, A.: Limiting the spread of misinformation in social networks. In: Proceedings of the 20th International Conference on World Wide Web, pp. 665–674. ACM (2011)
4. Chen, G.H., Nikolov, S., Shah, D.: A latent source model for nonparametric time series classification. In: Advances in Neural Information Processing Systems, pp. 1088–1096 (2013)
5. Chen, M., Zheng, Q.P., Boginski, V., Pasiliao, E.L.: Reinforcement learning in information cascades based on dynamic user behavior. In: Tagarelli, A., Tong, H. (eds.) CSoNet 2019. LNCS, vol. 11917, pp. 148–154. Springer, Cham (2019). https://doi.org/10.1007/978-3-030-34980-6_17
6. Domingos, P.: Mining social networks for viral marketing. IEEE Intell. Syst. 20(1), 80–82 (2005)
7. Fei, H., Jiang, R., Yang, Y., Luo, B., Huan, J.: Content based social behavior prediction: a multi-task learning approach. In: Proceedings of the 20th ACM International Conference on Information and Knowledge Management, pp. 995–1000. ACM (2011)
8. Goyal, A., Bonchi, F., Lakshmanan, L.V.: Learning influence probabilities in social networks. In: Proceedings of the third ACM International Conference on Web Search and Data Mining, pp. 241–250. ACM (2010)
9. Granovetter, M.: Threshold models of collective behavior. Am. J. Sociol. 83(6), 1420–1443 (1978)
10. Guille, A., Hacid, H.: A predictive model for the temporal dynamics of information diffusion in online social networks. In: Proceedings of the 21st International Conference on World Wide Web, pp. 1145–1152. ACM (2012)
11. Jiang, B., et al.: Retweeting behavior prediction based on one-class collaborative filtering in social networks. In: Proceedings of the 39th International ACM SIGIR Conference on Research and Development in Information Retrieval, pp. 977–980. ACM (2016)
12. Kempe, D., Kleinberg, J., Tardos, E.: Maximizing the spread of influence through a social network. In: Proceedings of the Ninth ACM SIGKDD International Conference on Knowledge Discovery and Data Mining, KDD 2003, pp. 137–146. ACM, New York (2003). https://doi.org/10.1145/956750.956769, http://doi.acm.org/10.1145/956750.956769
13. Lagnier, C., Denoyer, L., Gaussier, E., Gallinari, P.: Predicting information diffusion in social networks using content and user's profiles. In: Serdyukov, P., et al. (eds.) ECIR 2013. LNCS, vol. 7814, pp. 74–85. Springer, Heidelberg (2013). https://doi.org/10.1007/978-3-642-36973-5_7
14. Peng, H.K., Zhu, J., Piao, D., Yan, R., Zhang, Y.: Retweet modeling using conditional random fields. In: 2011 11th IEEE International Conference on Data Mining Workshops, pp. 336–343. IEEE (2011)

15. Qiang, Z., Pasiliao, E.L., Zheng, Q.P.: Model-based learning of information diffusion in social media networks. Appl. Netw. Sci. **4**(1), 1–16 (2019). https://doi.org/10.1007/s41109-019-0215-3
16. Reimers, N., Gurevych, I.: Making monolingual sentence embeddings multilingual using knowledge distillation. In: Proceedings of the 2020 Conference on Empirical Methods in Natural Language Processing. Association for Computational Linguistics (2020). https://arxiv.org/abs/2004.09813
17. Rodriguez, M.G., Balduzzi, D., Schölkopf, B.: Uncovering the temporal dynamics of diffusion networks. arXiv preprint arXiv:1105.0697 (2011)
18. Saito, K., Kimura, M., Ohara, K., Motoda, H.: Learning continuous-time information diffusion model for social behavioral data analysis. In: Zhou, Z.-H., Washio, T. (eds.) ACML 2009. LNCS (LNAI), vol. 5828, pp. 322–337. Springer, Heidelberg (2009). https://doi.org/10.1007/978-3-642-05224-8_25
19. Saito, K., Nakano, R., Kimura, M.: Prediction of information diffusion probabilities for independent cascade model. In: Lovrek, I., Howlett, R.J., Jain, L.C. (eds.) KES 2008. LNCS (LNAI), vol. 5179, pp. 67–75. Springer, Heidelberg (2008). https://doi.org/10.1007/978-3-540-85567-5_9
20. Saito, K., Ohara, K., Yamagishi, Y., Kimura, M., Motoda, H.: Learning diffusion probability based on node attributes in social networks. In: Kryszkiewicz, M., Rybinski, H., Skowron, A., Raś, Z.W. (eds.) ISMIS 2011. LNCS (LNAI), vol. 6804, pp. 153–162. Springer, Heidelberg (2011). https://doi.org/10.1007/978-3-642-21916-0_18
21. Shah, D., Zaman, T.: Detecting sources of computer viruses in networks: theory and experiment. SIGMETRICS Perform. Eval. Rev. **38**(1), 203–214 (2010). https://doi.org/10.1145/1811099.1811063, http://doi.acm.org/10.1145/1811099.1811063
22. Suh, B., Hong, L., Pirolli, P., Chi, E.H.: Want to be retweeted? Large scale analytics on factors impacting retweet in twitter network. In: 2010 IEEE Second International Conference on Social Computing, pp. 177–184. IEEE (2010)
23. Tsur, O., Rappoport, A.: What's in a hashtag?: content based prediction of the spread of ideas in microblogging communities. In: Proceedings of the fifth ACM International Conference on Web Search and Data Mining, pp. 643–652. ACM (2012)
24. Varshney, D., Kumar, S., Gupta, V.: Predicting information diffusion probabilities in social networks: a Bayesian networks based approach. Knowl.-Based Syst. **133**, 66–76 (2017)
25. Yun, G., Zheng, Q.P., Boginski, V., Pasiliao, E.L.: Information network cascading and network re-construction with bounded rational user behaviors. In: Tagarelli, A., Tong, H. (eds.) CSoNet 2019. LNCS, vol. 11917, pp. 351–362. Springer, Cham (2019). https://doi.org/10.1007/978-3-030-34980-6_37
26. Zhang, J., Tang, J., Li, J., Liu, Y., Xing, C.: Who influenced you? Predicting retweet via social influence locality. ACM Trans. Knowl. Discov. Data **9**(3), 25:1–25:26 (2015). https://doi.org/10.1145/2700398, http://doi.acm.org/10.1145/2700398
27. Zhang, M., Chen, Y.: Link prediction based on graph neural networks. arXiv preprint arXiv:1802.09691 (2018)
28. Zhu, J., Xiong, F., Piao, D., Liu, Y., Zhang, Y.: Statistically modeling the effectiveness of disaster information in social media. In: 2011 IEEE Global Humanitarian Technology Conference (GHTC), pp. 431–436. IEEE (2011)

Driving Factors of Polarization on Twitter During Protests Against COVID-19 Mitigation Measures in Vienna

Marcus Röckl[1], Maximilian Paul[1], Andrzej Jarynowski[2,3(✉)],
Alexander Semenov[4], and Vitaly Belik[3]

[1] Freie Universität Berlin, Berlin, Germany
[2] Interdisciplinary Research Institute, Wroclaw, Poland
a.jarynowski@fu-berlin.de
[3] System Modeling Group, Institute for Veterinary Epidemiology and Biostatistics,
Freie Universität Berlin, Berlin, Germany
vitaly.belik@fu-berlin.de
[4] Herbert Wertheim College of Engineering, University of Florida,
Gainesville, FL, USA
asemenov@ufl.edu

Abstract. We conduct the analysis of the Twitter discourse related to the anti-lockdown and anti-vaccination protests during the so-called 4th wave of COVID-19 infections in Austria (particularly in Vienna). We focus on predicting users' protest activity by leveraging machine learning methods and individual driving factors such as language features of users supporting/opposing Corona protests. For evaluation of our methods we utilize novel datasets, collected from discussions about a series of protests on Twitter (40488 tweets related to 20.11.2021; 7639 from 15.01.2022 – the two biggest protests as well as 192 from 22.01.2022; 8412 from 11.12.2021; 3945 from 11.02.2022). We clustered users via the Louvain community detection algorithm on a retweet network into pro- and anti-protest classes. We show that the number of users engaged in the discourse and the share of users classified as pro-protest are decreasing with time. We have created language-based classifiers for single tweets of the two protest sides – random forest, neural networks and a regression-based approach. To gain insights into language-related differences between clusters we also investigated variable importance for a word-list-based modeling approach.

Keywords: Social network analysis · Twitter · Polarization · Classification · COVID-19

1 Introduction

The aim of our analysis is to evaluate the reaction of Twitter users of German speaking Twitter in Austria to highly polarizing and dominating protest agenda

---

M. Röckl and M. Paul—These authors contributed equally.

T. N. Dinh and M. Li (Eds.): CSoNet 2022, LNCS 13831, pp. 15–26, 2023.
https://doi.org/10.1007/978-3-031-26303-3_2

topics in the last two years – COVID-19. We are going to detect and analyze the main common features of the COVID-19 discourse at national level by social media listening thus pursuing an infodemiological approach [9].

Social movement scholars have increasingly sought to understand the multi-topic dynamics of movements mobilization [8].

Generally, anti-lockdown protests (as in the case of Austrian protests) were commonly organized in many countries by many independent organizations [17,21]. Movements have made greater use of digital forms of action and combined them in part with analog ones. The advent of the Internet and social media platforms has changed the way social movements are communicating. For instance, the first letters of a location (e.g. W for Wien) combined with digits describing a date of a rally (e.g. 1501 for 15 January) in the hashtag form (e.g. #W1501) is widely in use since the Arab Springs [12]. Digital communities have given rise to new socio-cultural scenarios. E.g. in this article we will investigate behavioral and affective properties of these communities. The demand to respect individual freedom above all through ignoring basic health norms could lead to the so called pandemic riots [13] being highly confrontational, often involving clashes with police and governments.

1.1 Austrian Context

In our previous study [17] we observed that supporters of antivaxx/ "coronascepticism" in Germany mainly belong to far right as well as far left political spectrum being interconnected between each-others. The protest participants political sympathies in Austria seems to follow the same pattern [5]. The protest is concentrated on the right wing with FPÖ (30.2%) or ÖVP (20.2%) parties with still a significant representation of left wing such as the Greens (20.5%). The Austrian government strategically raised awareness for COVID-19 during the first wave of infections and the spikes in the related interest at the beginning of March 2020 in traditional and social media may be partially an effect of these practices [24]. The COVID-19 pandemic was largely controlled by governmental top-down interventions and Stringency Index in Austria was among the highest in Europe during first 2 years of pandemic [15]. Protests against these measures and a lack of compliance in some parts of the population have underlined the need to understand what drives this protest. It is unclear who shows health protective behavior (mask wearing, contact reductions, vaccine uptake) and under what conditions.

1.2 Coronasceptic Protests

Thousands of people across hundreds of German, Austrian or Swiss towns have taken to the street against the government's actions on COVID-19. These rallies gathered various kind of magical thinking categories such as QAnon followers, Querdenker, believers in alternative medicine, esoteric or folk religion communities [1,28]. The biggest street rally took place in the Austrian capital on 20.11.2021 shortly after the announcement of mandatory vaccination (on

Monday 22.11.2021 Austria went into a hard lockdown), according to the police, almost 40000 people took part[1]. The second biggest demonstration took place in Vienna on 15.01.2022 (when the Austrian parliament approved a mandatory vaccination order) with around 30000 protesters[2].

1.3 Anti-lockdown, Anti-segregation Attitude and Vaccine Hesitancy

Digital green passes such as 2G (geimpft–vaccinated or genesen–recovered) or 3G (geimpft–vaccinated or genesen–recovered or getestet–with a negative test result) were used in Austria as an efficient measure of separating potentially infectious citizens from the rest of the population. However, among many it was seen as the foundation for a system of social credits not commonly attributed to the Western civilization of freedom [7]. Therefore, one of the most used words of protest sympathizers in the debate surrounding the first protest was "Freiheit" (*freedom*).

1.4 Research Questions

Influential studies concerning the Corona protests that have been published already [10,14,20] in the German speaking world have some limitations since they use narrative or survey analysis and focus on individuals who have already participated in these protests. We contribute to this research by studying the individual mobilization potential via prediction of patterns of discussion for protesters and anti-protesters. We want to investigate: What are the behavioral and affective differential characteristics of individuals' willingness to be engaged against the anti-containment measures/or being against protests in the 3rd/4th wave of the Corona pandemic in Vienna?

2 Data and Methodology

In this study we retrospectively processed Twitter data in the contexts of major rallies against COVID-19-related mitigation measures in Vienna, Austria. An ethical committee approval (by the ethical board at Freie Universität Berlin) was given for the analysis of COVID-19 related discourse on the internet.

The first protest exhibits by far the greatest participation within the Twitter community as more than 2/3 of the tweets in our data referred to this first protest (Fig. 1). Around the first protest on November 20, 2021 we have 290 and 1213 active users (i.e. authoring tweets) taking the side of protesters and supporters of the measures, respectively, and are left with merely eight users of the protest and 276 of the compliant cluster being active during the last protest.

---

[1] https://www.tagesschau.de/ausland/proteste-wien-103.html. Accessed on 11.11. 2022.

[2] https://wien.orf.at/stories/3138745/. Accessed on 11.11.2022.

One reason for the low participation on the protest side is that the February 2022 protest was a car convoy driving though Vienna. As a consequence there were no speeches by politicians and little interaction between the two groups. The relatively high number of people of the compliant cluster might be in reaction to the noise disturbance caused by protesters honking their car horns.

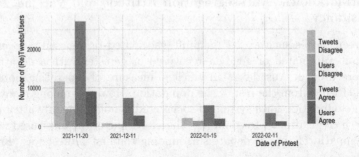

Fig. 1. Temporal development of tweets that were retweeted at least twice and the respective users where Agree means agree with measures. (Color figure online)

2.1 Clustering Method for Users Labelling

Retweet network consists of users being vertices or nodes and edges or links representing retweets (one user retweeted the other). We used *Spin Glass* and *Louvain* algorithm for clustering the retweet networks of all tweets relating to one of the four major protests and were retweeted at least twice; following the assumption in Ref. [6] where posts with less retweets were considered as noise.

2.2 Features for Classification

For the classification task of this paper, we aim to predict the cluster-memberships resulting from the aforementioned clustering of the retweet-network. Namely, tweets will be subject to a binary prediction of either falling into the protest or compliant group on the basis of lexical information regarding the content of the tweets, similarly to [6].

However, given the fact that in the present case only tweets that were part of the network creation were analyzed, we departed from the approach of [6] who used a BoW (bag of words) model on the users historical tweets to predict cluster membership of users. Accordingly, the present analysis was restricted to a corpus of only 8,513 original tweets (excluding 52,163 of the 60,676 collected tweets as they were retweets and did not contain original language contributions), rendering the creation of approaches such as the BoW unfeasible. Given our special interest in the semantic dimension of the tweets of the two clusters, we therefore use external libraries to obtain language features for a classification of single tweets and not users.

More specifically, we resort to **LIWC** *(Linguistic Inquiry and Word Count) dictionary*, a word-list approach that counts words in psychologically meaningful categories, see [29] and [4] as well as [23] for the German adaptation used in the present work and has been used in the context of Twitter-data previously (e.g. [26]). With this tool, we aim to analyze all tweets for each individual user in the network. The 97 categories developed for the German version of the **LIWC** dictionary include summary variables (e.g. *Words per Sentence* or *Dictionary Words*), Linguistic Dimensions (e.g. *Total pronouns* or *Common Adverbs*), Other Grammar (e.g. *Common verbs* or *Comparisons*) as well as Psychological Processes including Social processes (e.g. *Family*), Cognitive processes (e.g. *insight*), Perceptual processes (e.g. *See*) or Informal language (e.g. *swear words*) among many others. Furthermore, the **LIWC** list provides punctuation categories including *Periods, Commas, Colons, Semicolons, Question marks, Exclamation marks, Dashes, Quotation marks, Apostrophes, Parentheses (pairs) and Other punctuation* (including hashtags).

Additionally to **LIWC** we employ **fastText**, a supervised learning approach for text classification and sentiment analysis which computes a 300-dimensional vector for each word. These word specific vectors can then be averaged over sentences or – in our case – tweets and used for modelling [18]. More precisely, we make use of pre-trained word vectors for German language trained on Wikipedia-data using the **fastText** method made available by the Facebook-AI team[3]. As Ref. [11] points out, since meaning is abstracted to an arbitrary number of dimensions (300 dimensions in the present **fastText** case), embedding methods lack the interpretability that lexical approaches such as **LIWC** offer. However, since these word embedding techniques have been shown to perform favourably in similar contexts [2], we include them to enrich the prediction task and provide a comparison to the word list approach. Lastly, we acknowledge that a key problem of natural language processing and text mining of Twitter posts is their short length, where Ref. [4] suggests special caution in case of texts below 25–50 words which is the case for many tweets (Fig. 2).

User ID	Tweet
96850282...	Auch in Österreich wird zu...
96850282...	#w2011 Dieser Dude hat sc...
...	...
10790336...	Volle Solidarität mit allen d...

Fig. 2. Three truncated tweets of the data set. Their representation processed for analysis can be found in Fig. 3. User IDs and tweets have been truncated for privacy reasons.

Concerning the cleaning of the tweet-text before feature extraction, we decided on differing strategies for the two approaches. Since the **LIWC** list includes punctuation categories that count hashtags as well as e.g. question

---

[3] https://fasttext.cc/docs/en/crawl-vectors.html. Accessed on 11.11.2022.

UserID	Analytic	Authentic	...	function		1	2	...	300
96850282...	99.00	1.76	...	50.00		−0.0068	0.0130	...	−0.0079
96850282...	13.74	1.00	...	33.33		−0.0035	0.0193	...	−0.0290
⋮	⋮	⋮	⋱	⋮		⋮	⋮	⋱	⋮
10790336...	20.23	5.07	...	9.09		−0.0279	0.0082	...	−0.0224
	1	2	...	97		1	2	...	300

Fig. 3. Text in tweets processed as **LIWC** (*left*) and **fastText** (*right*) matrix with 8,513 rows and 97 (*LIWC*) and 300 (*fastText*) columns. User IDs have been truncated for privacy reasons.

marks or exclamation marks and other lexical categories such as word count, words per sentence and verb count, cleaning only consisted of the removal of URLs and user-name tags. That way, in addition to word-based features, we could also investigate the effect of simpler lexical features similarly to Ref. [6]. There lexical features (e.g. number of characters, verbs, question marks and exclamation marks) Twitter specific features (e.g. tweets that have hashtags or URLs) as well as users features (e.g. account age, tweet rate) were used in a first step to predict cluster membership before including the aforementioned BoW model. For the **fastText** approach, we followed the conventional practice of excluding punctuation and stop words as well as using stemming.

3   Results

In the Twitter discussion we can observe a clear visual polarization of two filter bubbles (Fig. 4). Most social movements are accompanied with opposite movements [25] (phenomenon of Hashtag hijacking [27] where opponents are discussing under the hashtag of the enemy). Accordingly, the border between corona protester/anti-corona protesters is blurred. The size of the protest supporter community is decreasing over time (Fig. 1).

3.1   Clustering

As stated in [3] Louvain algorithm reveals its particular strength for large networks due to short computation times and empirically reasonable results. In our case Spin Glass was outperformed by Louvain. By looking at Fig. 4 one may notice that the blue cluster in Fig. 4a has many orange sprinkles whereas the clusters in Fig. 4b look "cleaner". And indeed, by checking several representatives of each cluster manually, users of the compliant cluster whose posts were retweeted several hundred times were wrongly classified by Spin Glass. Nevertheless, one downside of Louvain is that the number of clusters can not be controlled directly which resulted in a total of four cluster of which two are negligible small (*green* and *red* in Fig. 4b).

As expected, the cluster of the users who agree with protective measures is by far larger than the cluster of the protesters and their sympathizers (Fig. 4). After discarding every tweet that has been retweeted once or less, we are left with a total of 58,059 tweets of which 43,105 were assigned to the members of the compliant cluster and the remaining 14,954 tweets were posted by members of the protest cluster. Another interesting finding is the ratio of verified accounts, as 0.381 % (57) of the accounts in the protest cluster and 1.513 % (652) in the compliant cluster are verified.

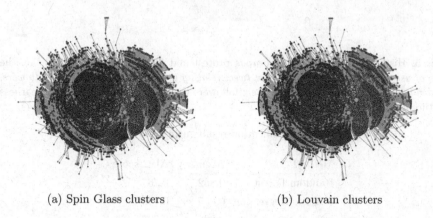

(a) Spin Glass clusters   (b) Louvain clusters

Fig. 4. Aggregated retweet network of tweets posted during the four major protest against protective measures against Coronavirus in Vienna; *blue* nodes are (re)tweets of users who agree with governmental measures; *orange* nodes are (re)tweets of users who disagree. (Color figure online)

Figure 5 shows a selection of user-data to illustrate exemplary differences between clusters. Among these, especially the age of the user account until first tweet related to the protest was posted shows significant differences for the distinct clusters as the mode of the accounts belonging to the protest clusters is around 0 (*days*). Note that we include only users that really authored their tweets. By also including users who merely retweeted other users posts, the discrepancy becomes even more distinct. A possible explanation is that these users may have created an account to participate in the Twitter debate.

3.2 Classification Results

In a first classification approach, we train a Random Forest Model (RF) and a Neural Network (NN) as well as a logistic regression (LR) on the **LIWC** data to predict class membership of 8,513 original tweets of which 7,021 (82.47%) fall into the anti-protest class.

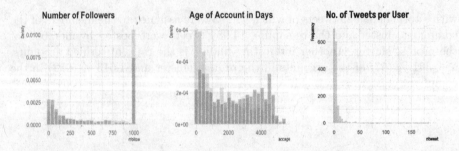

Fig. 5. Histograms of user data (*orange* protest and *blue* compliant cluster), i.e. the *age of user account in days until first tweet relating to the protest, number of followers*, and *number of original tweets per user*; followers were truncated at 1,000 for clearness of the plot. (Color figure online)

Table 1. Classification results using **LIWC**-data

Method	Accuracy	AUC-score
Random Forest	0.862	0.767
Logistic Regression	0.855	0.749
Neural Network	0.828	0.727

The Random Forest tuned for the data-set of **LIWC** variables uses 200 trees and a minimum of 3 samples per split. As it can be seen from the AUC-scores in Table 1, classification yields acceptable results where accuracy is slightly improved for RF when compared to LR. However, when examining the relatively high prediction accuracy one has to consider the unequal distribution of members from the two clusters (1,569 anti-protesters vs. 347 protesters) resulting in a baseline accuracy of 82.47%. We used fully connected neural network (NN), the first layer had 100 neurons, second had 50, next one had 12; we have used ReLU as activation for all layers and added dropout to prevent over-fitting. The network was trained for 20 epochs.

In a next step, we train the same three models (RF, LR, NN) on the **fastText** features instead of the **LIWC** variables.

The Random Forest tuned for the data-set of **fastText** variables uses 400 trees and a minimum of 2 samples per split. As we see from the results in Table 2, accuracy and AUC-score in RF and LR are slightly lower for the **fastText**-data while NN performs on par with how RF performed in a **LIWC** setting. However, when comparing these results, one has to keep in mind that the **LIWC**-dataset contains additional information in the form of lexical and punctuation categories that are absent in the **fastText**-dataset. In light of that, it seems plausible that from this additional information, the prediction model might learn characteristics that are relatively constant over several tweets of one individual (e.g. one individual might use a similar amount of hashtags over all of their

Table 2. Classification results using **fastText**-data

Method	Accuracy	AUC-score
Random Forest	0.846	0.758
Logistic Regression	0.840	0.744
Neural Network	0.852	0.765

tweets) and use it to predict an outcome for a different tweet of that one certain individual, therefore potentially improving accuracy in a non-meaningful way.

3.3 Variable Importance

When considering the variable importance of a Random Forest or any other prediction model, the SHAP (Shapley Additive Explanations) approach, which is based on the work of [22] has benefits such as providing consistent and more detailed explanations of variable influence. Using SHAP-values, we can obtain the marginal contributions of each predictor variable to the outcome in a consistent way.

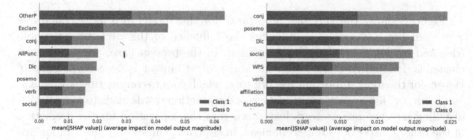

Fig. 6. Variable importance for **LIWC**-model with punctuation categories (*left*) and without punctuation categories (*right*). (Color figure online)

As has been described earlier, word-list based approaches such as **LIWC** have the advantage of providing meaningful categories that can be inspected in terms of variable importance to shed light on potential semantic differences between tweets of different clusters. Consequently, we investigated the variable importance of the **LIWC**-based Random Forest model from Table 1 to gain insight into semantic differences between clusters. As can be seen in the left plot of Fig. 6, which illustrates variable importance for the RF-model for **LIWC**, the category *OtherP*, which captures other punctuation and in case of the present data mostly hashtags, seems to be the most important predictor by a great margin, affirming the suspicion we voiced at the end of the classification section that the prediction model might use hashtags to correctly predict the outcome for specific users that occur repeatedly in the corpus.

To control for this effect in a setting where the main focus lies on semantic analysis, we propose a different RF model for meaningful variable importance for the **LIWC**-approach where tweets have been cleaned of hashtags before running **LIWC** and all punctuation categories (listed in data and methodology) have been excluded, putting more emphasis on actual semantic categories. This model performs slightly worse than its counterpart in Table 1 and comparable to the **fastText** models in Table 2 (**Accuracy:** 0.847 & **AUC:** 0.720).

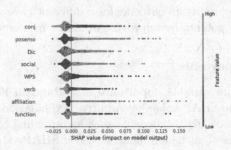

Fig. 7. Variable importance for RF **LIWC**-model without punctuation categories. (Color figure online)

Figure 7 shows the variable importance for this model in a more conclusive way than Fig. 6. Namely, it indicates the influence of the value of a variable (denoted by color) on the class prediction. In the present case, the pro-protest cluster is denoted as *class 1* and the anti-protest cluster is denoted as *class 0*. Hence, for the most important feature *conj.* which counts conjunction words (e.g. "and, but, or"), we can see that low use of conjunction words leads to a prediction of falling into the pro-protest cluster, whereas the separation for the prediction of the other cluster does not seem clear. In contrast, the second most important variable *posemo*, which counts positive emotion words where low use of positive emotion leads to a prediction of falling into the anti-protest cluster and high use of positive emotion leads to falling into the pro-protest cluster, shows clearer separation. This corresponds with the expectation that pro-protest tweets would speak positively of the current event and anti-protest tweets would not. Another variable that shows clear separation is *affiliation* (e.g. "ally, friend, social").

4   Conclusion

As the amount of COVID-19 discourse on German speaking Twitter continues to decrease, it is increasing on Telegram on the other hand [16,19]. We observe this pattern (decreasing share of protesters with time in protest discourse) in the analyzed Vienna protests (Fig. 1).

We have proposed different classifiers to detect positions in conflicted opinions on Twitter using language based metrics achieving accuracy above 0.83

and AUC-score above 0.73 and discussed them in terms of interpretability. We find results from a word-list based approach to be comparable to an embedding-based approach but discover decreased performance when restricting the word-list based approach to semantic categories (without punctuation categories). We use the interpretability-properties of the word-list based approach to identify the use of conjunction words and positive emotion words to be among the most important predictors for the class of tweets cluster. The use of transparent models could help to understand the dynamics of COVID-19 protests [30]. The major limitation of this work lies in the relatively low improvements in prediction accuracy compared to baseline, which also limits the meaningfulness of the word-list based variable-importance considerations. We point out the difficulty in using language models based on single-tweets for prediction that arises from the issues of short tweet-length and repeated contributions of users.

Acknowledgement. This study was partially funded by the Deutsche Forschungsge-meinschaft (DFG, German Research Foundation, project number 458528774).

References

1. Anti-lockdown activity: Germany country profile (2021). https://www.isdglobal.org/wp-content/uploads/2022/01/ISD-Anti-lockdown-Germany-briefing.pdf
2. Alessa, A., Faezipour, M., Alhassan, Z.: Sentiment and structure in word co-occurrence networks on twitter. In: IEEE International Conference on Healthcare Informatics (ICHI), pp. 366–367 (2018)
3. Blondel, V.D., Guillaume, J.L., Lambiotte, R., Lefebvreo, E.: Fast unfolding of communities in large networks. J. Stat. Mech. Theory Exp. **2008**, P10008 (2008)
4. Boyd, R.L., Ashokkumar, A., Seraj, S., Pennebaker, J.W.: The development and psychometric properties of LIWC-22. In: University of Texas at Austin (2022)
5. Brunner, M., Daniel, A., Knasmüller, F., Maile, F., Schadauer, A., Stern, V.: Corona-protest-report. Narrative-motive-einstellungen (2021)
6. Cossard, A., Morales, G.D.F., Kalimeri, K., Mejova, Y., Paolotti, D., Starnini, M.: Falling into the echo chamber: the Italian vaccination debate on Twitter. In: ICWSM (2020)
7. Di Sia, P.: Current perception of epidemic between traditional and social media: an Italian case study (2022)
8. Diani, M.: The Cement of Civil Society. Cambridge University Press, Cambridge (2015)
9. Eysenbach, G.: How to fight an infodemic: the four pillars of infodemic management. J. Med. Internet Res. **22**(6), e21820 (2020)
10. Frei, N., Schäfer, R., Nachtwey, O.: Die proteste gegen die corona-maßnahmen. Forschungsjournal Soziale Bewegungen **34**(2), 249–258 (2021)
11. Fudolig, M., Alshaabi, T., Arnold, M., Danforth, C., Dodds, P.: Sentiment and structure in word co-occurrence networks on Twitter. Appl. Netw. Sci. **7**(9), 1–27 (2022)
12. Gerbaudo, P.: Tweets and the Streets: Social Media and Contemporary Activism. Pluto Press (2012)
13. Gerbaudo, P.: The pandemic crowd. J. Int. Aff. **73**(2), 61–76 (2020)

14. Grande, E., Hutter, S., Hunger, S., Kanol, E.: Alles covidioten? Politische potenziale des corona-protests in deutschland. Technical report, WZB Discussion Paper (2021)
15. Hale, T., et al.: A global panel database of pandemic policies (Oxford COVID-19 government response tracker). Nat. Hum. Behav. 5(4), 529–538 (2021)
16. Hoseini, M., Melo, P., Benevenuto, F., Feldmann, A., Zannettou, S.: On the globalization of the QAnon conspiracy theory through telegram. arXiv preprint arXiv:2105.13020 (2021)
17. Jarynowski, A., Semenov, A., Belik, V.: Protest perspective against COVID-19 risk mitigation strategies on the German internet. In: Chellappan, S., Choo, K.-K.R., Phan, N.H. (eds.) CSoNet 2020. LNCS, vol. 12575, pp. 524–535. Springer, Cham (2020). https://doi.org/10.1007/978-3-030-66046-8_43
18. Joulin, A., Grave, E., Bojanowski, P., Mikolov, T.: Bag of tricks for efficient text classification. arXiv preprint arXiv:1607.01759 (2016)
19. Kemmesies, U., et al.: Motra-monitor (2022)
20. Koos, S.: Die "querdenker". Wer nimmt an Corona-Protesten teil und warum (2021)
21. Kowalewski, M.: Street protests in times of COVID-19: adjusting tactics and marching 'as usual'. Soc. Move. Stud. 20, 1–8 (2020)
22. Lundberg, S., Lee, S.I.: A unified approach to interpreting model predictions. In: 31st Conference on Neural Information Processing Systems (NIPS) (2017)
23. Meier, T., et al.: "LIWC auf deutsch": the development, psychometrics, and introduction of de- LIWC2015 (2018). https://osf.io/tfqzc/
24. Pellert, M., Lasser, J., Metzler, H., Garcia, D.: Dashboard of sentiment in Austrian social media during COVID-19. Front. Big Data 3, 32 (2020)
25. Płatek, D.: Przemoc skrajnej prawicy w polsce. Analiza strategicznego pola ruchu społecznego. Studia Socjologiczne 239, 123–153 (2020)
26. Pope, D., Griffith, J.: An analysis of online Twitter sentiment surrounding the European refugee crisis. In: KDIR (2016)
27. Rodak, O.: Hashtag hijacking and crowdsourcing transparency: social media affordances and the governance of farm animal protection. Agric. Hum. Values 37(2), 281–294 (2020)
28. Seiler, M.: From anti-mask to anti-state: anti-lockdown protests, conspiracy thinking and the risk of radicalization (2021)
29. Tausczik, Y.R., Pennebaker, J.W.: The psychological meaning of words: LIWC and computerized text analysis methods. J. Lang. Soc. Psychol. 29(1), 24–54 (2010)
30. van der Zwet, K., Barros, A.I., van Engers, T.M., Sloot, P.: Emergence of protests during the COVID-19 pandemic: quantitative models to explore the contributions of societal conditions. Human. Soc. Sci. Commun. 9(1), 1–11 (2022)

Categorizing Memes About the Ukraine Conflict

Keyu Chen[1], Ashley Feng[1], Rohan Aanegola[1], Koustuv Saha[2], Allie Wong[3],
Zach Schwitzky[3], Roy Ka-Wei Lee[4], Robin O'Hanlon[5],
Munmun De Choudhury[6], Frederick L. Altice[1], Kaveh Khoshnood[1],
and Navin Kumar[1,7(✉)]

[1] Yale University, New Haven, USA
navin183@gmail.com
[2] Microsoft Research, Montreal, Canada
[3] Limbik, New York, USA
[4] Singapore University of Technology and Design, Singapore, Singapore
[5] Memorial Sloan Kettering Cancer Center, New York, USA
[6] Georgia Tech, Atlanta, USA
[7] National University of Singapore, Singapore, Singapore

Abstract. The Russian disinformation campaign uses pro-Russia
memes to polarize Americans, and increase support for the Russian invasion of Ukraine. Thus, it is critical for governments and similar stakeholders to identify pro-Russia memes, countering them with evidence-based
information. Identifying broad meme themes is crucial for developing a
targeted and strategic counter response. There are also a range of pro-
Ukraine memes that bolster support for the Ukrainian cause. As such,
we need to identify pro-Ukraine memes and aid with their dissemination to augment global support for Ukraine. We address the indicated
issues through the following contributions: 1) Creation of an annotated
dataset of pro-Russia (N = 70) and pro-Ukraine (N = 121) memes regarding the Ukraine conflict; 2) Identification of broad themes within the
pro-Russia and pro-Ukraine meme categories. Broadly, our findings indicated that pro-Russia memes fall into thematic categories that seek to
undermine specific elements of US and their allies' policy and culture.
Pro-Ukraine memes are far more diffuse thematically, highlighting admiration for Ukraine's people and its leadership. Stakeholders may utilize
our findings to develop targeted strategies to mitigate Russian influence
operations - possibly reducing effects of the conflict.

Keywords: Ukraine · Social media · Memes

1 Introduction

The Russian Federation's *special military operation* in Ukraine began on February 24 2022 and, at the time of writing, shows no sign of abating. Social
media has become essential to understanding the Russian invasion of Ukraine
[4,18]. Within social media, memes have become a popular way of conveying

© The Author(s), under exclusive license to Springer Nature Switzerland AG 2023
T. N. Dinh and M. Li (Eds.): CSoNet 2022, LNCS 13831, pp. 27–38, 2023.
https://doi.org/10.1007/978-3-031-26303-3_3

information about the conflict. For example, there exist pro-Ukraine memes portraying Ukraine President Volodymyr Zelenskyy as an Avenger, and pro-Russia memes claiming that Ukraine belongs to Russia. The Russian disinformation campaign uses pro-Russia memes to polarize Americans, particularly those at the extreme ends of the political spectrum [2], and increase support for the Russian invasion of Ukraine. Thus, it is critical for governments and similar stakeholders to identify pro-Russia memes, countering them with evidence-based information and corresponding regulation of social media platforms. To effectively counter such memes and prevent their further dissemination, identifying broad themes is crucial. Only by understanding the thematic elements and cultural ramifications of these memes can effective counter strategies be successfully developed. For example, if pro-Russia memes focus on the existence of secret US-funded biolabs in Ukraine, interventions can provide evidence-based information that demonstrates the non-existence of such facilities.

Similarly, there are a range of pro-Ukraine memes that bolster support for the Ukrainian cause and may increase global aid for Ukrainians. As such we need to identify pro-Ukraine memes and aid with their dissemination, augmenting global support for Ukraine, possibly reducing effects of the conflict. There currently exists no annotated dataset of pro-Russia and pro-Ukraine memes, central to developing classifiers for these cases. We address the indicated issues through the following contributions: 1) Creation of an annotated dataset of pro-Russia ($N = 70$) and pro-Ukraine ($N = 121$) memes regarding the Ukraine conflict; and 2) Identification of broad themes within pro-Russia and pro-Ukraine categories.

2   Related Work

2.1   Ukraine-Related Memes

A qualitative content analysis was conducted of memes posted to the RuNet Memes Twitter account in 2014 and indicated that most memes fell into one of two categories: pro-Russia or pro-Ukraine. While the memes reference a given news story or event, they continued to be consumed and reproduced along similar thematic categories [28]. Another study analyzed memes created during anti-government protests in Ukraine (2013–2014), indicating that pro-government memes usually rely on simple emotional messages for propaganda/polarization purposes, whereas anti-government memes produce more nuanced statements used as a form of creative protest or as a coping mechanism [14]. Researchers studied memes created during the 2019 Ukrainian presidential election, examining the influence of Ukrainian politicians on the development of such memes. The researchers also detailed the role memes play in societal polarization. While past work detailed and thematically categorized Ukraine-related memes, previous research did not explore Ukraine memes in relation to the current conflict, or assemble datasets for developing pro-Russia or pro-Ukraine meme classifiers.

2.2   Mis/Disinformation in Memes

Existing research on identifying mis- or disinformation in memes is limited. One study focused on two widely circulating memes in the anti-vaccination movement, namely lists of vaccine ingredients containing mercury, and quotes attributed to Mahatma Gandhi [3]. The article analyzed both memes, and illustrated how the repurposed, often ironic use of visual tropes can either undermine or strengthen the accompanying claims, exploring how memes can function as vehicles for the spread of controversial health-related information. Another study used an experimental design to examine the credibility and persuasiveness of COVID-19-related memes [27]. Results indicated that memes with expert source attribution are more credible than those with nonexpert source attribution. A positive correlation between the credibility of a meme and its persuasiveness was observed. Overall, previous work provides an overview of mis/disinformation in memes, but does not detail possible disinformation within memes around the Ukraine conflict.

Complementarily, there exist several classifiers for offensive [23] or hateful [7,10,29] meme content. However, there are limited classifiers for disinformation within memes and no classifiers for disinformation in memes for the Ukraine conflict context. We thus provide an annotated dataset covering the initial memes produced during the conflict, with an emphasis on memes possibly resulting from disinformation campaigns, simultaneously aiding the development of Ukraine conflict-centric meme classifiers.

3   Data

We first assembled a list of keywords derived from related literature reviews [11,13,15,19] on the Ukraine conflict, such as *Ukraine invasion, Ukraine-Russia Conflict, Russian special military operation.* Two content experts then reviewed the keyword list independently. We selected only keywords which both experts approved. We then used these keywords to search for and collect all relevant memes in the following sites: memegine.com; knowyourmeme.com. We also used the keywords + *meme* e.g., Ukraine conflict meme, to obtain all memes within the first 20 pages in a Google image search.

4   Methods

Two content experts then independently coded (85% agreement) the resultant 1426 memes into three categories: pro-Russia; pro-Ukraine; irrelevant. A third content expert made the final decisions on coding disagreements. We selected content experts who had published at least five academic articles on mis/disinformation and/or Ukraine, defined broadly. Irrelevant memes were deleted to result in 70 pro-Russia and 121 pro-Ukraine memes. Examples of irrelevant memes included those which had too low resolution to determine the content, memes that were screenshots of tweets, and memes that were relating the conflict to anime content. We sought to develop multimodal classifiers to detect pro-Russia and pro-Ukraine memes, but were unable to do so due to limited data.

We then grouped memes within the pro-Russia and pro-Ukraine meme groups into broad themes, to determine if there were themes related to disinformation or pro-Ukraine sentiment. We first used DeepCluster. However, due to limited data, clusters were not meaningful and we categorized memes manually. Two content experts independently coded the pro-Russia and pro-Ukraine memes separately to identify broad themes. Coders first indicated if a meme was part of a broader theme, based on the overarching message of the meme. For example, if a meme was about the US preferring Ukrainian refugees to Syrian refugees, the coder would assign *refugee* as the theme. Memes were then organized based on their theme. Coders then reviewed thematic groups with two or fewer memes to see if these groups could be subsumed into larger thematic groups. Coders then compared thematic groups to develop themes common across coders (75% agreement). Any disagreement was resolved with a third content expert.

5   Results

5.1   Pro-Russia Memes

Coders identified five broad themes for pro-Russia memes: Russian Competence (n = 7); US and Allies Incompetence (n = 35); Azov Regiment (n = 6); Refugee (n = 12); Sexuality (n = 10). Overall, pro-Russia memes fall into specific thematic categories, e.g., targeting aspects of US and their allies' culture and beliefs. We provide an overview of these categories next.

Fig. 1. Russian competence memes

Russian Competence Memes (Fig. 1) frame Putin or Russian soldiers as highly competent or benevolent. Such memes may be linked to existing Russian media framing of Putin as a strongman [12]. These memes may be part of broader disinformation campaigns where Putin and the Russian Army are seen as able to take on all challenges as a global force for good [25].

Fig. 2. US and Allies incompetence memes

US and Allies Incompetence memes (Fig. 2) are about how the US and its allies are incompetent. Examples of such memes suggest that NATO and the UN are unable to respond to the conflict, or NATO or Ukrainian soldiers are poorly trained. A significant portion of memes centered on how inmates in Ukraine with combat experience were released from jail to help defend against Russia, implying that the Ukrainian government is disingenuous. These memes may suggest that the US, its allies, and Ukraine are ineffectual, building support for Russia's claim on Ukraine.

Azov Regiment Memes (Fig. 3) center on the Azov Regiment and their purported National Socialist leanings. The Azov Battalion, a regiment of the Ukrainian Army with roots in ultranationalist political groups, has been used by the Russian media since 2014 as an example of far-right support in Ukraine. The Russian media's portrayal of the group exaggerates the extent to which its members hold neo-Nazi views. Multiple expert assessments conclude the modern Azov Regiment is a fairly typical fighting unit, with little, if any, political bent [21]. Examples of such memes detail National Socialist imagery within the Azov Regiment and the Western media's supposed suppression of reports around the Azov Regiment's National Socialist leanings. Such memes may be part of larger influence campaigns to bolster Russia's claims of denazification and that the US is implicitly supporting a National Socialist state [8].

Memes in the Refugee theme (Fig. 4) focus on how the US and its allies supposedly prefer white Ukrainian refugees over Middle Eastern refugees [1,9]. Examples of such memes indicate that Europeans are Islamophobic and do not want to resettle Syrian and Afghan refugees, but are supportive of Ukrainians displaced by the conflict. These memes bolster Russian influence campaigns that the US and its allies are hypocritical and racist, casting Russia as a benevolent state that has a rightful claim over Ukraine [24].

Fig. 3. Azov regiment meme examples

Fig. 4. Refugee meme examples

Memes in the Sexuality theme (Fig. 5) center on how the US and its allies are too focused on progressive issues i.e., *woke*, to be effective supporters of Ukraine in the conflict [22]. Examples of such memes purport that the US army, despite not being deployed in the conflict, is comprised of numerous individuals who use they/them pronouns, which somehow renders the US army as ineffective. Such memes build on Russia's anti-LGBT+ policies [17], which hold Russia to be a paragon of conservatism, hypermasculinity, and white culture, in opposition to

Fig. 5. Sexuality meme examples

the supposedly weak and ineffectual US. These memes may be part of Russian disinformation campaigns that reaffirm anti-LGBT+ rhetoric and build support for Russia's position among right-leaning US individuals [20].

5.2 Pro-Ukraine Memes

Coders identified three broad themes for pro-Ukraine memes: Russian Deceit (n = 15); Russian Incompetence (n = 31); Ukrainian Fortitude (n = 75). Because of the diffuse nature of the Pro-Ukraine memes compared to the pro-Russia means, fewer broad themes emerged as durable patterns within the data. We provide an overview of these categories.

Memes in the Russian Deceit theme (Fig. 6) center on how Russia and/or Putin have engaged in underhanded tactics during the conflict. For example, some memes indicate that the Russian media obscures facts about Ukraine, and the Russian army is *peacefully* occupying parts of Ukraine. Such memes build on evidence of Russian disinformation campaigns, dispelling notions of Russia being a altruistic entity.

Memes in the Russian Incompetence theme (Fig. 7) center on how Russia and/or Putin are inept and are unable to organize their attempted occupation of Ukraine. Such memes denote how Putin has endangered the Russian economy and is not providing leadership to Russian troops. These memes echo reports of Russian economic collapse and ineffectual Russian military movements, providing a concise version of news reports regarding certain aspects of the conflict. We suggest that news organizations or similar stakeholders use memes to present the latest news on the conflict in a simple format, suitable for younger people or those with limited English language proficiency.

Fig. 6. Russian deceit meme examples

Fig. 7. Russian incompetence meme examples

Memes in the Ukrainian Fortitude theme (Fig. 8) are about the bravery and resourcefulness exhibited by Ukrainians and Zelenskyy. Such memes focus on sub-themes such as Ukrainian farmers appropriating Russian tanks, the Ghost of Kyiv (folk hero who destroyed numerous Russian aircraft), and Zelenskyy's

THE UKRAINIAN PEOPLE
AFTER DESTROYING
A RUSSIAN TANK

CAPTAIN UKRAINE

It ain't much, but it's honest work

Fig. 8. Ukrainian fortitude meme examples

courage in the face of Russian occupation. Such memes may boost morale among Ukrainians, aiding Ukrainian mental health and wellbeing during the conflict.

6 Discussion

6.1 Implications of Findings

Our goal was to categorize memes involving the Ukraine conflict into pro-Russia and pro-Ukraine categories, and then group memes into broad themes within these categories. The dataset curated can be used to build pro-Russia and pro-Ukraine meme classifiers. A strength of our work is our systematic annotation strategy. The systematic strategy we employed suggests the veracity of our findings and we hope that results can add to research and policy around limiting disinformation around the Ukraine conflict and bolstering support for Ukrainians. Results detailed that pro-Russia memes tend to fall into specific thematic categories that seek to undermine specific elements of US and their allies' policy and culture. Pro-Ukraine memes are far more diffuse thematically, highlighting admiration for Ukraine's people and its leadership. Our findings are supported by previous research, around Russian disinformation campaigns [24] and attempts by Ukrainians to build support for their cause and improve morale among Ukrainians affected by the conflict. However, previous work does not explore the range of memes around the conflict, instead centering on other forms of media, and does not explore how memes are used to build support for Russia or Ukraine. We expand on previous research, providing an overview of memes around the Ukraine conflict.

6.2 Recommendations

Key to mitigating pro-Russia memes, which are likely part of Russian intelligence operations, are targeted efforts that focus on the underlying themes of such memes. Pro-Ukraine memes with specific themes may counter pro-Russia memes more effectively than more generic pro-Ukraine memes detailed in our findings. For example, stakeholders can develop memes around how support for Ukrainian refugees is leading to debate around refugee policy [26], or memes indicating the National Socialist leanings of the Azov Regiment are exaggerated. In line with recent work [5,6], we suggest that where possible, Ukrainians affected by the conflict should be consulted around development of media-based efforts to counter pro-Russia influence campaigns, co-creating work [16]. For example, stakeholders developing memes to counter influence operations can discuss strategies with Ukrainians active in meme development, to ensure memes are effective, culturally responsive, and trauma informed. Such strategies may not only mitigate effects of influence operations, but also improve Ukrainian mental health and bolster public support for Ukrainians.

6.3 Limitations

Our findings relied on the validity of data collected with our search terms, and there may be memes which did not include our search terms. Our data may not be generalizable to non-English language memes around the Ukrainian conflict. We will include non-English terms in future work. We were unable to develop multimodal classifiers for pro-Russia or pro-Ukraine memes due to limited data and the limitations of computational tools. Given recent advancements in multimodal zero-shot and few-shot classification, we hope that future techniques will allow us to build meme classifiers with limited data, allowing us to respond swiftly to influence operations at the start of crises.

Acknowledgments. We thank the editor and reviewers for their comments. This study was funded with a grant from the The Whitney and Betty MacMillan Center for international and Area Studies at Yale, and Yale Fund for Lesbian and Gay Studies Research Awards.

References

1. Bayoumi, M.: They are 'civilised' and 'look like us': the racist coverage of Ukraine. The Guardian, 2 March 2022
2. Beskow, D.M., Carley, K.M.: Characterization and comparison of Russian and Chinese disinformation campaigns. In: Shu, K., Wang, S., Lee, D., Liu, H. (eds.) Disinformation, Misinformation, and Fake News in Social Media. LNSN, pp. 63–81. Springer, Cham (2020). https://doi.org/10.1007/978-3-030-42699-6_4
3. Buts, J.: Memes of Gandhi and mercury in anti-vaccination discourse. Media Commun. 8(2), 353–363 (2020)

4. Chen, E., Ferrara, E.: Tweets in time of conflict: a public dataset tracking the Twitter discourse on the war between Ukraine and Russia. arXiv preprint arXiv:2203.07488 (2022)
5. Chen, K., et al.: Partisan US news media representations of syrian refugees. arXiv preprint arXiv:2206.09024 (2022)
6. Chen, K., et al.: How is vaping framed on online knowledge dissemination platforms? arXiv preprint arXiv:2206.10594 (2022)
7. Deshpande, T., Mani, N.: An interpretable approach to hateful meme detection. In: Proceedings of the 2021 International Conference on Multimodal Interaction, pp. 723–727 (2021)
8. Kushnir, O.: Seven truths of Russian neo-imperialism: unceasing expansion. In: Forum for Ukrainian Studies. Canadian Institute for Ukrainian Studies (2022)
9. Kuttab, D.: Palestine and ukraine: Exposing the double standard. Palestine-Israel J. Polit. Econ. Cult. **27**(1/2), 99–105 (2022)
10. Lee, R.K.W., Cao, R., Fan, Z., Jiang, J., Chong, W.H.: Disentangling hate in online memes. In: Proceedings of the 29th ACM International Conference on Multimedia, pp. 5138–5147 (2021)
11. Ludvigsson, J.F., Loboda, A.: Systematic review of health and disease in Ukrainian children highlights poor child health and challenges for those treating refugees. Acta Paediatr. **111**(7), 1341–1353 (2022)
12. Lyudmila, S., Anastasia, L., Ilya, S.: Media representation of the image of the Russian political leader in western online media (on the material daily news and der spiegel). Int. J. Media Inf. Lit. **6**(2), 396–405 (2021)
13. Magill, P., Rees, W.: UK defence policy after Ukraine: revisiting the integrated review. Survival **64**(3), 87–102 (2022)
14. Makhortykh, M., González Aguilar, J.M.: Memory, politics and emotions: internet memes and protests in Venezuela and Ukraine. Continuum **34**(3), 342–362 (2020)
15. Mbah, R.E., Wasum, D.F.: Russian-Ukraine 2022 war: a review of the economic impact of Russian-Ukraine crisis on the USA, UK, Canada, and Europe. Adv. Soc. Sci. Res. J. **9**(3), 144–153 (2022)
16. Mitchell, D.: Making Foreign Policy: Presidential Management of the Decision-Making Process. Routledge, London (2019)
17. Moss, K.: Russia's queer science, or how anti-LGBT scholarship is made. Russ. Rev. **80**(1), 17–36 (2021)
18. Park, C.Y., Mendelsohn, J., Field, A., Tsvetkov, Y.: Voynaslov: A data set of Russian social media activity during the 2022 Ukraine-Russia war. arXiv preprint arXiv:2205.12382 (2022)
19. Patel, S.S., Moncayo, O.E., Conroy, K.M., Jordan, D., Erickson, T.B.: The landscape of disinformation on health crisis communication during the COVID-19 pandemic in Ukraine: hybrid warfare tactics, fake media news and review of evidence. JCOM J. Sci. Commun. **19**(5) (2020)
20. Robinson, P.: Russia's emergence as an international conservative power. Russia Glob. Aff. **18**(1), 10–37 (2020)
21. Smart, C.: How the Russian media spread false claims about Ukrainian nazis. The New York Times (2022). https://www.nytimes.com/interactive/2022/07/02/world/europe/ukraine-nazis-russia-media.html
22. Sohl, B.: Discolored revolutions: information warfare in Russia's grand strategy. Wash. Q. **45**(1), 97–111 (2022)

23. Suryawanshi, S., Chakravarthi, B.R., Arcan, M., Buitelaar, P.: Multimodal meme dataset (MultiOFF) for identifying offensive content in image and text. In: Proceedings of the Second Workshop on Trolling, Aggression and Cyberbullying, pp. 32–41 (2020)
24. Švedkauskas, Ž., Sirikupt, C., Salzer, M.: Russia's disinformation campaigns are targeting African Americans. Washington Post, 24 July 2020
25. Tolz, V., Teper, Y.: Broadcasting agitainment: a new media strategy of Putin's third presidency. Post-Sov. Aff. **34**(4), 213–227 (2018)
26. Venturi, E., Vallianatou, A.I.: Ukraine exposes Europe's double standards for refugees. Expert Comment Chatham House (2022)
27. Wasike, B.: Memes, memes, everywhere, nor any meme to trust: examining the credibility and persuasiveness of COVID-19-related memes. J. Comput.-Mediat. Commun. **27**(2), zmab024 (2022)
28. Wiggins, B.E.: Crimea river: directionality in memes from the Russia-Ukraine conflict. Int. J. Commun. **10**, 35 (2016)
29. Zhu, J., Lee, R.K.W., Chong, W.H.: Multimodal zero-shot hateful meme detection. In: 14th ACM Web Science Conference 2022, pp. 382–389 (2022)

Analyzing Scientometric Indicators of Journals and Chief Editors: A Case Study in Artificial Intelligence (AI) Domain

Salim Sazzed[✉]

Old Dominion University, Norfolk, VA 23529, USA
ssazz001@odu.edu

Abstract. This study investigates the relationship between the rankings of artificial intelligence (AI) journals and the chief editor's scholarly reputations by mining various scientometric data. The associations between these two types of entities are studied with respect to the top AI journals (selected based on Google Scholar ranking) and journals from various quartiles (based on Scimago quartile ranking). Three quantitative reputation metrics (i.e., citation count, h-index, and i10-index) of editor-in-chief (EiC) and four journal ranking metrics (i.e., h5-index, h5-median, the impact factor (IF), and Scimago Journal Rank (SJR)) of journals are considered to find any relationships. To determine the correlation between various pairs of scholarly metrics of EiC and top AI journals, we employ the Spearman and Kendall correlation coefficients. Furthermore, we investigate whether machine learning (ML) classifiers can predict the SJR and IF of journals utilizing EIC's scholarly reputation metrics. It is observed that the comparative rankings (based on various metrics) of top AI journals do not correlate with the EiC's scholarly achievements. The high prediction errors of ML classifiers indicate that the EiC's scholarly indices are not comprehensive enough to build a good model for predicting the IF or SJR of top AI journals. Nevertheless, when AI journals of various qualities are analyzed, we observe that Q1 journals usually have EiCs with a much higher number of citations and h-index compared to the EiCs from the journals from the bottom two quarterlies (Q3 and Q4). The Mann-Whitney U test indicates the differences between the scholarly metrics of EiCs of Q1 journals and journals from Q3 and Q4 are significant. The results imply that while selecting the EiC of a journal, scientometric indices should be considered prudently.

1 Introduction

The research related to the assessment of scientific impact and contributions of various entities such as articles, authors, journals, and institutions has become a key research area due to its importance for scientific knowledge dissemination [6]. The journal's reputation has high significance in scientific research as the initial impression regarding a newly published article is often influenced by

T. N. Dinh and M. Li (Eds.): CSoNet 2022, LNCS 13831, pp. 39–50, 2023.
https://doi.org/10.1007/978-3-031-26303-3_4

the recognition of the publishing journal [17]; though, researchers also warned against such kind of conclusion [13]. Based on Walter [23], scientometric metrics such as total citations, IF, and h-index along with some other criteria can quantitatively evaluate the rank of a journal. The Editor-in-Chief (EiC) usually plays the utmost role in final decision making considering the feedback of reviewers [9,20]. Scholarly journals hope to appoint academically excellent scholars as editors who suit the journal's orientations and development [9].

As the EiC plays a key role in the journal's academic impact, in this work, we investigate the relationship between the scholarly reputation indices of EiC and the journal's citation metrics. In particular, we try to provide insight into the below research questions-

- RQ1: To what extent does the scholarly reputation of the EiC correlate with the scientometric indices of top AI journals (based on Google Scholar)?
- RQ2: Can ML classifiers predict the SJR and IF of top AI journals correctly from the EiC's scholarly reputation?
- RQ3: Are there any noticeable differences in the EiC's scholarly metrics in journals from the top quartile (Q1) and bottom quartiles (Q3 or Q4)?

We select the top Artificial Intelligence (AI) journals based on the Google Scholar (GS) ranking. Note that the GS top AI journal list substantially overlaps with the Scopus-based Scimago top journal list. To find journals of various qualities, we use the Scimago journal ranking, which ranks journals based on the quarterlies they represent for a particular research area. As the journal quantitative metrics, we consider the h5-index, h5-median, impact factor, and Scimago Journal Rank (SJR) score. As the scholarly metrics of the EiC, we utilize the total citation counts, h-index, and i-10 index. Spearman's ρ and Kendall's τ correlation coefficients are utilized to find the presence of any correlation. Besides, we employ ML classifiers to predict the quantitative metrics (i.e., IF and SJR) of the journals. We find none of the Spearman and Kendall correlation coefficients show any correlation between the journal and EiC's scientometric measures among the top AI journals. Also, we observe LR and SVM show high RMSE and R-squared errors that indicate the limitation of EiC's scholarly metrics for estimating the relative rankings of top AI journals. Yet, when we compare journals with diverse qualities (quartile rank), we observe that the EiC scholarly indices of journals of the top quartile (Q1) are much higher compared to Q3 and Q4 journals.

1.1 Motivations and Contributions

Existing works primarily considered the scholarly reputation of entire editorial boards to find any relationship with the journal metrics. Since the editorial board usually consist of editors and associated editors with different levels of scholarly achievements, it is hard to characterize the importance of the EiC's scholarly reputation exclusively for the journal ranking. Moreover, until now, no study investigated the relationship between AI journal rankings and the quantitative scholarly metrics of the editorial boards. The main contributions of this study can be summarized as follows-

- In this paper, we investigate the correlation between the EiC's scholarly metrics and the various scientometric indices of top AI journals and AI journals of various qualities.
- We train ML models utilizing EiC reputation metrics as input features to predict the SJR and IF of top AI journals.
- We reveal that the scholarly indices of EiC do not correlate well with the comparative ranking of the top AI journals.
- We show that, however, the scholarly reputation of the EiC is reflected in the journal quartile rank in most cases.

2   Related Work

A number of studies investigated various aspects of the journal editorial board. Zdenek and Lososova [25] analyzed the productivity (i.e., publication output) of editorial board members of agricultural economics and policy journals. Hardin et al. [8] analyzed the research productivity of leading financial journals. A number of attributes of the editorial boards, such as geography [7], gender [16], and institutional affiliation [5], have been investigated to find any relation with the rankings of the journal. Bedeian et al. [2], in their study, considered various metrics such as normalized total articles, adjusted quality index, and h-index score to determine the scientific achievement of editorial board members. Their results advocated that the scholarly metrics of the researchers should be given high importance while selecting editors. A similar study was conducted by Lowe and Van Fleet [11] for the top accounting journals. The authors observed no correlation between the impact factors and the scholarly achievement of the editorial board members.

Pagel et al. [18] found for the Anesthesia journals, a positive correlation exists between the median h-index of editorial boards and the journal's impact factor. A similar study was performed by Kay et al. [10] for the top sports medicine journals; The authors measured the correlations considering various aspects such as sex, country, degree, and faculty status.

The association between the EiC's reputation indicators and the ranking of the top Bioinformatics journal has been investigated in [21]. The author found no positive correlation between the EiC indices and the relative ranks of top journals. Asnafi et al. [1] analyzed the h-indices of editorial board members of Radiology journals to find whether the hypothesis 'editorial board members of the Radiology journals with higher impact factors (IF) have higher h-indices' is true. They noticed that various quantitative metrics such as h-indices, total publications, and citations (both total and average) of editorial board members of the above-median journals are higher than the below-median journal. Mendonça et al. [15] found an association between the research productivity (h-index) of editorial boards and the performance of top African studies journals. Valderrama et al. [22] performed regression analysis to determine the Dentistry, Oral Surgery, and Medicine journals rankings from various metrics. Mazov and Gureev [14] conducted a literature review focusing on the publications that studied various demographic aspects (e.g., geographic, linguistic, and gender distributions) of the editorial board of the journals.

For economics journals, Wu et al. [24] noticed a significant positive relation between institution-based diversity and the impact factor for non-US-based journals.

3 Data Description

3.1 Journal Selection

This study concentrates on two research questions; the first task performs a correlation analysis of the journal and EiC scientometric data for the top AI journals (based on GS ranking). The second task analyzes statistical differences of the EiCs scholarly metrics in journals of various quarterlies such as Q1, Q3, and Q4 based on SCGImago ranking. Thus, we collect two sets of data.

Top AI Journals. The top AI journals are selected from Google Scholar (GS) publication ranking. GS lists the top 20 publication venues (conferences and journals) of a research domain based on the h5-index. Among the top 20 AI venues, 13 journals are included[1]. We find no GS profile of EiC for two journals; thus, we exclude them.

AI Journals from Various Quartiles. The AI journals belonging to various quartiles (i.e., Q1, Q3, and Q4) are collected from the Scimago website[2] in August 2021. We notice some AI journals are also listed in different categories, such as *Computer Science Applications*, with higher quartile ranks. We exclude these journals having a higher quartile rank for any non-AI category, as they do not exclusively represent a particular quartile. Similar to GS ranking, we exclude journals for which we do not find the GS profile of EiC. Finally, the Q1 list contains 16 journals, while Q3 and Q4 lists contain 7 and 6 journals, respectively[3]. Note that only a limited number of Q3 (only 7) and Q4 (only 6) journals are available for the comparison (based on Scimago ranking) since most of the Q3 or Q4 AI journals have better quartile ranks in other research areas, which means they do not represent Q3 or Q4 category exclusively.

3.2 Quantitative Metrics of EiC and Journals

Scholarly Reputation Indices of EiC: In this study, the following three scholarly metrics of the EiC are studied as indicators of the scholarly reputation-

- total citation: The total citation reflects the number of times a researcher's works have been cited by other researchers.

---

[1] https://github.com/sazzadcsedu/AIJournalList.
[2] https://www.scimagojr.com/.
[3] https://github.com/sazzadcsedu/AIJournalList.

– h-index: h-index of a researcher refers that the scholar has h number of articles with at least h number of citations for each.
– i10-index: i10-index of a scholar refers to the number of publications the scholar has with minimum of 10 citations.

For journals having more than one EiC (e.g., some journals may have two or three EiCs), we take the average of the values of their scholarly indices (i.e., citations, h5 index, and i10 index). Table 1 shows statistics of various scholarly indices of EiC.

Table 1. Statistics of the scholarly metrics of EiC in top AI journals

Metric	Scholarly Index		
	citation	h-index	i10-index
Mean	25165.19	63.57	243.09
Median	14814	63	199
STD	28753.40	27.23	159.92

Journal Ranking Metrics: The journal performances are analyzed based on the following quantitative metrics,

– h5-index: h5-index is the h-index of journal calculated based on the recent 5 years.
– h5-median: It is the median value of h5-index.
– Impact factor (IF): The journal IF represents the yearly average number of citations based on the total number of citations and the articles published in the preceding two years.
– SJR index: ISJR takes the reputation of the source of citation into account in addition to the number of citations a journal receives. Besides, SJR considers the last 3 years instead of 2 years to calculate the weighted average for the current year.

3.3 Data Collection

In this paper, we refer EiC to the scholar who leads a team of editors and associate editors of a journal. Using the author search option of GS[4], we obtain EiC's scholarly metrics. However, it is not uncommon to have multiple researchers with the same name. Thus, we manually check those author profiles to find their affiliation and identify the author with the correct affiliation. Then we collect the total number of citations, h-index, and i10-index of the EiC.

---

[4] https://scholar.google.com/.

We retrieve the h5-index and h5-median of the journals from GS. The IF is obtained from the journal website, while the SJR is extracted from the Scimago website[5]. Table 2 shows various scientometric indices of top AI journals.

Table 2. Scientometric indices of GS-based top AI journals

Metric	Scientometric Index			
	h5-index	h5-median	SJR	IF
Mean	81.36	117.36	1.71	5.22
Median	85	121	1.494	5.452
STD	21.59	29.28	0.94	2.31

4 Correlation Analysis

We utilize two correlation measures, the Spearman rank correlation coefficient (ρ) and Kendall rank correlation coefficient (τ), to compute the association between various EiC's scholarly reputation metrics and the journal's scientometric indices.

4.1 Spearman Rank-Order Correlation Coefficient

The Spearman rank-order correlation coefficient computes the monotonicity between two sets of observations. The Spearman's ρ values could be within the range of $[-1, +1]$. Let X and Y two variables with n number of observations of $X_1, ..., X_n$ and $Y_1, ...Y_n$, respectively. The ranks of $X_1, ..., X_n$ are represented by $R_x(X_1, ..., X_n)$ and ranks of $Y_1, ...Y_n$ are represented by $R_y(Y_1, ..., Y_n)$. Spearman's ρ is defined as,

$$\rho = 1 - \frac{6\sum r_i^2}{n(n^2 - 1)} \tag{1}$$

where, r_i is the difference between the ranks of corresponding observations of each set and equals to $R_x(X_i) - R_y(Y_i)$.

4.2 Kendall Rank Correlation Coefficient

Similar to the Spearman, Kendall's τ coefficient computes the correlation between two sets of variables with an output value within the range between $[-1, +1]$, where $+1$ and -1 represent maximum correlation in the positive and negative directions, respectively (0 means no correlation at all). However, unlike Spearman, Kendall's τ is a non-parametric measure as it is not dependent on the distributions of the variable(s).

---

[5] https://www.scimagojr.com/.

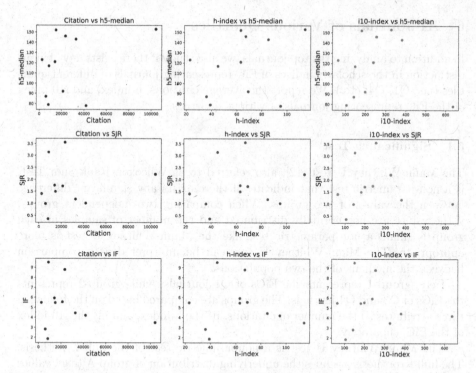

Fig. 1. Plots of various metrics of top AI journals and corresponding EiCs scholarly indices

4.3   Pairs Used for Correlation Analysis

We compute the correlation values for nine pairs of quantitative metrics, $(x_1, y_1), (x_2, y_2), \ldots, (x_9, y_9)$, where x_i represents a quantitative metric of the EiC and y_i refers to a quantitative metric of the journal, and $i = 1 \ldots 9$. The nine pairs are: (citation, IF), (citation, SJR), (citation, h5-median), (h-index, IF), (h-index, SJR),(h-index, h5-median), (i10-index, IF), (i10-index, SJR), (i10-index, h5-median).

5   Regression Analysis

We investigate whether incorporating multiple attributes can yield an effective predictive model. Utilizing various EiC scholarly indices (e.g., citation, h-index, and i10 index) as input features, we train the ML models. We employ linear regression (LR) and support vector regression (SVR) for predicting SJR and IF. The default parameter settings of LR and SVR model of scikit-learn library [19] is used. As the performance indicators of the ML classifiers, we consider both Root Mean Squared Error (RMSE) and R-squared (denoted as R^2) errors. The Leave-one-out cross-validation (LOOCV) is used to assess the performance of various ML classifiers.

6   AI Journals of Various Quartiles

In addition to analyzing the top journals, we also study if there exists any obvious distinction in the scholarly indices of EiC representing journals of different quartiles (i.e., Q1, Q4, etc.). We report the average citations, h-index, and i10-index of the EiC representing journals of various quartiles.

6.1   Significance Test

The Mann-Whitney U test [12], also referred to as Wilcoxon Rank Sum Test, is a non-parametric test that indicates if there exists any significant difference between the values of two groups. When comparing two independent groups, if the values are not normally distributed and the number of samples in each group is small, a non-parametric test like the Mann-Whitney U test is more appropriate. The Mann-Whitney U test can be interpreted as a comparison between the medians of the two populations.

Here, group-1 represents the EiCs of Q1 journals, while group-2 represents the EiCs of Q3 and Q4 journals. The groups are compared based on the following three attributes: i) the number of citations, ii) the h-index, and iii) the i10-index of the EiC, separately.

The Mann-Whitney U test is a nonparametric test of the null hypothesis. The null hypothesis assumes the underlying distribution of group A (e.g., values from group-1 representing EiC of Q1 journals) is the same as the distribution of group B (e.g., values from group-2 representing EiC of Q3 and Q4 journals).

The Mann-Whitney U test assumes the following conditions:

- All the observations from both groups are independent of each other.
- The responses are at least ordinal.
- Under the null hypothesis H_0, the distributions of both populations are equal.
- The alternative hypothesis H_1 is that the distributions are not equal.

7   Results

We observe that none of the nine scientometric pairs (where one entity represents an EiC metric and the other entity represents a journal metric) shows any kind of correlation for the top AI journals (Table 3). The highest correlation coefficient is found between EiC's citation and the journal's h5-median by the Spearman coefficient, which is 0.327. Nevertheless, the high p-value of 0.325 associated with it indicates that the coefficient score is not significant.

Table 3. Correlation scores (with p-values) between various indices of top AI Journal and EiC's scholarly metrics

Correlation coefficient	Metric of EiC	Journal Metric		
		h5-median	IF	SJR
Spearman	citation	0.327 (0.325)	−0.027 (0.936)	0.309 (0.355)
	h-index	0.264 (0.432)	−0.031(0.925)	0.282 (0.400)
	i10-index	0.200 (0.555)	0.045 (0.894)	0.245 (0.466)
Kendall	citation	0.236 (0.358)	−0.127 (0.648)	0.163 (0.542)
	h-index	0.183 (0.434)	−0.146 (0.532)	0.146 (0.532)
	i10-index	0.2 (0.44)	0.054 (0.879)	0.155 (0.445)

Table 4. Performances of regression models for predicting SJR and IF of top AI journals

Journal Metric	Feature	LR		SVR	
		RMSE	R^2	RMSE	R^2
SJR	citation, h-index	1.543	−1.925	0.980	−0.181
	citation, i10-index	1.596	−2.131	0.980	−0.180
	h-index, i10-index	1.060	−0.381	0.957	−0.126
	citation,h-index,i10-index	1.079	−0.431	0.9874	−0.197
IF	citation, h-index	3.274	−1.196	2.253	−0.040
	citation, i10-index	4.039	−2.342	2.253	−0.040
	h-index, i10-index	3.093	−0.960	2.245	−0.033
	citation, h-index, i10-index	2.525	−0.306	2.256	−0.042

Table 5. Statistics of various scientometric indices of EiC of journals from various quartiles

Journal Quartile Rank	EiC Scholarly Metric		
	citation	h-index	i10-index
	avg./med./min./max	avg./med./min./max	avg./med./min./max
Q1	36603.59/18612/1444/220690	73.75/66/21/186	304.94/219/39/636
Q3	6946.333/6746/1102/11554	34.66/37/16/39	94.11/91/26/148
Q4	7272.71/2627/313/31075	31.14/25/8/78	112.29/68/6/359

Both LR and SVR fail to create a predictive model for SJR and IF calculation. ML classifiers yield high prediction errors (i.e., RMSE values) regardless of the input features used for training (Table 4). Particularly, for the IF prediction, both LR and SVR provide an RMSE value between 2.25 to 4.03, which is around 1 to 2 standard deviations (std.) away from the true value. Furthermore, a negative R^2 value is observed which demonstrates that the data is not comprehensive enough to fit the ML model.

Table 5 shows various statistics such as average (avg.), median (med.), minimum (min.), and maximum (max.) values of the EiC's scholarly metrics of journals from various quartiles. The Mann-Whitney U test indicates a significant difference between both groups based on all three measures. The p-values below 0.05 are observed for those measures, which suggests both groups are significantly different with respect to various measures.

8   Discussion

As seen by Table 3 and 4, for the top AI journals, no positive correlation exists between the scientometric indices of EiC and the journal rankings. When two top AI journals are compared, the journal with the higher IF or SJR does not necessarily have EiC with higher citations or h-index. Note that the GS top AI journal list is very similar to the Scopus-based Scimago top AI journal list (Q1). Scopus is a well-regarded abstract and citation database. Since the top AI journals considered here are listed in both databases (GS and Scoups), they are true representatives of top AI journals.

Several factors may contribute to the above-mentioned observations. For example, the journal citation count depends on a number of factors, such as publication models (e.g., subscription-based, open access) and article type (e.g., research article, review article). Usually, review articles get a higher number of citations than the other type, thus, may boost the citation count of the journals. The journals having a much broader scope than domain-specific journals often get more citations. Besides, the publication model (e.g., open access), number of issues per year, and number of researchers in particular fields may influence a journal's citation count [3, 4].

Similarly, the research areas and venues of publication of the EiC can have an impact on the total number of citations accumulated (assuming a similar number of years in a research career). Publications on hot research topics often bring more citations than the already saturated research topics.

More importantly, the publishers (e.g., Elsevier, Springer, ACM, IEEE) may have preferable measures for recruiting editors for the journals. The quantitative metrics of EiC may not be the only deciding factor; Rather, publishers may look for researchers who can serve the journal and community best.[6].

The data representing the scholarly reputation of the EiCs (e.g., citation, h-index) reveal that EiCs of the top-ranked AI journals are leading researchers in their research domains. We find the citation counts of EiCs range from around 2500 to over 100000, with an average and median citation count of 25165 and 14814, respectively.

Although the limitations of the quantitative metrics for the granular level analysis are revealed when top AI journals are considered, we find that when journals from different quartiles are investigated, the quantitative metrics align

---

[6] https://www.elsevier.com/connect/how-do-publishers-choose-editors-and-how-do-they-work-together.

with the assumption that the higher-ranked journals have EiCs with better scientific reputations than EiCs of journals from the bottom quartiles. For example, when the top (Q1) and bottom quartiles (Q3 and Q4) are considered, we see substantial differences in the average and median values of various scholarly metrics of the EiC (shown in Table 5). For example, the median citations of EiC of Q1 journals are 18612, while for Q3 and Q4 journals, it is much lower, 6746 and 2627, respectively. The other scientometric indices, such as the h-index, also suggest that the EiC of Q1 journals are more established researchers than the EiC of Q3 or Q4 journals. The results indicate that although scientometric indices have limitations, they are still useful for assessing the overall quality of journals in a broader scope.

9 Summary

This study analyzes the connection between various scientometric indices of the EiC and AI journals. We do not notice any positive correlation between the scholarly indices of EiC and the ranking of top AI journals considering the IF, h5-median, or SJR. We train multiple ML models employing various EiC scientometric indices as input features; however, we discover that ML models cannot estimate the SJR and IF satisfactorily. The high prediction errors of ML models suggest that the EiC's scholarly metrics are inadequate indicators of top AI journal rankings. However, when journals from various qualities are investigated, it is observed that, in general, scholarly metrics are significantly higher for the EiCs of Q1 journals than the EiCs of Q3 and Q4 journals.

References

1. Asnafi, S., Gunderson, T., McDonald, R.J., Kallmes, D.F.: Association of h-index of editorial board members and impact factor among radiology journals. Acad. Radiol. **24**(2), 119–123 (2017)
2. Bedeian, A.G., Van Fleet, D.D., Hyman III, H.H.: Scientific achievement and editorial board membership (2009)
3. Bornmann, L., Daniel, H.D.: What do citation counts measure? A review of studies on citing behavior. J. Doc. **64**(1), 45–80 (2008)
4. Bornmann, L., Marx, W., Gasparyan, A.Y., Kitas, G.D.: Diversity, value and limitations of the journal impact factor and alternative metrics. Rheumatol. Int. **32**(7), 1861–1867 (2012)
5. Burgess, T.F., Shaw, N.E.: Editorial board membership of management and business journals: a social network analysis study of the financial times 40. Br. J. Manag. **21**(3), 627–648 (2010)
6. Cai, L., et al.: Scholarly impact assessment: a survey of citation weighting solutions. Scientometrics **118**(2), 453–478 (2019)
7. García-Carpintero, E., Granadino, B., Plaza, L.: The representation of nationalities on the editorial boards of international journals and the promotion of the scientific output of the same countries. Scientometrics **84**(3), 799–811 (2010)

8. Hardin, W.G., Liano, K., Chan, K.C., Fok, R.C.: Finance editorial board membership and research productivity. Rev. Quant. Financ. Account. **31**(3), 225–240 (2008)
9. Herteliu, C., Ausloos, M., Ileanu, B.V., Rotundo, G., Andrei, T.: Quantitative and qualitative analysis of editor behavior through potentially coercive citations. Publications **5**(2), 15 (2017)
10. Kay, J., et al.: The h-index of editorial board members correlates positively with the impact factor of sports medicine journals. Orthop. J. Sports Med. **5**(3), 2325967117694024 (2017)
11. Lowe, D.J., Van Fleet, D.D.: Scholarly achievement and accounting journal editorial board membership. J. Account. Educ. **27**(4), 197–209 (2009)
12. Mann, H.B., Whitney, D.R.: On a test of whether one of two random variables is stochastically larger than the other. Ann. Math. Stat. 50–60 (1947)
13. Marks, M.S., Marsh, M., Schroer, T.A., Stevens, T.H.: Misuse of journal impact factors in scientific assessment (2013)
14. Mazov, N., Gureev, V.: The editorial boards of scientific journals as a subject of scientometric research: a literature review. Sci. Tech. Inf. Process. **43**(3), 144–153 (2016)
15. Mendonça, S., Pereira, J., Ferreira, M.E.: Gatekeeping african studies: what does "editormetrics" indicate about journal governance? Scientometrics **117**(3), 1513–1534 (2018)
16. Metz, I., Harzing, A.W., Zyphur, M.J.: Of journal editors and editorial boards: who are the trailblazers in increasing editorial board gender equality? Br. J. Manag. **27**(4), 712–726 (2016)
17. Mingers, J., Yang, L.: Evaluating journal quality: a review of journal citation indicators and ranking in business and management. Eur. J. Oper. Res. **257**(1), 323–337 (2017)
18. Pagel, P., Hudetz, J.: An analysis of scholarly productivity in united states academic anaesthesiologists by citation bibliometrics. Anaesthesia **66**(10), 873–878 (2011)
19. Pedregosa, F., et al.: Scikit-learn: machine learning in python. J. Mach. Learn. Res. **12**, 2825–2830 (2011)
20. Primack, R.B., et al.: Are scientific editors reliable gatekeepers of the publication process? (2019)
21. Sazzed, S.: Association between the rankings of top bioinformatics and medical informatics journals and the scholarly reputations of chief editors. Publications **9**(3) (2021). https://doi.org/10.3390/publications9030042, https://www.mdpi.com/2304-6775/9/3/42
22. Valderrama, P., Escabias, M., Jiménez-Contreras, E., Rodríguez-Archilla, A., Valderrama, M.J.: Proposal of a stochastic model to determine the bibliometric variables influencing the quality of a journal: application to the field of dentistry. Scientometrics **115**(2), 1087–1095 (2018)
23. Walters, W.H.: Citation-based journal rankings: Key questions, metrics, and data sources. IEEE Access **5**, 22036–22053 (2017)
24. Wu, D., Lu, X., Li, J., Li, J.: Does the institutional diversity of editorial boards increase journal quality? The case economics field. Scientometrics **124**(2), 1579–1597 (2020)
25. Zdeněk, R., Lososová, J.: An analysis of editorial board members' publication output in agricultural economics and policy journals. Scientometrics **117**(1), 563–578 (2018)

Link Prediction of Complex Networks Based on Local Path and Closeness Centrality

Min Li[1,3] , Shuming Zhou[2]([📧]) , and Gaolin Chen[1,2]

[1] College of Computer and Cyber Security, Fujian Normal University,
Fuzhou, Fujian 350117, People's Republic of China
[2] College of Mathematics and Informatics, Fujian Normal University,
Fuzhou, Fujian 350117, People's Republic of China
zhoushuming@fjnu.edu.cn
[3] Concord University College, Fujian Normal University,
Fuzhou, Fujian 350117, People's Republic of China

Abstract. Due to evolving nature of complex network, link prediction plays a crucial role in exploring likelihood of new relationships among nodes. There exist a great number of techniques to apply the similarity-based metrics for estimating proximity of vertices in the network. In this work, a novel similarity-based metric based on local path and closeness centrality is proposed, and the Local Path and Centrality based Parameterized Algorithm (LPCPA) is suggested. The proposed method is a new variant of the well-known index of Common Neighbor and Centrality based Parameterized Algorithm (CCPA). Extensive experiments are conducted on thirteen real networks originating from diverse domains. The experimental results indicate that the proposed index improves the prediction accuracy measured by AUC and has achieved a competitive result on Precision compared to the existing state-of-the-art link prediction methods.

Keywords: Link prediction · Local paths · Closeness centrality · Complex networks

1 Introduction

Link prediction in complex networks aims at estimating the existence likelihood of a link between two nodes. Finding missing links in an observed network or predicting the future links is the main task of link prediction issue. So far, this problem has attracted growing interests from diverse disciplines. Some of its

Supported by the National Natural Science Foundation of China (Nos. 61977016 and 61572010), Natural Science Foundation of Fujian Province (Nos. 2020J01164, 2017J01738) and Education and Scientific Research Project for Young and Middle-aged Teachers of Fujian Province (No. JAT191119).

T. N. Dinh and M. Li (Eds.): CSoNet 2022, LNCS 13831, pp. 51–63, 2023.
https://doi.org/10.1007/978-3-031-26303-3_5

applications include protein-protein interaction in biological networks [1], trade relation finding in international trade [2], recommender systems [3] for friends, products, movies, etc. on different online platforms, and so on.

Recently, many link prediction methods have been implemented under different backgrounds. These methods can be roughly divided into several groups, such as similarity-based methods [4], probabilistic models [5], maximum likelihood methods [6], embedding-based methods [7], matrix factorization-based methods [8], learning-based methods [9], etc. As a major approach for link prediction, similarity-based methods have received considerable attention from researchers due to their simplicity and effectiveness. The structure-based similarity methods always assign similarity scores to node pairs with the aid of the topology features of networks, which can be further classified into three categories: local methods such as the common neighbors [10], global methods such as Katz index (Katz [19]), quasi-local methods such as local path index (LP) [20]. Local similarity metrics are widely used due to their simplicity, low complexity and competitive performance. However, their performance is not very robust for various networks. Global similarity metrics may present more effective and higher performance, while they are usually time-consuming and only suitable for small networks. Quasi-local metrics achieve a tradeoff between local and global metrics, which extract both local and global topology information from the networks. They are believed to be a good choice for link prediction problem. Thus, how to design a novel quasi-local method by combing the local and global structure information of the network is an important issue deserved to study, which is still a major challenge in link prediction problem.

Motivated by the above discussions, in this paper, we propose a new quasi-local method named "Local Path and Centrality based Parameterized Algorithm" (LPCPA), which employs the number of 3-hop paths based on "Common neighbor and Centrality based Parameterized Algorithm" (CCPA). On the one hand, we investigate the prediction accuracy effect of choosing β which controls the proportion of the number of 3-hop paths. On the other hand, we compare the proposed method with state-of-the-art methods and show the superiority of the proposed method in terms of the metric of AUC.

The rest of the paper is organized as follows. First, we briefly review the relevant studies on link prediction in Sect. 2, and then describe the proposed method in Sect. 3. Next, experimental results are discussed in Sect. 4. Finally, Sect. 5 concludes our work.

2   Related Works

Inspired by the practical signifcance of link prediction, many kinds of link prediction algorithms have been suggested for undirected networks. In this section, a briefly review of structure-based similarity link prediction methods is presented. The most commonly used network structure includes common neighbors, paths, and triangles. These methods can be divided into three categories: local, quasi-local and global.

In local methods, more attention is paid to the direct neighbor information. The CN (Common Neighbors) index [10] is the simplest one, which calculates the similarity between two nodes by counting the number of their shared neighbors. The Jaccard index (Jaccard) [11] is a normalization of CN index. The AA (Adamic-Adar) [12] and RA (Resource Allocation) [13] emphasize that the contribution of a shared neighbor to the similarity is inversely proportional to its degree. Obviously, CN, AA and RA don't take the degree of two end nodes into consideration. The PA (Preferential Attachment) index [14] defines similarity between two nodes as the product of their degrees. However, the LHNL (Leicht-Holme-Newman Local Index) [15], Salton (Salton Index) [16], Sørensen (Sørensen Index) [17], HPI (Hub Promoted Index) [18], and HDI (Hub Depressed Index) [18] harbor the idea that the degree of two end nodes have negative contribution to the similarity between them. The global similarity-based indices, such as Katz Index [19], consider the topological structure information of all possible paths between non-adjacent nodes. However, global methods have lower accuracy and higher computational complexity. Quasi-local methods seek to take the advantages of both local and global methods to obtain a nice trade-off between performance and computational complexity. These metrics are comparable to local indices and their time complexity is low in comparison to the global indices. Local Path (LP) [20] falls into this category. The comparison baselines are illustrated in Table 1.

For a detailed survey of link prediction techniques, the readers are referred to [21].

3   Proposed Framework

3.1   LPCPA Similarity

Recently, the fact that 3-hop-based similarity indices perform better than 2-hop-based indices has been revealed. Pech et al. [22] assume that the likelihood of a link is a linear sum of all its neighbors' contribution and provides an analytical theory to show that the number of 3-hop paths more accurately predicts missing links than the number of 2-hop paths. Zhou et al. [23] implement extensive experimental comparisons between 2-hop-based and 3-hop-based similarity indices on 137 real networks. The statistical results show that 3-hop-based indices and 2-hop-based indices are competitive on precision, while 3-hop-based indices outperform 2-hop-based indices on ROC-AUC. Besides, Ahmad et al. [24] propose a new link prediction algorithm named "Common neighbor and Centrality based Parameterized Algorithm" (CCPA), which utilizes both the number of common neighbor and the shortest-path distance between two nodes. Motivated by the idea above, in this paper, a new strategy, namely Local Path and Centrality based Parameterized Algorithm (LPCPA), is proposed.

Definition 1. *(LPCPA Similarity)*
The similarity score, S_{xy}^{LPCPA}, between two nodes x and y is defined as:

$$S_{xy}^{LPCPA} = \alpha(A_{xy}^2 + \beta A_{xy}^3) + (1 - \alpha)\frac{N}{d_{xy}}, \tag{1}$$

where A is the adjacency matrix of the network, $d(x, y)$ is the length of the shortest path between x and y, and α and β are free parameters. Here, Dijkstra algorithm is used to compute $d(x, y)$.

3.2   Link Prediction Algorithm Based on LPCPA

Algorithm 1. Link prediction algorithm based on LPCPA

Input: $G(V, E)$, α, β.
Output: AUC and Precision.
1: Divide the original network G into trainning set G^T and test set G^P.
2: Compute the adjacency matrix A of G^T;
3: Trainning set size $N = |G^T|$;
4: **for** each non-observation edge (x, y) in G^T **do**
5:     Compute the score of 2 path length A_{xy}^2;
6:     Compute the score of 3 path length A_{xy}^3;
7:     Compute the shortest path distance d_{xy};
8:     Obtain the S_{xy}^{LPCPA} according to Eq. (4);
9: **end for**
10: Sort all the non-observation edges in descending order according their similarity
    scores;
11: Insert edges from the ordered list into G^T;
12: Compute AUC and Precision;

Consider an undirected network $G(V, E)$ without multiple links or self-connections, where V is the set of nodes and E is the set of links. The set of non-existent links is $U - E$, where U denotes all possible links in this network. To estimate the accuracy of the prediction algorithm, E is randomly divided into two sets, i.e., a training set E^T and a probe set E^P. There is no intersection between the two sets, that is, $E^T \cup E^P = E$, $E^T \cap E^P = \emptyset$. The training set E^T is used to calculate node similarity for training purpose, while the probe set E^P is treated as the set of missing links to validate the accuracy of predictors. In this work, E^T occupies 90% of the overall data set, while E^P occupies the rest 10%, and 100 independent implementations are performed to get the average results. The LPCPA index takes 3-hop paths into consideration based on CCPA. Algorithm 1 illustrates the pseudo-code of link prediction algorithm based on LPCPA. Figure 1 shows the overall framework of the solution for Algorithm 1.

As Fig. 2 shows, three simple networks are illustrated to interpret the proposed method. Our main task is to calculate the similarity score of a non-observation edge (x, y). According to Eq. (1), LPCPA involves 2-hop paths, 3-hop paths and the shortest paths. Hence, in this example, in terms of A and B, G_1

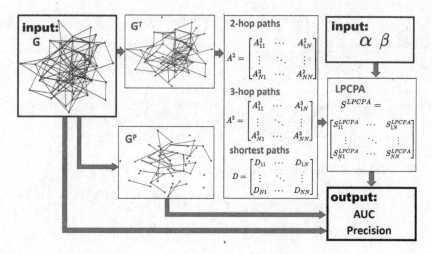

Fig. 1. A overall framework of the solution for the proposed link prediction method.

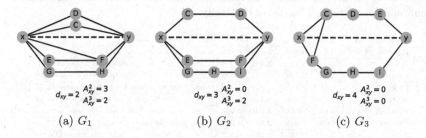

(a) G_1 (b) G_2 (c) G_3

Fig. 2. Three simple networks.

has 2-hop and 3-hop paths, G_2 only has 3-hop path, while G_3 has no 2-hop and 3-hop paths. In Fig. 2, A_{xy}^2, A_{xy}^3 and d_{xy} represents the number of 2-hop, 3-hop, the shortest paths between x and y, respectively.

Table 2 shows the different similarity scores between x and y for G_1. Here, We set ($\alpha = 0.5$) for CCPA and ($\alpha = 0.5$, $\beta = 0.1$) for LPCPA.

4   Experimental Analysis

4.1   Datasets

Table 3 shows the basic topology characters of these networks. Most of these datasets can be downloaded from the websites of http://konect.uni-koblenz.de/ and http://networkrepository.com/networks.php.

Two evaluation metrics, AUC (area under the receiver operating characteristic curve) [21] and Precision [25], are utilized for evaluating the performance of the proposed methods in comparison with baseline methods. In the algorithmic implementation, a missing link and a non-existent link are randomly selected

Table 1. Comparison baselines.

No	Similarity	Scores	No	Similarity	Scores
1	Jaccard	$S_{xy}^{Jaccard} = \frac{|\Gamma(x) \cap \Gamma(y)|}{|\Gamma(x) \cup \Gamma(y)|}$	7	Sørensen	$S_{xy}^{Srensen} = \frac{2|\Gamma(x) \cap \Gamma(y)|}{k_x + k_y}$
2	AA	$S_{xy}^{AA} = \sum_{z \in \Gamma(x) \cap \Gamma(y)} \frac{1}{\log |k_z|}$	8	HPI	$S_{xy}^{HPI} = \frac{|\Gamma(x) \cap \Gamma(y)|}{\min(k_x, k_y)}$
3	RA	$S_{xy}^{RA} = \sum_{z \in \Gamma(x) \cap \Gamma(y)} \frac{1}{k_z}$	9	HDI	$S_{xy}^{HDI} = \frac{|\Gamma(x) \cap \Gamma(y)|}{\max(k_x, k_y)}$
4	PA	$S_{xy}^{PA} = k_x \cdot k_y$	10	CCPA	$S_{xy}^{CCPA} = \alpha \cdot (A_{xy}^2) + (1 - \alpha) \frac{N}{d_{xy}}$
5	LHNL	$S_{xy}^{LHNL} = \frac{|\Gamma(x) \cap \Gamma(y)|}{k_x \cdot k_y}$	11	LP	$S_{xy}^{LP} = A^2 + \epsilon A^3$
6	Salton	$S_{xy}^{Salton} = \frac{|\Gamma(x) \cap \Gamma(y)|}{\sqrt{k_x \cdot k_y}}$	12	Katz	$S_{xy}^{Katz} = (I - \gamma A)^{-1} - I$

Table 2. The different similarity scores between x and y on G_1.

Network	CN	RA	AA	Jaccard	PA	Salton	Sorenson	HDI	HPI	LHNL	CCPA	LPCPA
G_1	3	1.33	3.8	0.5	20	0.67	0.67	0.6	0.75	0.15	3.5	3.6
G_2	0	0	0	0	9	0	0	0	0	0	1.5	1.6
G_3	0	0	0	0	4	0	0	0	0	0	1.12	1.12

Table 3. The basic characters of the real networks, where $\langle k \rangle$ denotes the average degree, $\langle d \rangle$ is the average shortest distance, $\langle C \rangle$ indicates the average clustering coefficient, and $\langle \rho \rangle$ represents the network density.

No	Networks	$|V|$	$|E|$	$\langle k \rangle$	$\langle d \rangle$	$\langle C \rangle$	$\langle \rho \rangle$	Description
1	Karate	34	78	4.588	2.408	0.571	0.139	Zachary's Karate Club network [26]
2	Dolphin	62	159	5.129	3.357	0.259	0.084	A bottlenose dolphins network [27]
3	Lesmis	77	254	6.597	2.641	0.573	0.087	A characters network in a novel [28]
4	Polbook	105	441	8.4	3.079	0.488	0.081	Books about US politics network [29]
5	Word	112	425	7.589	2.536	0.173	0.068	An adjacency network in a novel [30]
6	Football	115	613	10.661	2.508	0.403	0.094	An American football game network [31]
7	Physician	117	465	7.949	2.587	0.219	0.069	An network among physician [32]
8	Jazz	198	2742	27.697	2.235	0.617	0.141	A network of jazz musicians [33]
9	Celegan	297	2148	14.465	2.455	0.292	0.049	A neural network [34]
10	USAir	332	2126	12.807	2.738	0.625	0.039	A network of flights betweeen airports [35]
11	Netsci	379	914	4.823	6.042	0.741	0.013	A collaboration network [30]
12	Email	1133	5451	9.622	3.606	0.22	0.009	An email communication network [36]
13	Yeast	2225	6610	5.942	4.378	0.138	0.003	A protein-protein interaction network [37]

Table 4. Average AUC of the LPCPA on selected real networks with $\alpha = 0.5$ and different values of β.

β	AUC												
	Karate	Word	Dolphin	Football	Polbook	Physician	Email	Celegan	Yeast	Jazz	USAir	Netsci	Lesmis
-0.01	0.6531	0.6435	0.7802	0.8431	0.8965	0.7043	0.866	0.8268	0.7792	**0.9627**	**0.9277**	**0.9334**	**0.899**
0	0.6953	0.6814	0.7852	0.8504	0.8979	0.7166	0.8794	0.8453	0.7918	0.9532	0.926	0.9314	0.8948
0.01	0.7375	0.7194	**0.7902**	**0.8578**	**0.8993**	**0.729**	**0.8927**	0.8614	0.8045	0.9458	0.9233	0.9294	0.8906
0.03	0.7375	0.7198	**0.7902**	**0.8578**	**0.8993**	**0.729**	**0.8927**	**0.8623**	0.8044	0.9359	0.9211	0.9294	0.8906
0.05	0.7375	0.7216	**0.7902**	**0.8578**	0.8991	**0.729**	0.8926	0.862	0.8044	0.9297	0.919	0.9294	0.8903
0.07	0.7375	0.7233	**0.7902**	**0.8578**	0.8986	**0.729**	0.8925	0.8612	0.8045	0.9255	0.9176	0.9294	0.8896
0.09	0.7376	0.7245	**0.7902**	**0.8578**	0.8979	0.7288	0.8923	0.8602	0.8045	0.9226	0.9166	0.9293	0.8889
0.1	0.7378	0.7253	0.7901	0.8577	0.8974	0.7286	0.8922	0.8596	**0.8046**	0.9213	0.9161	0.9293	0.8884
0.3	0.7501	0.7294	0.788	0.8563	0.8916	0.7252	0.8913	0.8535	**0.8046**	0.912	0.9127	0.928	0.883
0.5	0.7545	0.7302	0.786	0.8553	0.8889	0.7226	0.8908	0.8509	**0.8046**	0.9093	0.9117	0.9269	0.8805
0.7	0.7556	0.7307	0.785	0.8546	0.8874	0.7212	0.8906	0.8496	0.8043	0.908	0.9112	0.9264	0.8792
0.9	**0.7562**	**0.7316**	0.785	0.8546	0.8873	0.7212	0.8906	0.8493	0.8045	0.9071	0.9111	0.9264	0.8789

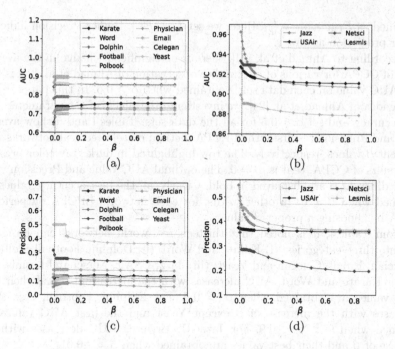

Fig. 3. Comparison of the AUC and the Precision of the *LPCPA* on selected real networks with $\alpha = 0.5$ and different values of β.

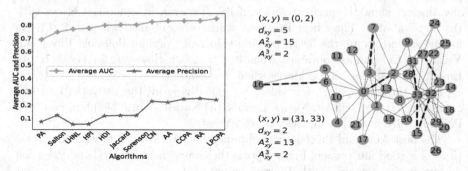

Fig. 4. Average of AUC and Precision on all the networks.

Fig. 5. An example for Karate network to show the case that LPCPA outperforms CCPA.

and their similarity scores are compared at each time. If there are n independent performed comparisons, such that there are n' times that the missing link has a higher score than the non-existent link, and n'' times that both of them have equal scores, then AUC is defined as $AUC = \frac{n' + 0.5n''}{n}$. All the non-observed links are ranked in descending order according to their similarity scores. Among the top-L non-observed links, L_r links belong to missing links, then the precision is defined as $Precision = \frac{L_r}{L}$. Here, we set $L = |E^P|$. Higher Precision indicates higher prediction accuracy.

According to Ahmad et al. [24], there is no significant change in the average AUC of CCPA for various of α. Another observation of Ahmad et al. [24] is that best AUC value on each data set is obtained when $\alpha \geq 0.5$. In this paper, based on the idea of Ahmad et al. [24], we investigate the effects of the parameter β on the accuracy and set $\alpha = 0.5$ for all the data sets. Tables 4 and 5 show average AUC and Precision results of the LPCPA method on selected real networks with different β values, respectively. The row highlighted in dark gray color presents the results of CCPA, that is, $\beta = 0$. The optimal AUC value and Precision value under different β are displayed in bold. Clearly, all the best accuracy values are obtained when $\beta \neq 0$. In other words, for each network, LPCPA outperforms CCPA by choosing a proper β value.

From Table 4, it is noted that thirteen real-world networks can be classified into three categories: (i) Karate and Word; (ii) Dolphin, Football, Polbook, Physician, Email, Celegan and Yeast; (iii) Jazz, USAir, Netsci and Lesmis.

For Karate and Word, AUC increases with the increase of β and their best AUC values are obtained when $\beta = 0.9$. For Dolphin, Football, etc., AUC decreases with the increase of β except Yeast and the best AUC values are obtained when $0 < \beta < 0.9$. For Jazz, USAir, etc., AUC decreases with the increase of β and their best values are obtained when $\beta = -0.01$.

As shown in Table 5, in terms of the third groups (Jazz, USAir, etc.,), the performance of Precision is consistent with AUC. For Karate and Word, Precision doesn't show the perfect monotonicity. However, in general, it grows with the increase of β. Their best Precision values are obtained when $\beta > 0.1$. As for the second group, the best Precision values of Dolphin, Polbook, Physician, Yeast and Celegan are obtained when $0 < \beta < 0.9$. However, for Football and Email, the best Precision achieves when $\beta = -0.01$.

From the view point of β value, we can divide all the networks into two categories: (i) Jazz, USAir, Netsci, Lesmis; (ii) Karate, Word, Dolphin, Football, Polbook, Physician, Email, Celegan and Yeast.

The best AUC and Precision are obtained for (i) when $\beta < 0$, while $\beta > 0$ for (ii). For a good illustration, Fig. 3 reports the comparisons of AUC and Precision on two groups networks with different values of β.

4.2   Comparison with Classical Methods

To validate the accuracy of the proposed index, we compare it with other common-neighbor-based indices including CN, RA, AA, Jaccard, PA, Salton, Sorenson, HDI, HPI, LHNL, CCPA. Note that, here, the values of CCPA and LPCPA are the best results from Tables 4 and 5 with $\alpha = 0.5$ and $-0.01 \leq \beta < 1$.

Table 6 presents the accuracy results measured by AUC and Precision. In terms of the AUC, LPCPA index has the top two best performance on ten

Table 5. Average Precision of the LPCPA on selected real networks with $\alpha = 0.5$ and different values of β.

β	Karate	Word	Dolphin	Football	Polbook	Physician	Email	Celegan	Yeast	Jazz	USAir	Netsci	Lesmis
-0.01	0.0929	0.056	0.1253	**0.3177**	0.1791	0.0835	**0.1482**	0.0925	0.0875	**0.5488**	**0.3741**	**0.3916**	0.512
0	0.1186	0.0624	0.13	0.2848	0.1768	0.0789	0.1423	0.0973	0.0902	0.5018	0.3708	0.3549	0.4852
0.01	0.1329	0.0702	0.1353	0.259	**0.1807**	0.0822	0.1362	0.1019	0.0945	0.4683	0.3686	0.2855	0.4568
0.03	0.1329	0.0745	0.1353	0.259	0.1805	0.0822	0.1356	0.1102	0.1006	0.4295	0.3674	0.2855	0.4556
0.05	0.1329	0.08	0.1353	0.259	**0.1807**	**0.0824**	0.1308	0.1131	0.1008	0.4072	0.3663	0.2843	0.4432
0.07	0.1329	0.0843	**0.136**	0.2589	0.1805	0.0822	0.1271	0.1168	**0.102**	0.3947	0.365	0.2823	0.4304
0.09	0.1329	0.0857	**0.136**	0.2579	**0.1807**	0.0811	0.1241	0.1172	0.1013	0.3868	0.3642	0.278	0.4204
0.1	0.1329	0.0881	0.1353	0.258	0.1798	0.0809	0.1225	0.1179	0.1001	0.384	0.3638	0.276	0.414
0.3	**0.1414**	0.0988	0.1267	0.2472	0.1766	0.0724	0.1087	0.1205	0.0986	0.3635	0.3618	0.2364	0.3812
0.5	0.1314	**0.1002**	0.1233	0.2415	0.1761	0.0663	0.1047	0.1212	0.0983	0.3583	0.3611	0.2116	0.3684
0.7	0.1329	**0.1002**	0.12	0.2387	0.1727	0.0602	0.1028	**0.1214**	0.0968	0.3561	0.3608	0.2005	0.3672
0.9	0.1329	0.0998	0.1187	0.2387	0.1725	0.0596	0.102	0.1213	0.0984	0.3548	0.3608	0.1978	0.3656

Table 6. Average AUC and Precision of the algorithms on selected data sets.

Network	AUC											
	CN	RA	AA	Jaccard	PA	Salton	Sorenson	HDI	HPI	LHNL	CCPA	LPCPA
Karate	0.6918	**0.7409**	0.7304	0.6002	0.699	0.6272	0.6002	0.5861	0.6804	0.5897	0.6953	**0.7562**
Word	0.6702	0.6662	0.6689	0.6114	**0.7501**	0.6141	0.6114	0.6097	0.6085	0.575	0.6814	**0.7316**
Dolphin	0.7711	0.7734	**0.7745**	0.7707	0.6292	0.7659	0.7707	0.7724	0.7035	0.7528	0.7852	**0.7902**
Football	0.846	0.8457	0.8457	0.8572	0.2665	0.8569	0.8572	0.8564	0.8553	**0.8588**	0.8504	0.8578
Polbook	0.8855	0.8951	0.8937	0.8717	0.673	0.8791	0.8717	0.8602	0.8863	0.8398	**0.8979**	0.8993
Physician	0.709	0.7118	0.7124	0.7054	0.6035	0.7074	0.7054	0.7015	0.7054	0.6944	**0.7166**	0.729
Email	0.8433	0.8449	0.8453	0.8411	0.7793	0.0536	0.8411	0.8409	0.7985	0.8342	**0.8794**	0.8927
Celegan	0.8446	**0.8653**	0.861	0.7902	0.7531	0.7969	0.7902	0.7799	0.7906	0.724	0.8453	0.8614
Yeast	0.7061	0.7065	0.7067	0.7033	0.7767	0.7042	0.7033	0.7033	0.6049	0.7029	**0.7918**	0.8046
Jazz	0.9537	**0.9697**	0.9607	0.9595	0.7673	0.9641	0.9595	0.9502	0.9404	0.9015	0.9532	0.9627
USAir	0.935	**0.9524**	0.9463	0.898	0.8853	0.909	0.898	0.8916	0.8245	0.7686	0.926	0.9277
Netsci	0.9507	**0.9543**	0.9541	0.9486	0.6267	0.9502	0.9486	0.9473	0.9124	0.9459	0.9314	0.9334
lesmis	0.9142	**0.9227**	0.9224	0.8837	0.7855	0.887	0.8837	0.8814	0.7911	0.8237	0.8948	0.899
Network	Precision											
	CN	RA	AA	Jaccard	PA	Salton	Sorenson	HDI	HPI	LHNL	CCPA	LPCPA
Karate	0.1186	**0.1272**	0.1157	0.0014	0.0672	0.0014	0.0014	0.0043	0.0314	0.0029	0.1186	**0.1329**
Word	0.0657	0.059	0.0671	0.0031	**0.0928**	0.0026	0.0031	0.0038	0.0079	0	0.0624	**0.0998**
Dolphin	**0.1327**	0.108	0.12	0.0987	0.022	0.0873	0.0987	0.1007	0.01	0.014	0.13	**0.1353**
Football	0.2867	0.2807	0.282	0.3469	0	0.3462	0.3469	0.3408	**0.3495**	**0.416**	0.2848	0.3177
Polbook	0.1798	**0.208**	**0.207**	0.1157	0.0468	0.1166	0.1157	0.1052	0.1207	0.0664	0.1768	0.1791
Physician	0.0789	0.0759	**0.0848**	0.0548	0.0252	0.0578	0.0548	0.0515	0.0522	0.0222	0.0789	**0.0835**
Email	0.1418	0.1399	**0.1547**	0.0501	0.0184	0.0536	0.0501	0.0629	0.0009	0.0037	0.1423	**0.1482**
Celegan	0.0973	0.1016	**0.1073**	0.0214	0.0585	0.021	0.0214	0.0223	0.0032	0	0.0973	**0.1214**
Yeast	0.0901	0.075	**0.0927**	0.0001	0.0099	0.0003	0.0001	0.0001	0.0006	0.0001	0.0902	**0.1001**
Jazz	0.5021	**0.5383**	0.5217	0.5127	0.1295	0.5281	0.5127	0.4645	0.1275	0.1008	0.5018	**0.5488**
USAir	0.3712	**0.4558**	0.3914	0.0731	0.3203	0.0553	0.0731	0.0843	0.0039	0.0075	0.3708	0.3741
Netsci	0.3531	**0.5643**	0.5357	0.2594	0.014	0.2638	0.2594	0.2577	0.0075	0.0755	0.3549	0.3916
lesmis	0.4852	**0.5464**	0.5352	0.0472	0.1052	0.0584	0.0472	0.0452	0.024	0.0024	0.4852	0.512

out of thirteen networks. RA outperforms LPCPA on Celegan, Jazz, USAir, NetSci and Lesmis. According to Precision, LPCPA index has the top two best performance on eight out of thirteen networks. Figure 4 depicts the average AUC

Table 7. Comparisons of LPCPA and other metrics including global index Katz ($\gamma = 0.01$) and quasi-local index LP ($\epsilon = 0.01$).

Method	AUC												
	Karate	Word	Dolphin	Football	Polbook	Physician	Email	Celegan	Yeast	Jazz	USAir	Netsci	lesmis
Katz	0.733	0.719	0.791	**0.858**	0.899	0.7279	0.894	0.859	0.806	0.941	0.922	0.929	0.89
LP	0.742	0.721	**0.794**	0.858	**0.900**	0.729	**0.899**	**0.862**	**0.823**	0.946	0.926	**0.954**	0.897
LPCPA	**0.756**	**0.732**	0.79	0.858	0.8993	0.728	0.893	0.861	0.805	**0.963**	**0.928**	0.933	**0.899**

Method	AUC												
	Karate	Word	Dolphin	Football	Polbook	Physician	Email	Celegan	Yeast	Jazz	USAir	Netsci	lesmis
Katz	0.129	0.07	0.131	0.258	0.1802	0.0813	0.136	0.102	0.096	0.443	0.365	0.285	0.456
LP	**0.133**	0.071	**0.136**	0.259	**0.1807**	0.0822	0.136	0.102	0.098	0.468	0.369	0.286	0.456
LPCPA	**0.133**	**0.099**	0.135	**0.318**	0.1791	**0.084**	**0.148**	**0.121**	**0.100**	**0.549**	**0.374**	**0.392**	**0.512**

and Precision of each algorithm on all networks, in which LPCPA performs best on AUC.

Table 7 shows comparisons of LPCPA and other metrics including global index Katz and quasi-local index LP. In terms of AUC, LP performs best, while LPCPA performs best according to Precision.

Figure 5 shows a case for Karate network that LPCPA outperforms CCPA. The dashed edges denotes the probe set E^P including edges (3, 7), (0, 2), (6, 16), (31, 33), (15, 32), (15, 33), (23, 27). All the non-observed links are ranked in descending order according their similarity score. The top-L ($L = 7$) edges for CCPA are (2, 33), (0, 2), (1, 33), (13, 19), (8, 13), (7, 13), (4, 5). Clearly, only edge (0, 2) falls into E^P. However, according to LPCPA, top-L edges includes (2, 33), (0, 2), (1, 8), (13, 32), (1, 33), (31, 33), (2, 19) and two edges (0, 2) and (31, 33) belong to the missing ones. That is, both CCPA and LPCPA can identify the missing link (0, 2), while only LPCPA finds (31, 33). As shown in Fig. 7, although the number of 2-hop paths between nodes 31 and 33 is small, more 3-hop paths between them should obtain a higher similarity of the link.

4.3   Effects of the Average Clustering Coefficient of the Network

We have some interesting observations that can be made from the results of Tables 4, 5 and 6, where the values of Jazz, USAir, Netsci and Lesmis are highlighted in light gray color.

From the view point of topological statistical properties, their average clustering coefficients are top-four highest among all the networks.

On the one hand, based on Tables 4 and 5, for Jazz, USAir, Netsci and Lesmis, the best AUC and Precision are obtained when $\beta = -0.01$. As we have mentioned above, β controls the proportion of 3-hop-based paths. How to determine β to ensure that LPCPA outperforms CCPA? With respect to the networks with higher average clustering coefficient, $\beta = -0.01$, whereas, for the networks that have lower average clustering coefficient, $\beta > 0$. On the other hand, based on Tables 6, for Jazz, USAir, Netsci and Lesmis, in terms of AUC and Precision, RA has a better performance than LPCPA in most cases. In other words, compared with benchmark algorithms, the LPCPA index is particularly suitable for the networks with lower average clustering coefficient and density.

5   Conclusion

In this work, we propose a new prediction method LPCPA by introducing 3-hop paths into the well-known method CCPA. To evaluate the performance of the presented method, extensive experiments are implemented on thirteen real networks under the metric of AUC and Precision. The proposed method has better performance than CCPA under the metric of AUC and Precision, and it is superior to the baselines in most of the cases. With respect to the networks with higher average clustering coefficient, $\beta = -0.01$, whereas, $\beta > 0$ for the networks that have lower average clustering coefficient. In the future, we will employ the proposed method on weighted, directed and also bipartite networks.

Acknowledgements. This work was partly supported by the National Natural Science Foundation of China (Nos. 61977016 and 61572010), Natural Science Foundation of Fujian Province (Nos. 2020J01164, 2017J01738) and Education and Scientific Research Project for Young and Middle-aged Teachers of Fujian Province (No. JAT191119).

References

1. Wang, Y.B., et al.: Predicting protein-protein interactions from protein sequences by a stacked sparse autoencoder deep neural network. Mol. BioSyst. **13**(7), 1336–1344 (2017)
2. Liu, S., Dong, Z., Ding, C., Wang, T., Zhang, Y.: Do you need cobalt ore? Estimating potential trade relations through link prediction. Resour. Policy **66**, 101632 (2020)
3. Lü, L., Medo, M., Yeung, C.H., Zhang, Y.-C., Zhang, Z.-K., Zhou, T.: Recommender systems. Phys. Rep. **519**(1), 1–49 (2012)
4. Aziz, F., Gul, H., Uddin, I., Gkoutos, G.V.: Path-based extensions of local link prediction methods for complex networks. Sci. Rep. **10**, 19848 (2020)
5. Das, S., Das, S.K.: A probabilistic link prediction model in time-varying social networks. In: IEEE International Conference on Communications(ICC), pp. 1–6. IEEE (2017). https://doi.org/10.1109/ICC.2017.7996909
6. Pan, L., Zhou, T., Lü, L., Hu, C.-K.: Predicting missing links and identifying spurious links via likelihood analysis. Sci. Rep. **6**, 22955 (2016)
7. Grover, A., Leskovec, J.: node2vec: scalable feature learning for networks. In: Proceedings of the 22nd ACM SIGKDD International Conference on Knowledge Discovery and Data Mining (KDD 2016), pp. 855–864 ACM (2016). https://doi.org/10.1145/2939672.2939754
8. Chen, G.-F., Xu, C., Wang, J.-Y., Feng, J.-W., Feng, J.-Q.: Graph regularization weighted nonnegative matrix factorization for link prediction in weighted complex network. Neurocomputing **369**, 50–60 (2019)
9. Chen, B.L., Chen, L., Li, B.: A fast algorithm for predicting links to nodes of interest. Inf. Sci. **329**, 552–567 (2016)
10. Newman, M.E.J.: Clustering and preferential attachment in growing networks. Phys. Rev. E **64**(2), 025102 (2001)
11. Jaccard, P.: Distribution de la flore alpine dans le bassin des Dranses et dans quelques régions voisines. Bulletin de la Societe Vaudoise des Sciences Naturelles **37**(140), 241–272 (1901)

12. Adamic, L.A., Adar, E.: Friends and neighbors on the web. Soc. Netw. **25**(3), 211–230 (2003)
13. Zhou, T., Lü, L., Zhang, Y.-C.: Predicting missing links via local information. Eur. Phys. J. B **71**(4), 623–630 (2009)
14. Barabási, A.L., Jeong, H., Néda, Z., Ravasz, E., Schubert, A., Vicsek, T.: Evolution of the social network of scientific collaborations. Phys. A **311**, 590–614 (2002)
15. Leicht, E.A., Holme, P., Newman, M.E.J.: Vertex similarity in networks. Phys. Rev. E **73**, 026120 (2006)
16. Salton, G., McGill, M.J.: Introduction to Modern Information Retrieval. McGraw Hill Inc, New York, NY, USA (1986)
17. Sørensen, T.: A method of establishing groups of equal amplitude in plant sociology based on similarity of species and its application to analyses of the vegetation on Danish commons. Kongelige Danske Videnskabernes Selskab **5**(4), 1–34 (1948)
18. Ravasz, E., Somera, A.L., Mongru, D.A., Oltvai, Z.N., Barabási, A.-L.: Hierarchical organization of modularity in metabolic networks. Science **297**(5586), 1551–1555 (2002)
19. Katz, L.: A new status index derived from sociometric analysis. Psychometrika **18**(1), 39–43 (1953)
20. Lü, L., Jin, C.-H., Zhou, T.: Similarity index based on local paths for link prediction of complex networks. Phys. Rev. E **80**, 046122 (2009)
21. Lü, L., Zhou, T.: Link prediction in complex networks: a survey. Phys. A **390**, 1150–1170 (2011)
22. Pech, R., Hao, D., Lee, Y.-L., Yuan, Y., Zhou, T.: Link prediction via linear optimization. Phys. A **528**, 121319 (2019)
23. Zhou, T., Lee, Y.-L., Wang, G.-N.: Experimental analyses on 2-hop-based and 3-hop-based link prediction algorithms. Phys. A **564**, 125532 (2021)
24. Ahmad, I., Akhtar, M.U., Noor, S., Shahnaz, A.: Missing link prediction using common neighbor and centrality based parameterized algorithm. Sci. Rep. **10**, 364 (2020)
25. Herlocker, J.L., Konstan, J.A., Terveen, L.G., Riedl, J.T.: Evaluating collaborative filtering recommender systems. ACM Trans. Inf. Syst. **22**(1), 5–53 (2004)
26. Zachary, W.W.: An information flow model for conflict and fission in small groups. J. Anthropol. Res. **33**(4), 452–473 (1977)
27. Lusseau, D., Schneider, K., Boisseau, O.J., et al.: The bottlenose dolphin community of doubtful sound features a large proportion of long-lasting associations. Behav. Ecol. Sociobiol. **54**(4), 396–405 (2003)
28. Knuth, D.E.: The stanford GraphBase: a platform for combinatorial algorithms. In: The Fourth Annual ACM-SIAM Symposium on Discrete Algorithms (SODA), pp. 41–43. Philadelphia, PA, USA (1993)
29. Newman, M.E.J.: Modularity and community structure in networks. Proc. Natl. Acad. Sci. **103**(23), 8577–8582 (2006)
30. Newman, M.E.J.: Finding community structure in networks using the eigenvectors of matrices. Phys. Rev. E **74**, 036104 (2006)
31. Girvan, M., Newman, M.E.J.: Community structure in social and biological networks. Proc. Natl. Acad. Sci. **99**(12), 7821–7826 (2002)
32. Coleman, J., Katz, E., Menzel, H.: The diffusion of an innovation among physicians. Sociometry **20**(4), 253–270 (1957)
33. Gleiser, P.M., Danon, L.: Community structure in jazz. Adv. Complex Syst. **6**(4), 565–573 (2003)

34. White, J.G., Southgate, E., Thomson, J.N., Brenner, S.: The structure of the nervous system of the nematode caenorhabditis elegans. Philos. Trans. R. Soc. Lond. **314**(1165), 1–340 (1986)
35. Batagelj, V., Mrvar, A.: Pajek. In: Alhajj, R., Rokne, J. (eds.) Encyclopedia of Social Network Analysis and Mining, pp. 1245–1256. Springer, New York, NY (2014). https://doi.org/10.1007/978-1-4614-6170-8_310
36. Guimera, R., Danon, L., Diaz-Guilera, A., Giralt, F., Arenas, A.: Self-similar community structure in a network of human interactions. Phys. Rev. E **68**, 065103 (2003)
37. Batagelj, V., Mrvar, A., Pajek datasets. https://vlado.fmf.uni-lj.si/pub/networks/data/

The Influence of Color on Prices of Abstract Paintings

Maksim Borisov[2], Valeria Kolycheva[2(✉)], Alexander Semenov[1(✉)],
and Dmitry Grigoriev[2(✉)]

[1] University of Florida, Gainesville, FL 32611, USA
asemenov@ufl.edu
[2] Center of Econometrics and Business Analytics, Saint Petersburg State University,
Saint Petersburg, Russia
v.kolycheva@spbu.ru, gridmer@mail.ru

Abstract. Determination of price of an artwork is a fundamental problem in cultural economics. In this work we investigate what impact visual characteristics of a painting have on its price. We construct a number of visual features in CIELAB color space measuring complexity of the painting, its points of interest using Discrete symmetry transform, segmentation-based features using Felzenszwalb segmentation and Regions adjacency graph merging, local color features from segmented image, features based on Itten and Kandinsky theories, and utilize mixed-effects model with authors bias as fixed effect to study impact of these features on the painting price. We analyze the influence of the color on the example of the most complex art style - abstractionism, created by Kandinsky, for which the color is the primary basis. We use Itten's theory - the most recognized color theory in art history, from which the largest number of subtheories was born. For this day it is taken as the base for teaching artists. We utilize novel dataset of 3,885 paintings collected from Christie's and Sotheby's and find that color harmony has a little explanatory power, color complexity metrics are impact price negatively and color diversity explains price well.

Keywords: Paintings · abstractionism · Color theory · Itten's color wheel · Pricing · Mixed-effects

1 Introduction

The reaction of people to a particular color can be determined by the cultural environment (in different cultures the same colors can symbolize different concepts) and the personal preferences of a particular person. For example, red in France is the color of aristocrats, as it is known. But the French led the Egyptians to complete bewilderment, as in Egypt, red is a symbol of mourning.

In art literature, the relationship between the price of a painting and its attributes such as size [12], material and signature, sale conditions such as

T. N. Dinh and M. Li (Eds.): CSoNet 2022, LNCS 13831, pp. 64–68, 2023.
https://doi.org/10.1007/978-3-031-26303-3_6

year, salesroom and sale location has been studied with the hedonic model since the 70's (e.g., [1,2,14]). The painting size is normally found to be a significant explanatory variable for price.

In the past years, quantitative studies of color in art have been conducted multiple times (such as [9,11], and [5]). Their work suggest that by analysing art one can obtain results useful for understanding human perception [7].

The connection between color composition of an artwork and its price is still a new research area. Article [6] concludes that the selling price is not predictive of preference, while shared preferences may to some extent be predictable based on image statistics [6]. Paper [13] finds significant evidence that contrastive paintings (those with a high diversity of colors) carry premium price than equivalent monochrome artworks. Empirical results presented in [10] present significant evidence of intense colors fetching a premium over equivalent artworks which are less intense in color. Furthermore, darkness carries a premium over lightness.

In the scope of this article we will answer a research question: How do color metrics, such as color harmony according to Itten, color complexity and others, influence the abstract painting price?

We construct a number of visual features measuring complexity of the painting, its points of interest, segmentation-based features, local color features, and six color harmony features based on Itten and Kandinsky theories, and utilize mixed-effects model to study impact of these features on the painting price. We analyze the influence of the color on the example of the most complex art style - abstractionism, by created Kandinsky, for which the color is the primary basis. We use Itten's theory - the most recognized color theory in art history, from which the largest number of subtheories was born. For this day it is taken as the base for teaching artists. We utilize novel dataset of 3,885 paintings collected from Christie's and Sotheby's and find that color harmony has some explanatory power, color complexity metrics are insignificant and color diversity explains price well.

Our study makes several contributions to the current literature.

1. We apply Itten color theory to price-harmony relationship.
2. We study the effect of visual complexity of the paintings on their price. Visual complexity is operationalized as a number of features constructed from the digital image content.
3. We utilize the mixed-effects model to take into account the effect of the author, visual features, and other characteristics.

2   Datasets and Methods

We analyze 3885 paintings collected from Sotheby's and Christies auction houses. The main style of paintings is abstract. Our dataset includes paintings from such authors as Franz Kline, Mark Rothko, Cy Twombly, Alexander Calder, Joan Miró, Joan Mitchell, Lucio Fontana, Philip Guston, Richard Diebenkorn, Wassily Kandinsky, Willem de Kooning, Zao Wou-ki, Henry Moore. The minimum number of paintings of a single author is 114, and the maximum is 655. The selected images belong to high price category, with a mean normalized price totaling around 11.9 million dollars. Author birth years range is from 1866 to 1928, and the most frequent material is paper. Ninety percent of paintings have a signature, and about half of all paintings were obtained from Sotheby's. The most used color in the paintings is green.

Majority of studies use the hedonic model to find price determinants [10,13]. In our case, we have multiple authors, which implies a different price range for each of them; this may result in heteroscedasticity, making the hedonic model unsuitable. In order to account for the heteróscedasticity, we employ the linear mixed model.

3 Discussion and Conclusions

The present study was designed to determine the effect of visual characteristics on abstract painting price. This section summarizes our findings and contributions.

One interesting finding is significance of signature on an abstract painting. When comparing our results to those of older studies, it must be pointed out that presence of signature has unstable explanatory power [4].

This study supports evidence from previous observations about insignificance of auction house [10]. Although, certain limitations of this study could be addressed, such as moderate sample and focus on abstract paintings.

Although the present results clearly support insignificance of the painter's birth date similar to [4], it is appropriate to recognize this limitation. That is, in our data, this feature ranges from 1866 to 1928, with the median being 1899, so its insignificance could be caused by small variation.

A strong relationship between size of a painting and its price has been reported in the literature ([4,13,14]). Our results are in agreement with the reported findings.

Further on, reviewing the literature, we have found data supporting the correlation between the number of exhibitions of a painting, the number of mentions in the literature, the number of owners and the price [4]. Our findings highlight a strong positive significance of these predictors, which represent the cultural importance of a painting.

The current study found clear support for the significance of canvas. This finding broadly supports the work of other researchers in this area ([4,8,13]). Lithography and paper are found to be inferior in terms of price to others, since lithography is a printing process, so copies of such artwork could be easily produced, and paper demands special preservation conditions, as such paintings are fragile. In addition, sketches are usually depicted on the paper, while the completed idea is transferred by the artist to the canvas. Oil on canvas is the most costly technique that requires special efforts and skill from the artist. There are no two absolutely identical canvases, even if they are author's copies.

Our findings on color distribution at least hint that in abstract paintings blue color might be superior to others. In case of Picasso, similar result was obtained by Stepanova [13]. The main limitation is the lack of paintings to make a stronger conclusion.

Overall findings about color harmony are in accordance with previous research [3]. That is, color harmony has low significance. However, our results go beyond previous reports, showing that color harmony can hold explanatory power, e.g. analogue triad.

From the results, it is clear that color complexity metrics has low explanatory power. For example, lines variance and points of interest are found to be insignificant. Moreover, higher color complexity measure is associated with lesser price. That is, painting has too complicated color distribution, therefore it becomes less attractive for a potential buyer.

The present study has confirmed the findings about diversity of color composition, that is, contrast in hue, contrast in saturation, and number of segments are positively significant ([3,13]). On the other hand, contrast in value represents color diversity poorly. This may be the reason why it was not found to be significant.

References

1. Buelens, N., Ginsburgh, V.: Revisiting baumol's 'art as floating crap game'. Eur. Econ. Rev. **37**(7), 1351–1371 (1993)
2. Chanel, O.: Is art market behaviour predictable? Eur. Econ. Rev. **39**(3–4), 519–527 (1995)
3. Charlin, V., Cifuentes, A.: A general framework to study the price-color relationship in paintings with an application to mark rothko rectangular series. Color Res. Appl. **46**(1), 168–182 (2021)
4. Cinefra, J., Garay, U., Mibelli, C., Pérez, E.: The determinants of art prices: an analysis of joan miró. Academia Revista Latinoamericana de Administración (2019)
5. Farrell, L., Fry, J.M., Fry, T.R.: Gender differences in hammer prices for Australian indigenous art. J. Cult. Econ. **45**(1), 1–12 (2021)
6. Graham, D.J., Friedenberg, J.D., McCandless, C.H., Rockmore, D.N.: Preference for art: similarity, statistics, and selling price. In: Human Vision and Electronic Imaging XV, vol. 7527, p. 75271A. International Society for Optics and Photonics (2010)
7. Graham, D.J., Redies, C.: Statistical regularities in art: relations with visual coding and perception. Vis. Res. **50**(16), 1503–1509 (2010)
8. Ju, L., Tu, Z., Xue, C.: Art pricing with computer graphic techniques. arXiv preprint arXiv:1910.03800 (2019)
9. Montagner, C., Linhares, J.M., Vilarigues, M., Melo, M.J., Nascimento, S.M.: Supporting history of art with colorimetry: the paintings of amadeo de souza-cardoso. Color Res. Appl. **43**(3), 304–310 (2018)
10. Pownall, R.A., Graddy, K.: Pricing color intensity and lightness in contemporary art auctions. Res. Econ. **70**(3), 412–420 (2016)
11. Romero, J., Gómez-Robledo, L., Nieves, J.: Computational color analysis of paintings for different artists of the xvi and xvii centuries. Color Res. Appl. **43**(3), 296–303 (2018)

12. Schönfeld, S., Reinstaller, A.: The effects of gallery and artist reputation on prices in the primary market for art: a note. J. Cult. Econ. **31**(2), 143–153 (2007)
13. Stepanova, E.: The impact of color palettes on the prices of paintings. Empir. Econ. **56**(2), 755–773 (2019)
14. Worthington, A.C., Higgs, H.: Art as an investment: risk, return and portfolio diversification in major painting markets. Account. Financ. **44**(2), 257–271 (2004)

ELA: A Time-Series Forecasting Model for Liner Shipping Based on EMD-LSTM and Attention

Jiadong Chen, Xiaofeng Gao[✉], and Guihai Chen

MoE Key Lab of Artificial Intelligence, Department of Computer Science and Engineering, Shanghai Jiao Tong University, Shanghai, China
chenjiadong998@sjtu.edu.cn, {gao-xf,gchen}@cs.sjtu.edu.cn

Abstract. The development of modern information technology has led to the digital and intelligent transformation of many traditional industries. In order to face the challenge of difficult transition, liner shipping industry is in urgent need of prediction methods for strategy adjustment, and much research remains to be done in this area. In our paper, we put forward a time-series forecasting model named ELA based on modified EMD-LSTM (Empirical Mode Decomposition and Long Short-Term Memory), AR (Autoregressive) and the attention mechanism. Tailored for handling the market characteristics of the liner shipping industry, our method effectively reduces the negative impact of noise on model training and learns the short-term dependencies and long-term trend in time series. Experiments on real-world liner shipping datasets show the powerful prediction performance of ELA.

Keywords: Liner shipping · Time-series forecasting · Digital transformation

1 Introduction

Liner shipping plays a significant role in the development of human society. This service is provided by shipping companies, and the goods are allocated according to the number of shipping slots left in the remaining voyage period. Liner shipping is cheap and carries large quantities of goods. According to World Shipping Council [1], it provides cost-efficient and effective transportation for almost everything we need in our life. Nowadays, more than 50% of the value of goods by sea are transported by liner shipping.

Unlike shipping by chartering, which means that the shipping company charters a ship for the customer and arranges the transportation schedule according to the charter contract, liner shipping targets at a wider range of customers. As a result, statistics such as cargo type and container fare can vary a lot. Given the rapidly changing market, it is important for the transportation company to predict the trends of shipment revenue.

What's special with the liner shipping data is that it contains uncorrelated noise and fluctuations, and has a unique market lagging feature. To be concrete,

T. N. Dinh and M. Li (Eds.): CSoNet 2022, LNCS 13831, pp. 69–80, 2023.
https://doi.org/10.1007/978-3-031-26303-3_7

the customers of the liner shipping industry are mainly companies or corporations, who have relatively steady and regular demand and shipment behavior. Nevertheless, when summing the demands and orders up, the regular features are not available any more, along with large amount of noise involved in the data time-series, which brings difficulty to our prediction. What's more, the lagging influence involved in the liner shipping market means that the current market value is not only influenced by the temporal adjacent market condition, but also heavily affected by the market condition in the long term.

Currently, there are many methods to manage time-series prediction. EMD (Empirical Mode Decomposition) is an excellent time-frequency analysis method. It decomposes the original signal with large fluctuations into several sub-signals, called IMFs (Intrinsic Mode Functions), which is more regular and with much less noise [2–4]. Besides, Vanilla LSTM [5] (Long Short-Term Memory) is designed to fix the gradient vanishing problem of RNN (Recurrent Neural Networks), and is capable of selectively forgetting or memorizing historical information in the sequential chain. However, LSTM cannot perform well when encountering time series of even longer span with uncorrelated noise and fluctuations [2,6–8], so it cannot be directly applied to the liner shipping industry. Many scholars have combined LSTM and EMD method to effectively reduce the impact of noise on model training and have achieved good prediction ability [2,9–12], but the lagging problem still remains unsolved.

To address this, attention mechanism can be applied. Experiments have shown that when combining attention mechanism with neural network, the previous memory can be assigned detailed weights to prevent essential and critical information from vanishing in the long term, so that the global context features can be exploited [13–15]. Hence, we add attention mechanism to the EMD-LSTM model to well utilize the data from long-span ago, which may be diluted and forgotten in the LSTM cell chains. Based on these two methods, the two main problems (fluctuations and lagging condition) involved in prediction tasks of liner shipping industry can be addressed.

Therefore, we design the ELA model based on the characteristics of liner shipping, which combines the EMD algorithm, LSTM neural network, AR (Autoregressive) and the attention mechanism. The EMD method is used to decompose the time-series and dilute its noise. Then, we divide the sub-signals into high-frequency and low-frequency components, which we use a modified LSTM layer and a classical AR layer to model respectively. The modified LSTM layer refers to the addition of attention mechanism in Vanilla LSTM module to adjust the weights of outputs from each LSTM cell. This modification extracts the important temporal patterns from historical observations and strengthens the ability of our model when processing long time series. Finally, we apply our model to predict the daily average container fare of shipping lines, which proves to have an excellent forecasting ability. By accomplishing a forecasting model with fewer training parameters, lower computational costs and higher prediction accuracy, we provide our solution to help traditional liner shipping industry moving towards intelligence.

The contribution of our work is four-fold: We propose a novel model to solve the problem of high volatility of the input signal by decomposing the liner shipping data into several sub-signals (IMFs) by EMD and predicting the trend of IMFs by LSTM. We address the disadvantages of classical EMD-LSTM method by dividing the sub-signals into high-frequency and low-frequency components and modeling them separately. To better solve the lagging problem involved in liner shipping, we add the attention mechanism into the Vanilla LSTM module to learn the recurring patterns of data over a long term. We conduct experiments on real-world liner shipping datasets, demonstrating the superiority of our model in terms of the prediction performance.

2 Related Work

The application of machine learning and deep learning gradually expands to various industries and fields, including liner shipping. For example, some researchers implement various machine learning models to predict the delay of container vessels and conclude that SVM (Support Vector Machine) with polynomial kernel is the best regression model [16]. Despite their rich model selection and comparison, they do not take into account the effect of time. As the shipping market changes rapidly, it is important to take into account not only the past information itself, but also discover how the information has changed over time.

A lot of researchers adopt LSTM to avoid gradient vanishing problem of RNNs, and achieve good results in time-series forecasting problems in all walks of life [14,17]. There are plenty of deep learning methods proposed for time-series forecasting, but many of them place little emphasis on directly grasping the useful time features from time signal itself. EMD-LSTM model which combines EMD for signal decomposition and LSTM, however, is one of the most useful methods for time-series prediction. For example, some researchers state that after applying EMD method, the original complicated time-series of stock price can be decomposed into several smoother, more stable and more regular sub-signals, which can be well predicted by LSTM neural networks [9]. Also, in Boning Zhang's research, he applies the EMD-LSTM neural network model to foreign exchange rate time-series, showing "superior forecasting capacity" [10]. In addition, Huiting Zheng et al. claim in their research that the combination of EMD methods and LSTM neural network performed better than using single separated LSTM neural networks when predicting electric load [11].

Apart from the EMD method, some scholars also integrate attention mechanism into the traditional neural networks due to its ability to capture long-term dependencies between historical states. Zhihui Lin's team apply self-attention mechanism to ConvLSTM and demonstrate its powerful ability to obtain global spatial context for spatio-temporal prediction [13]. According to Bin Li and Yuqing He [18], an attention mechanism oriented hybrid CNN-RNN is implemented to realize the automatic operation of container terminal handling systems. In their research, they take the time factor into consideration and try to fit the entire time-series rather than one data point. However, they treat the time factor as a sequence of system calling numbers, not real time. Asmaa Fahim's team combines LSTM with attention mechanism to predict the education reform in Morocco, and shows that it is able to assign more detailed and various weights to each data input, thus preventing the loss of essential and critical information [6]. Le Yan's team states in their research that the function of attention mechanism is to deal with uncorrelated time-series signals with noise, which pays attention to absolute action factors of stream [8]. Hence, attention mechanism with its great ability to extract critical information from historical data, have been proved to be effective when dealing with signals with noise.

Based on these, our work modifies and combines the EMD-LSTM model and attention mechanism so as to enhance the ability to deal with long-term correlation and large fluctuation problems involved in liner shipping area. Then, we apply the model to forecast the container fare in the shipping industry, which turns out to be relatively accurate.

3   Methodology

In this section, we describe the details of the proposed ELA model. Figure 1 shows the overall structure of the ELA model.

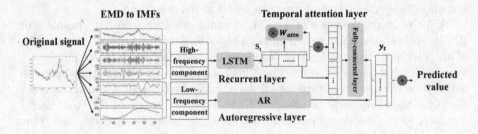

Fig. 1. The overall structure of the ELA model.

3.1   Input Reconstruction Through EMD

The common prediction objectives in liner shipping industry are container fare and port revenue. We denote the training series as $X = \{x_1, x_2, \cdots, x_n\} \in \mathbb{R}^n$, where x_i is the observation at timestep i and n is the size of the training set.

Considering the ever-changing market, it is natural that the liner shipping data is influenced by both short-term periodical fluctuations and long-term market trend. Hence, it is difficult to fully capture these underlying patterns when directly predicting the original variable. However, based on the assumption that the purchasing behavior of each individual company is regular, we can decompose the raw data into relatively regular signals with less noise involved, and make it easier for forecasting module to train and predict.

Empirical Mode Decomposition (EMD) is a useful time-frequency analysis method for time-series data. It can decompose the original signal into several Intrinsic Mode Functions (IMFs) and residuals [3]. An IMF is constrained to the following two criteria [4]: The number of extreme points and zero-crossings must be equal or the difference does not exceed one. The average value of the upper wrapping line (formed by local maximum points) and the lower wrapping line (formed by local minimum points) of one signal should be zero.

Because of the above properties, IMF components involve less noise and contain local characteristics of different time scales compared with the original signal, which provides more information for further analysis. Therefore, the original time series X is processed by EMD through the following process [9].

1. Fitting the local maximum and minimum points with cubic spline interpolation function to form the upper wrapping line and the lower wrapping line of X, denoted as U and L, respectively.
2. Form the average of U and L, denoted as A. Then, we obtain a candidate signal $C = X - A$.
3. Judge whether C satisfies the criteria of an IMF signal. If C complies with IMF criteria, C becomes one of the IMF decomposition of the original signal, denoted as Imf_i. Let $X - C$ be the new X and repeat the above three steps. Otherwise, let C be the new X and repeat the above three steps.
4. If there's no more IMF decomposition can be exploited from the original signal, we stop the EMD and get the residual signal Res.

Then, the original signal is decomposed into several IMF signals and one residual signal. After EMD, we turn a univariate time-series forecasting problem into a multivariate one. In a general time-series forecasting problem, the whole training series needs to be reconstructed into pairs of historical sequence and prediction label. Assume a fixed-size window is sliding over the decomposed series and cut them into snippets. For example, let X_t denotes the observations at timestep t, and w denotes the window size, we have: $X_t = \{Imf_1, \cdots, Imf_m, Res\} \in \mathbb{R}^{(m+1) \times w}, Imf_i = \{imf_i^1, \cdots, imf_i^w\} \in \mathbb{R}^w, Res = \{res^1, \cdots, res^w\} \in \mathbb{R}^w$.

Our goal is to predict: $x_{t+1} = \{imf_1^{t+1}, imf_2^{t+1}, \cdots, imf_m^{t+1}, res^{t+1}\} \in \mathbb{R}^{m+1}$.

The typical practice for classical EMD-LSTM is building and training a LSTM module for each sub-signal individually, and adding the predictions of sub-signals together to form the final result. This method has three disadvantages: (1) The number of decomposed IMFs of various datasets is different. The more sub-models we build, the more effort we need to pay on adjusting parameters for each model and the training time also multiplies; (2) Rigidly separating the prediction of sub-signals may fail to utilize the information from interactions between different signals. (3) Since the original signal is the sum of sub-signals, the prediction error of sub-models is likely to accumulate, which may become a bottleneck of improving the overall performance.

We divide the sub-signals into high frequency and low frequency components to address the above-mentioned disadvantages. From the theoretical knowledge of EMD, we learn that the IMF components are sorted from high frequency to low frequency, and their trends represent the inherent operating law of the original series. Every IMF is judged sequentially whether its mean value deviates significantly from zero through T-test. Assume that Imf_{d+1} is the first IMF whose mean value deviates from zero, then Imf_1 to Imf_d are concatenated to construct a high-frequency component, and the rest sub-signals are concatenated as well to form a low-frequency component. We denote the high-frequency component as H_t and the low-frequency component as L_t.

Generally speaking, high-frequency component contains distinct periodic patterns and are responsive to market fluctuations, while low-frequency component is less sensitive to short-term irregular events and reflects the long-term potential development direction. Therefore, we can use different strategies to predict the two components respectively in order to achieve better performance, and they will be discussed in detail in the following sections.

3.2 High-Frequency Component Prediction

Recurrent Layer. Long short-term memory neural network is one of the most useful deep learning methods when addressing time series with cyclical correlations, especially when common prediction objectives for liner shipping are rather influenced by the historical temporal patterns.

Instead of merely using hidden states h_t to transmit information between cells, LSTM also has cell states C_t which are able to store and renew information through the control of three gates, namely the input gate, the forget gate and the output gate. Based on this mechanism, the gradient of weight will not vanish quickly through the cell chain [9]. These three gates are all activated by the sigmoid activation function σ.

If we denote the input of the current cell state as x_t, then the function of the forget gate can be illustrated as follows. h_{t-1} is the previous hidden state, which will be combined with the current input and be processed by weight W_f and bias b_f. After the activation function is applied, the coefficient f_t determines how much information of the previous cell state C_{t-1} will be memorized. The input gate performs the following tasks. $\widetilde{C_t}$ records how much the input information will be added into the current cell state C_t, which will be updated after this gate. The output gate uses the renewed current cell state C_t and the gate weight o_t to produce the current output value y_t, also as the current hidden state to be transmitted to the next cell. We use LSTM layers to extract recurrent patterns in the high-frequency component H_t. The output of the LSTM layer is a hidden state matrix $S_t = \{s_1, s_2, \cdots, s_w\} \in \mathbb{R}^{f \times w}$, which is composed of outputs of all timesteps, and f denotes the number of hidden state features.

Temporal Attention Layer. The classical LSTM model utilizes a fully-connected layer to transform the last hidden state into a vector which is the same size of the prediction objective as output. In order to better extracting long-term correlation in time series, we add a temporal attention layer before forming the output. It can identify critical historical information from the long-run and better deal with the lagging conditions involved in the liner shipping market.

The attention mechanism allocates additional weight to each hidden state along the cell chain, which enables the information stored in the previous cell states be directly used [19]. It measures the similarity of the last hidden state and all previous hidden states through a scoring function. The common choice of scoring function includes dot-product, bilinear transformation and multilayer perceptron. Then, the similarity value is normalized into $(0, 1)$ by softmax, and the sum of all weights equals to 1. $\alpha_i = Softmax(Scoring(s_w, s_i))$, $i = 1, 2, \cdots, w$ The context vector is the weighted combination of all hidden states. $v_t = \sum_{i=1}^{w} \alpha_i s_i \in \mathbb{R}^f$ At last, we concatenate the context vector and the last hidden state and apply a fully-connected transformation to it so that it matches the dimension of high-frequency component, and get the prediction y_t^H. $y_t^H = W_H [v_t || s_t] + b_H \in \mathbb{R}^d$.

3.3 Low-Frequency Component Prediction

Compare the waveform of low-frequency component with high-frequency component, the former one is smoother and steadier. The low-frequency component usually shows the scale change and global trend of the original data, with high-frequency component showing great volatility. In liner shipping datasets, the scale of input series constantly changes in a non-periodic manner due to the influence of economy, society and politics condition. Many researches about time series forecasting have demonstrated the important role of autoregression in reacting to the scale change of input [20,21]. Hence, we use a classical Autoregressive (AR) layer to model the low-frequency component. It fits the prediction value through linear combination of historical observations. The input of the AR layer is L_t. The prediction of low-frequency component is: $y_t^L = W_A L_t + b_A \in \mathbb{R}^{m-d+1}$, where $W_A \in \mathbb{R}^w$ and $b_A \in \mathbb{R}$ are the coefficients of the AR layer. In our model, all low-frequency sub-signals share the same autoregressive parameters.

Finally, we concatenate the prediction of high-frequency and low-frequency components and obtain the whole prediction result. The objective value can be reconstructed by sum up the value of all sub-signals.

At the end of this section, we recapitulate the main ideas of the ELA model.

1. **Decompose and reconstruct by frequency.** The raw signal from liner shipping market includes large fluctuation due to varying purchasing behavior of multiple companies. Therefore, we decompose the original time series into several sub-signals (IMFs and residual) through EMD algorithm. Furthermore, we divide them into a high-frequency component and a low-frequency component through zero mean test. This approach can combine sub-signals with similar properties, making it convenient to apply suitable time-series forecasting method to each component.
2. **Make one-step ahead prediction.** For the high-frequency component, we have hidden states as outputs from LSTM. For the low-frequency component, we use autoregressive to generate prediction value.
3. **Grasp critical information from history.** Since there is lagging effect involved in liner shipping market, we attach a temporal attention layer after

the LSTM layer to add additional weight for each hidden states, so that we can exploit important information from relatively earlier timesteps.

4. **Back from sub-signals to the original.** We add the predicted values of all the sub-signals together to form the final prediction.

4   Experiment

In this section, we conduct extensive experiments on three real-world datasets collected from an anonymous liner shipping company. The experiment results of four baseline methods and the ablation study shows the state-of-the-art performance of the proposed ELA model.

4.1   Data Description

We collect our datasets from Shanghai PANASIA Shipping Co., Ltd., which is an affiliated enterprise of Cosco Container Lines Co., Ltd., COSCON. The raw data comes from three databases: SVVD (Shipping Line, Vessel Name, Voyage Number, Direction), container route and container fare. After data preprocessing, we obtain the year-round univariate time-series of container fare, and conduct experiments on port pairs Yingkou-Qinzhou, Yingkou-Nansha and Xingang-Nansha.

The data we collected is from 2019/1/2 to 2020/6/30, 547 entries in total. All datasets are split into training set (60%), validation set (20%) and test set (20%) in the chronological order.

4.2   Experiment Settings

Baselines. We choose several models widely used in time series forecasting and mentioned in the related work as baselines in model evaluation, including AR, LSTM, LSTM with attention (LSTM-A), and the classical EMD-LSTM.

Evaluation Metrics. We use three conventional evaluation metrics, Mean absolute error (MAE), Mean square error (MSE), R square (R^2).

Experiment Details. We conduct grid search over all tunable hyperparameters on the validation set for each method and dataset. All methods share the same window size $w = 30$, which means that we use observations of one month to predict the following day. The number of hidden state features in LSTM ranges in $\{2^2, 2^3, \cdots, 2^6\}$. The default batch size is 16. We use the Adam [22] algorithm to optimize the parameters in our model, and the learning rate is 0.001. During model training, we adopt the early-stopping technique to prevent model from over-fitting and the patience ranges from 10 to 50.

4.3   Results and Analysis

Table 1 shows the prediction results of five methods for port pairs Yingkou-Qinzhou, Yingkou-Nansha and Xingang-Nansha, respectively. The best evaluation metrics are highlighted in boldface.

Table 1. Prediction results of five methods on three datasets.

Dataset	Metric	AR	LSTM	LSTM-A	EMD-LSTM	ELA
Yingkou-Qinzhou	MAE	131.4476	85.7295	85.6918	64.2104	**54.0238**
	MSE	24796.48	12816.39	11902.21	7582.25	**4475.61**
	R^2	0.7267	0.8588	0.8688	0.9164	**0.9507**
Yingkou-Nansha	MAE	262.1004	230.4381	226.1365	214.5834	**211.4957**
	MSE	121522.9	93439.75	91413.19	88529.28	**76968.37**
	R^2	0.3612	0.5088	0.5195	0.5346	**0.5983**
Xingang-Nansha	MAE	93.4423	72.5317	76.1381	**50.9401**	51.9982
	MSE	14263.07	8627.76	9362.76	4295.67	**4126.04**
	R^2	0.7751	0.8639	0.8524	0.9323	**0.9349**

Comparing the former three baselines with the latter two models that adopt EMD, time-series decomposition contributes a lot in increasing accuracy, showing its ability to reducing intricate noise and enhancing feature learning. According to results of Yingkou-Qinzhou and Yingkou-Nansha, our model achieves the lowest MAE, MSE and the highest R^2 than the baseline models. Comparing the results of ELA with EMD-LSTM and LSTM with LSTM-A, the attention mechanism reveals positive effect on learning useful patterns in history. On the third dataset Xingang-Nansha, the performance of LSTM-A is worser than LSTM, and the MAE of EMD-LSTM is smaller than that of ELA. Figure 2 shows the EMD result of dataset Xingang-Nansha and can explain the reason behind this situation. There is an abnormal huge fluctuation around February, 2020, which indicates that there are unusual patterns exist in the validation set. Since our model is evaluated on the validation set during training, this phenomenon may have a negative impact on training. The attention mechanism cannot discover similar patterns from previous series, causing the baselines with or without attention to have a comparable performance.

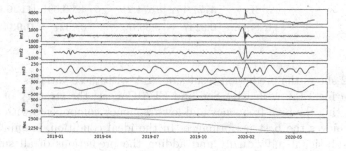

Fig. 2. The EMD result of dataset Xingang-Nansha.

4.4 Ablation Study

In this section, we carry out an ablation study to analyze the improvement of our model compared with classical EMD-LSTM. In Sect. 3.1, we have discussed that our model aims to deal with disadvantages of the classical EMD-LSTM through the following design: (1) Unlike training a LSTM for each IMF independently, we divide the IMFs into high- and low-frequency components and adopt different learning strategies (LSTM and AR) according to their characteristics; (2) We add the attention mechanism to the Vanilla LSTM in order to improve the prediction accuracy by memorizing temporal patterns in history. Based on these perspectives, we build several models for ablation study to verify and interpret our modification.

- **ELA-Sep** adds an attention layer to the classical EMD-LSTM, and trains a model for each sub-signals separately.
- **ELA-Concat** also adds an attention layer to the classical EMD-LSTM, but it concatenates all the sub-signals together to form one multivariate input to the LSTM layer.
- **EL** is a model which makes division of sub-signals as ELA but does not employ the attention mechanism.

Table 2. Results of ablation study.

Dataset	Metric	ELA-Sep	ELA-Concat	EL	ELA
Yingkou-Qinzhou	MAE	57.8114	80.9037	56.1789	**54.0238**
	MSE	5311.18	10443.71	4770.09	**4475.61**
	R-Square	0.9415	0.8849	0.9474	**0.9507**
Yingkou-Nansha	MAE	228.2582	229.8834	213.8965	**211.4957**
	MSE	97069.11	88753.08	79067.31	**76968.37**
	R-Square	0.4898	0.5335	0.5844	**0.5983**
Xingang-Nansha	MAE	54.6448	62.6586	**51.7282**	51.9982
	MSE	4771.46	7374.21	4268.09	**4126.04**
	R-Square	0.9247	0.8837	0.9327	**0.9349**

Table 2 shows the prediction results of ELA with these variants.

Comparing ELA with ELA-Sep and ELA-Concat, the strategy of dividing and predicting the high-frequency and low-frequency components with two sub-models achieves the best performance. In the ideal condition, training models for each sub-signal individually and adding the predictions of all sub-signals together to acquire the original signal is a fine-grained approach. However, the prediction error of models will accumulate and become a bottleneck to optimizing the overall performance. Additionally, we need to tailor the parameters for each sub-model, which takes huge time and effort in practice. On the other hand, concatenating all sub-signals together and forecasting the matrix as a whole also fails to generate satisfactory result. This is because concatenation merges separate signals again, which makes it difficult for the neural network to learn underlying features from high-dimensional input. Therefore, we conclude that our sub-signal reconstruction method successfully makes a tradeoff in maintaining accuracy and reducing training cost.

5 Conclusion

Based on the characteristics of data in liner shipping, our study constructs an ELA model and applies it to the container fare prediction for liner shipping companies. In terms of high volatility, EMD algorithm decomposes the original signal with noise into several regular and much more easily-predicted sub-signals, which we divide into two categories by frequency. Then, we use LSTM neural network to process the high-frequency sub-signals and obtain the hidden state features. To confirm the effectiveness of our model, we collect real-world transaction records from our cooperated shipping company and design appropriate experiments to test its forecasting ability. Results show that for the chosen three port pairs, ELA model has the lowest MAE, MSE and the highest R^2 among other baselines and ablation models on the whole.

Acknowledgement. This work was supported by the National Key R&D Program of China [2020YFB1707900]; the National Natural Science Foundation of China [62272302]; and the COSCO Shipping Container Transport Co., Ltd's Shipping Line Intelligent Pricing Model Construction Project [IT2020076].

References

1. World shipping council homepage. https://www.worldshipping.org/
2. Guo, R., Wang, Y., Zhang, H., Zhang, G.: Remaining useful life prediction for rolling bearings using EMD-RISI-LSTM. In: IEEE Transactions on Instrumentation and Measurement, vol. 70, pp. 1–12. IEEE (2021)
3. Stallone, A., Cicone, A., Materassi, M.: New insights and best practices for the successful use of Empirical Mode Decomposition, Iterative Filtering and derived algorithms. Sci. Rep. **10**(1), 1–15 (2020)
4. Ryan, M.: Decomposing signal using Empirical Mode Decomposition. Towards data science (2019). https://towardsdatascience.com/decomposing-signal-using-empirical-mode-decomposition-algorithm-explanation-for-dummy-93a93304c541
5. Hochreiter, S., Schmidhuber, J.: Long short-term memory. Neural Comput. **9**(8), 1735–1780 (1997)
6. Fahim, A., Tan, Q., Mazzi, M., Sahabuddin, M., Naz, B., Ullah Bazai, S.: Hybrid LSTM self-attention mechanism model for forecasting the reform of scientific research in Morocco. Comput. Intell. Neurosci. **2021**, 6689204 (2021)
7. Chen, X., Li, B., Wang, J., Zhao, Y., Xiong, Y.: Integrating EMD with multivariate LSTM for time series QoS prediction. In: 2020 IEEE International Conference on Web Services (ICWS), pp. 58–65. IEEE (2020)

8. Yan, L., Chen, C., Hang, T., Hu, Y.: A stream prediction model based on attention-LSTM. Earth Sci. Inf. **14**(2), 723–733 (2021). https://doi.org/10.1007/s12145-021-00571-z

9. Yujun, Y., Yimei, Y., Jianhua, X.: A hybrid prediction method for stock price using LSTM and ensemble EMD. Complexity **2020**, 6431712 (2020)

10. Zhang, B.: Foreign exchange rates forecasting with an EMD-LSTM neural networks model. J. Phys: Conf. Ser. **1053**(1), 012005 (2018)

11. Zheng, H., Yuan, J., Chen, L.: Short-term load forecasting using EMD-LSTM neural networks with a Xgboost algorithm for feature importance evaluation. Energies **10**(8), 1168 (2017)

12. Bedi, J., Toshniwal, D.: Data decomposition based learning for load time-series forecasting. In: Koprinska, I., et al. (eds.) ECML PKDD 2020. CCIS, vol. 1323, pp. 62–74. Springer, Cham (2020). https://doi.org/10.1007/978-3-030-65965-3_5

13. Lin, Z., Li, M., Zheng, Z., Cheng, Y., Yuan, C.: Self-attention ConvLSTM for spatiotemporal prediction. AAAI Conf. Artif Intell. **34**(07), 11531–11538 (2020)

14. Tang, G., Lei, J., Shao, C., Xiong, H., Cao, W., Men, S.: Short-term prediction in vessel heave motion based on improved LSTM model. IEEE Access **9**, 58067–58078 (2021)

15. Le, H.V., Murata, T., Iguchi, M.: Can eruptions be predicted? Short-term prediction of volcanic eruptions via attention-based long short-term memory. In: AAAI Conference on Artificial Intelligence (AAAI), vol. 34, pp. 13320–13325 (2020)

16. Viellechner, A., Spinler, S.: Novel data analytics meets conventional container shipping: predicting delays by comparing various machine learning algorithms. In: Hawaii International Conference on System Sciences (HICSS), pp. 1–10 (2020)

17. Huang, X., et al.: LSTM based sentiment analysis for cryptocurrency prediction. In: Jensen, C.S., et al. (eds.) DASFAA 2021. LNCS, vol. 12683, pp. 617–621. Springer, Cham (2021). https://doi.org/10.1007/978-3-030-73200-4_47

18. Li, B., He, Y.: An attention mechanism oriented hybrid CNN-RNN deep learning architecture of container terminal liner handling conditions prediction. Comput. Intell. Neurosci. **2021** (2021)

19. Agrawal, N.: Understanding attention mechanism: natural language processing (2020). https://medium.com/analytics-vidhya/https-medium-com-understanding-attention-mechanism-natural-language-processing-9744ab6aed6a

20. Lai, G., Chang, W.C., Yang, Y., Liu, H.: Modeling long-and short-term temporal patterns with deep neural networks. In: The 41st International ACM SIGIR Conference on Research & Development in Information Retrieval, pp. 95–104 (2018)

21. Huang, S., Wang, D., Wu, X., Tang, A.: DSANet: dual self-attention network for multivariate time series forecasting. In: Proceedings of the 28th ACM International Conference on Information and Knowledge Management, pp. 2129–2132 (2019)

22. Kingma, D.P., Ba, J.: Adam: a method for stochastic optimization. Comput. Sci. arXiv preprint arXiv:1412.6980 (2014)

Knowledge Transfer via Word Alignment and Its Application to Vietnamese POS Tagging

Hao D. Do[✉]

FPT University, Ho Chi Minh City, Vietnam
haodd3@fpt.edu.vn

Abstract. It is not difficult to build a linguistic tagger with a large annotated corpus. Labeled data becomes a big problem with low-resource languages such as Vietnamese. Due to the development and investment in research, there is no large and high-accuracy annotated corpus. This paper proposes a transfer learning strategy to build a high-quality tagger for Vietnamese using a bilingual corpus Vietnamese-English. Particularly, We inherit the strength of a POS tagger in English, which is constructed from a large-scale corpus, then transfer the knowledge to the Vietnamese POS tagger via word alignment. Experimental results show that the proposed method achieves high accuracy with 94.97%. The transfer strategy depends mostly on the source tagger's accuracy and the word alignment process's performance, so the proposed strategy can be extended and applied to another low-resource language as long as there is a large bilingual corpus with a rich resource language.

Keywords: Transfer learning · POS tagging · Bilingual corpus · Hidden Markov model · Joint probability

1 Introduction

In many fields of artificial intelligence, especially natural language processing, annotated data is one of the essential elements of the research. Data is the foundation on which we can build and evaluate models. Without data, we only can build rule-based models, which are effort and time-consuming. Besides, rule-based models can not solve the popular exceptions in natural languages. If we have a large enough annotated corpus, the model is accurate and can reduce errors when used in real-life applications.

In developing countries, the need for natural language processing applications is uncompromising. For example, in Vietnamese, users currently need many applications which can do many tasks such as plagiarism checking, sentiment analysis, spelling checking, or grammar checking. The studies are not going well because there is no annotated data for fundamental tasks, including POS, NER, or NP chunking. The complex applications below can not be completed without basic tasks.

T. N. Dinh and M. Li (Eds.): CSoNet 2022, LNCS 13831, pp. 81–92, 2023.
https://doi.org/10.1007/978-3-031-26303-3_8

Transfer learning provides a powerful approach for inheriting the achievement from a rich resource such as English to improve the learning model in another language. Our research focuses on building a model for Vietnamese POS tagger with the support of a high-quality English tagger via word alignment. Firstly, we used a large English-Vietnamese bilingual corpus from Ted talk to build a word alignment corpus. After that, we built a Hidden Markov Model (HMM) for Vietnamese tagger from a small annotated dataset and used it to tag Vietnamese sentences in the bilingual corpus. Then, on the English side, we used the Stanford tagger to label the sentences and computed the tag-correlation matrix of two tag sets. Next, we built our proposed model, which was the combination between an HMM and a bilingual projecting model, to tag again all sentences on the Vietnamese side in the bilingual corpus. Finally, we built the second Vietnamese POS tagger after merging the tagged Vietnamese sentences in the bilingual corpus with the original labeled sentence. Experiments demonstrate that the proposed method gains a better accuracy than current Vietnamese sequence tagging research.

The main contributions of this work include:

- Proposing a strategy to transfer the achievement of a rich resource language to a low resource language via a word alignment.
- Solving some limits of the previous projecting methods, such as cross alignment or out of alignment.
- Building a high-quality model for Vietnamese POS tagger.

Our paper is structured as follows: after introducing the research in Sect. 1, we present some related works in Sect. 2. In this section, we analyze the advantages and disadvantages of the previous models. Next, in Sect. 3, we present our proposed models. We focus on how to design the transfer strategy and how we resolve the problems, which are complicated with the previous models. After that, we do some experiments in Sect. 4. The paper ends with Sect. 5, presenting the conclusions and future works.

2   Related Works

2.1   POS Tagging for Monolingual Corpus

Sequence labeling is a sub-class of the classification problem. The sequence is an ordered list of objects. Each object is in a relationship with the others and influential together. With the POS task, the tag of one word depends on itself and many words in its surroundings or context. The first and the most simple model for labeling a sequence is Hidden Markov Model. TnT tagger, proposed by Thorsen [1], which used a second order Hidden Markov Model, was a former state-of-the-art model for POS tagging. One year later, a similar model was introduced by Thede [19]. Maximum Entropy and Maximum Entropy Markov Model were used for POS tagging by Toutanova [20,21] and Dennis [5]. These two models can solve some problems of previous models, so their results are better. Conditional Random Field was applied for POS tagging by Sun [18] and Huang [7], and then these models became the next state of the art.

Sequence labeling is a good approach but not perfect. The models use almost linguistic information to assign tags, so the results are very high. Besides, they do not require any particular data except a large annotated corpus. So with a rich resource language like English, it is easy to understand why these models are state of the art. But the best choices do not mean that they do not have any disadvantages. First, because they need a large annotated corpus, applying these models to low-resource language is difficult. Second, each language contains its exceptions, which are unlimited and uncountable. This leads to the fact that a limited corpus is never enough. Finally and clearly, we can improve labeling tasks with external information, which is not done by the sequence labeling approach. For these reasons, we need a better model for low-resource language.

Focusing on Vietnamese POS tagging, there are many pieces of research by domestic researchers. Excepting Dien [6], almost other researchers constructed the sequence labeling models. Huyen and Luong [10,11] used a probabilistic model while Hieu [14] implemented flexible Conditional Random Field for tagging. On the other hand, Minh proposed a Support Vector Machine for labeling [8]. Generally, these researches provided a high accuracy tagger, but not high enough for real applications.

2.2   Projecting Approach for Tagging

Different from models using internal information below, the bilingual projecting model completely depends on external information. First introduced by Yarowsky [24,25] and extended some years later by Dien [6], this model was the only choice for low-resource language. These studies used a bilingual corpus, including a rich-resource language and the language they wanted to label.

This approach is based on two foundations. The first is the excellent result from rich resource language. With this type, many researchers and projects build a large annotated corpus and beautiful labeling models. The satisfactory results from the rich side of the bilingual corpus are the information sources for transferring to the other side. The second is the result of word alignment when the bilingual corpus is large enough. With high accuracy from the source language and good alignment, this approach can label low-resource language with a favorable outcome. When the language is too low-resource, projecting is the best choice. It helps to build a large annotated corpus with high accuracy without annotated training data. But there is one disadvantage of this model. Because this approach is completely basing on external information, so we can not improve this model actively. The accumulation of errors from labeling in the source language and word alignment is the limitation of this approach.

2.3   Bilingual Learning for Sequence Labeling

Different from the bilingual projecting model, bilingual learning labels two words in two languages simultaneously. Besides, this approach is unsupervised learning. It means that we do not use annotated data for training. One of the first successful applications of bilingual learning is Word Sense Disambiguation by Brown [2]. The linguistic studies say that the ambiguation of different languages is different, so this leads us to a comment that we can use this word in this language to disambiguate its corresponding word in the other language.

In recent years, many applications of bilingual learning can be listed as POS tagging by Snyder [16,17] and Naseem [9], Name entity recognition by Wang [22,23], parsing by Burkett [3,4]. The same of these researches are combining two predictors of two languages with word alignment. The chosen labels are the balance between occurrence probability and co-occurrent probability.

This approach is more general and broader in application than the bilingual projecting model. In corpus, we can use any pair of languages, not only a rich resource and a low-resource language. On the other hand, this is pure supervised learning, so we do not need any annotated data. Because using two different languages, each language will disambiguate the other. These are all disadvantages of this approach.

There are two problems with all research in this method. First, the accuracy is not too high because this is a pure unsupervised learning method. We can improve it by using semi-supervised models. Second, bilingual learning requires a sentence pair not containing any cross-alignment. Snyder said that his research had been only applied for a pair of languages with under 5% cross alignment [16,17]. For Vietnamese-English bilingual data, we must transform this model into a more suitable model.

2.4 Transfer Learning and Its Potential

Transfer learning is a class of effective methods to deal with a difficult problem using an achievement from another resource. To solve a problem with machine learning approach, there are two must-have aspects, including data and model. In the world of transfer learning, the resource for transfer can be the data or model from another task. In this context, data could be a part, not a whole, of the original data. Similarly, the model could be the whole model, or a part of the model, for example, some beginning layers of a neural network.

In transfer learning, the similarity of two tasks is the most important. Obviously, it is challenging to deal with a hard problem when transferring irrelevant information. On the other hand, the transfer model or transfer strategy is important with two problems with some aspects that are similar. It is the key element in deciding whether the destination problem is solved or not. In this research, English and Vietnamese are not closely similar to English with French (in term of vocabulary) or Vietnamese with Chinese, but they also share some functions and structures of a language so that they could be seen as the same in some aspects. The remaining is how to design a transfer strategy to enhance the current tagger for Vietnamese.

3 Transfer Learning in English-Vietnamese Bilingual Corpus

3.1 Bilingual Learning

Let us start with the bilingual learning model proposed by Snyder [16,17] for POS tagging, which is illustrated in Fig. 1.

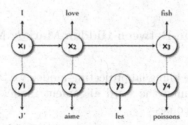

(a) Two HMM are combined via word alignment

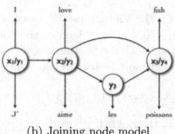

(b) Joining node model

Fig. 1. Bilingual learning for POS tagging

This is the combination of two Hidden Markov Models via word alignment. They constructed one Hidden Markov Model for predicting states with each language in the bilingual corpus. Then the states which are aligned will be tagged at the same time. The prediction begins from the first word of each sentence and finishes at the end of each sentence. The model is given by the formula as follows:

$$P(x_i, y_i | x_{i-1}, x_{i-2}, y_{i-1}, y_{i-2}) \propto P(x_i | x_{i-1}, x_{i-2}) P(y_i | y_{i-1}, y_{i-2}) P(x_i, y_i) \quad (1)$$

with x_n, y_m denote the tags for the labels of nth, m^{th} words in the two languages.

Because two Hidden Markov Models tag in the same direction, the two sequences of states or the sequences of tags are in the same direction too. But in the Vietnamese-English bilingual corpus, these orders are not the same.

In the example in Fig. 2, we can not tag state x_4 and y_5 because we need to know the tag of y_4 and x_5 before. So with the bilingual containing a high cross-alignment ratio like Vietnamese-English, we must solve it first.

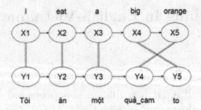

Fig. 2. Two sentences with cross alignment

3.2 The Combination Between Hidden Markov Model and Bilingual Learning

Inheriting the idea of the bilingual projecting model, we tagged the rich resource language first after running the word alignment model. This means that the objective probability becomes:

$$P(v_j|v_{j-1}, w_{v_j}, e(w_{v_j})) \tag{2}$$

with v_j denotes the label for the Vietnamese j^{th} word, w_{v_j} denotes the jth word in the Vietnamese sentence, and $e(w_{v_j})$ denotes the corresponding label of the English word which is aligned with the Vietnamese j^{th} word. On the other hand, the multiplication of two rightmost factors in the formula 1 is approximated by:

$$P(e_i|e_{i-1}, e_{i-2})P(v_i, e_i) \propto P(v_i|e(w_{v_j})) \tag{3}$$

with e_j denotes the label for the English j^{th} word. In this formula, the probability $P(v_i|e(w_{v_j}))$ present the tag-correlation information of two languages. Because tag sets of two language are limit, $P(v_i|e(w_{v_j}))$ can be presented as a matrix called tag-correlation matrix. This is the main agent to connect two languages, including containing rich information from a success model. So this matrix becomes the key element in the transfer process.

From this time to the end of our research, the sentences of rich resource language are a list of discrete states instead of a sequence of words or states (Fig. 3a. So there is not any cross-alignment line in our model. We can predict the label of one word by three pieces of information, including itself, the states before, and the corresponding aligned tag via the tag-correlation matrix (Fig. 3).

Our model is the combination of a model using internal information and a model using external information. There are three models based on how we combine these two models. If the combination operator is addition, there are two model corresponding with normal addition and weighted addition. The remaining model is created when the combining operator is multiplication. We describe our three models in detail in the next subsections.

The First Model. In our first model (Fig. 4), we maximize the probability:

$$P(v_j|v_{j-1}, w_{v_j}, e(w_{v_j})) \tag{4}$$

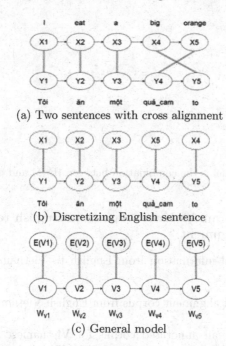

(a) Two sentences with cross alignment

(b) Discretizing English sentence

(c) General model

Fig. 3. The combination between HMM and bilingual learning model

To do this, we must choose the best probability $P(v_j|v_{j-1}, w_{v_j})$ and $P(v_j|e(w_{v_j}))$ and then merge them together. They are the Hidden Markov Model and bilingual projecting model. This means that we must find tag v_j such that:

$$v_j = ArgMaxP(v_j|v_{j-1}, w_{v_j}, e(w_{v_j})) \propto ArgMaxP(v_j|v_{j-1}, w_{v_j}) + P(v_j|e(w_{v_j}))$$
(5)

Tag v_j satisfies the above optimization equation is the balance between the model using internal information and the model using external information.

The Second Model. This model is a more general form of the first model. In the previous model, we want to find the optimal choice between two elements, but we do not present their different impacts on the model. So in the second model, we use a weight λ to highlight the role of one element in the model. This means that we need to find the tag v_j such that:

$$v_j \propto ArgMax\left(P(v_j|v_{j-1}, w_{v_j})\right)\lambda + \left(P(v_j|e(w_{v_j}))\right)(1 - \lambda)$$
(6)

The Third Model. Unlike the previous models, the third model uses multiplying to combine two probabilities instead of adding.

$$v_j \propto ArgMax\left(P(v_j|v_{j-1}, w_{v_j})\right)\left(P(v_j|e(w_{v_j}))\right)$$
(7)

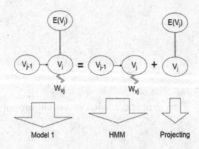

Fig. 4. The first model is the combination between HMM and projecting model

3.3   Strategy to Transfer Knowledge from English to Vietnamese via Word Alignment

To transfer the useful information from English to Vietnamese, our proposed strategy is as follows:

– Step 1: Build word alignment corpus from English-Vietnamese bilingual corpus.
– Step 2: From a small annotated corpus for Vietnamese named Corpus-V, build an HMM for POS tagger, named V-HMM, then use V-HMM to tag Vietnamese sentences in the bilingual corpus
– Step 3: Use a high-quality English POS tagger to tag English sentences in the bilingual corpus.
– Step 4: Compute the tag-correlation matrix, then ignore the tagged in Vietnamese sentences and apply the proposed models to tag the Vietnamese sentences in the bilingual corpus again.
– Step 5: Merge the original Corpus-V (in step 2) with all Vietnamese sentences in the bilingual corpus (after step 4), then use the merge corpus to build a new POS tagger for Vietnamese.

4   Experiments

4.1   Tools and Data Used in Experiments

In the experiment, we used three tools listed in Table 1 and two datasets listed in Table 2. To build the HMM for the Vietnamese POS tagger, we reused the TnT tagger. On the English side, we used the Stanford tagger to label the English sentence. Finally, we run GIZA++ [13] for word alignment process.

The small dataset used to build the Vietnamese POS tagger in this research was NIIVTB [12], which contains over 10k sentences. We also used TED2020 corpus [15] and then extracted over 300k pairs of bilingual English-Vietnamese sentences.

Table 1. List of tools used in this work

#	Name	Purpose
1	TnT tagger	Computing $P(v_j|v_{j-1}, w_{v_j})$
2	Stanford tagger	Computing $P(e_j|e_{j-1}, w_{e_j})$
3	GIZA++	Aligning English - Vietnamese

Table 2. List of datasets used in this work

#	Name	Size	Train/Dev/Test	Purpose
1	NIIVTB	10.431 sentences	8.431/1.000/1.000	Building HMM
2	TED2020	300.000+ sentences	All for training	Aligning English - Vietnamese

4.2   Training

The training process for our models includes two main phases. The first phase focuses on building the tag-correlation matrix as follows:

- Step 1: From bilingual corpus, running GIZA++ for word alignment.
- Step 2: Build TnT Tagger for Vietnamese with NIIVTB, then tag all Vietnamese sentences in the bilingual corpus.
- Step 3: Tag all English sentences in bilingual corpus with Stanford Tagger.
- Step 4a: Compute $P(v_j|e(w_{v_j}))$.

The result of this process is the tag-correlation matrix P, which stores the constraints between two languages via POS labels. Matrix P is the main parameter for our three models. It will help us construct a more powerful Vietnamese tagger. The second phase is the transfer step, which corresponds with the remaining of step 4 and step 5 in Subsect. 3.3. This phase could be separated into sub-steps as follow:

- Step 4b: Run the proposed models using $P(v_j|e(w_{v_j}))$ to tag all Vietnamese sentences in the bilingual corpus again.
- Step 5a: Merge the Vietnamese tagged sentences in step 4b with NIIVTB
- Step 5b: Use the merged corpus to build a new POS tagger for Vietnamese.

4.3   Experimental Results

We used our proposed models in the first experiment to test with 1.000 annotated samples in NIIVTB. The tags generated by the models were compared with the ground truth to compute total accuracy. Table 3 shows the accuracy of our different model configurations.

As can be seen in Table 3, all of our proposed models are better than the baseline model. The highest accuracy 94.97% comes from the third model, which uses multiplication as the combination operator. This is because the multiplication shows the mutual contribution of two factors better than the addition operator.

Table 3. Accuracy of our three proposed models

Model	Hyperparameter	Combination operator	Accuracy (%)
Baseline (HMM)	–	No	85.27
Model 1	–	Addition	88.54
Model 2	$\lambda = 0.2$	Addition	87.91
Model 2	$\lambda = 0.4$	Addition	88.40
Model 2	$\lambda = 0.6$	Addition	92.16
Model 2	$\lambda = 0.8$	Addition	91.08
Model 3	–	Multiplication	94.97

The result in this experiment also shows model 3 improves 2.81% than the highest accuracy of these two remaining models. Among the four instants of model 2, which uses a weighted addition operator, $\lambda = 0.6$ yields the best results with 92.16%. Model 2 gains better accuracy than model 1 because model 1 is only a special case of model 2, corresponding with $\lambda = 0.5$. The contributions of different elements are different, so a flexible λ can achieve higher accuracy.

In the second experiment, we compared our best model - model 3, with another well-known model for Vietnamese POS tagger.

Table 4. Results for Vietnamese POS Tagging using different methods

Authors	Data	Learning	Model	Accuracy (%)
Huyen et al.	Monolingual	Supervised	Probabilistic	93.04
Hieu et al.	Monolingual	Supervised	CRF	91.76
Chau et al.	Monolingual	Supervised	Probabilistic	89.54
Q. Minh et al.	Monolingual	Supervised	SVM	94.89
L. Minh et al.	Monolingual	Supervised	SVM	94.10
Phuong et al.	Monolingual	Supervised	Probabilistic	91.85
Phong et al.	Monolingual	Supervised	Probabilistic	94.50
Dien et al.	Bilingual	Projection	Projection	94.60
This research	Monolingual + bilingual	Transfer	Projection + HMM	94.97

Table 4 shows that the model in our research gains a good result, as well as the very high accuracies in the past. Tagger using SVM, researched by Q. Minh et al., yields a high accuracy with 94.89%. Our model gains a better numerical result, but this is not a clear improvement.

Via two experiments, our best model, which uses a multiplication operator to combine two probabilities, can perform as well as many other works. The difference here is we use less annotated data than the other. Our model can achieve a good result because the model inherits the strength of a rich resource language with transfer learning via word alignment in the bilingual corpus.

5 Conclusions

This paper presents a strategy to transfer knowledge from English, a rich resource language, to Vietnamese, a low-resource language, via word alignment in the bilingual corpus. The useful information is encoded and transferred from a highly accurate tagger in English into the Vietnamese POS tagger via the tag-correlation matrix. The joint probability, represented by the multiplication operator, performs better than the addition operator and gains the highest accuracy.

Because English and Vietnamese are similar in some aspects, the transfer strategy works well in this research. Via word alignment, information for the English tags is used to correct the Vietnamese tags. Currently, there are some bilingual or multilingual corpus such as news or Ted talk, so the strategy proposed in this work can be extended and used for any pair of a rich resource language and a low resource language.

Acknowledgement. Hao D. Do was funded by Vingroup JSC and supported by the Ph.D. Scholarship Programme of Vingroup Innovation Foundation (VINIF), Institute of Big Data, code VINIF.2021.TS.120. The author would like to thank the Computational Linguistics Center (Lab C44), University of Science - VNUHCM, for their support.

References

1. Brants, T.: TnT: a statistical part-of-speech tagger. ANLP (2002). https://doi.org/10.3115/974147.974178
2. Brown, P.F., Pietra, S.A.D., Pietra, V.J.D., Mercer, R.L.: Word-sense disambiguation using statistical methods. In: 29th Annual Meeting of the Association for Computational Linguistics, pp. 264–270 (1991). https://doi.org/10.3115/981344.981378
3. Burkett, D., Klein, D.: Two languages are better than one (for syntactic parsing). In: Proceedings of EMNLP, pp. 877–886 (2008). https://doi.org/10.3115/1613715.1613828
4. Burkett, D., Petrov, S., Blitzer, J., Klein, D.: Learning better monolingual models with unannotated bilingual text. In: Proceedings of CoNLL, pp. 46–54 (2010)
5. Denis, P., Sagot, B.: Coupling an annotated corpus and a morphosyntactic lexicon for state-of-the-art POS tagging with less human effort. In: PACLIC 23 - Proceedings of the 23rd Pacific Asia Conference on Language, Information and Computation, vol. 1, pp. 110–119 (2009)
6. Dien, D., Kiem, H.: POS-tagger for English-Vietnamese bilingual corpus. In: Proceedings of the HLT-NAACL 2003 Workshop on Building and Using Parallel Texts: Data Driven Machine Translation and Beyond, pp. 88–95 (2003). https://doi.org/10.3115/1118905.1118921
7. Huang, Z., Xu, W., Yu, K.: Bidirectional LSTM-CRF models for sequence tagging. arXiv preprint arXiv:1508.01991 (2015)
8. Minh, N., Xuan Bach, N., Nguyen, V.C., Minh, P.Q.N., Shimazu, A.: A semi-supervised learning method for Vietnamese part-of-speech tagging. In: Proceedings - 2nd International Conference on Knowledge and Systems Engineering (KSE 2010), pp. 141–146 (2010). https://doi.org/10.1109/KSE.2010.35

9. Naseem, T., Snyder, B., Eisenstein, J., Barzilay, R.: Multilingual part-of-speech tagging two unsupervised approaches. J. Artif. Intell. Res. **36**, 341–385 (2009). https://doi.org/10.1613/jair.2843

10. Nguyen, H., Romary, L., Rossignol, M., Vũ, X.: A lexicon for Vietnamese language processing. Lang. Resour. Eval. **40**, 291–309 (2006). https://doi.org/10.1007/s10579-007-9034-8

11. Nguyen, H., Vu, X., Phuong, L.H.: A case study of the probabilistic tagger QTAG for tagging Vietnamese texts. In: Proceedings of 10th TALN (2003)

12. Nguyen, Q.T., Miyao, Y., Le, H.T.T., Nguyen, N.T.H.: Ensuring annotation consistency and accuracy for Vietnamese treebank. Lang. Resour. Eval. **52**(1), 269–315 (2017). https://doi.org/10.1007/s10579-017-9398-3

13. Och, F.J., Ney, H.: A systematic comparison of various statistical alignment models. Comput. Linguist. **29**(1), 19–51 (2003)

14. Phan, X.H., Le-Minh, N., Inoguchi, Y.: Co-training of conditional random fields for segmenting sequence data. In: Proceedings of Symposium on Data/Text Mining from Large Databases (IFSR) (2022)

15. Reimers, N., Gurevych, I.: Making monolingual sentence embeddings multilingual using knowledge distillation. In: Proceedings of the 2020 Conference on Empirical Methods in Natural Language Processing, pp. 4512–4525 (2020)

16. Snyder, B., Naseem, T., Eisenstein, J., Barzilay, R.: Unsupervised multilingual learning for POS tagging. In: Proceedings of EMNLP, pp. 1041–1050 (2008). https://doi.org/10.3115/1613715.1613851

17. Snyder, B., Naseem, T., Eisenstein, J., Barzilay, R.: Adding more languages improves unsupervised multilingual part-of-speech tagging: a Bayesian nonparametric approach. Association for Computational Linguistics, pp. 83–91 (2009)

18. Sun, X.: Structure regularization for structured prediction. Adv. Neural Inf. Process. Syst. **3** (2014)

19. Thede, S., Harper, M.: A second-order hidden Markov model for part-of-speech tagging. In: Proceedings of the 37th Annual Meeting of the Association for Computational Linguistics, pp. 175–182 (2002). https://doi.org/10.3115/1034678.1034712

20. Toutanova, K., Klein, D., Manning, C., Singer, Y.: Feature-rich part-of-speech tagging with a cyclic dependency network. In: Proceedings of the 2003 Conference of the North American Chapter of the Association for Computational Linguistics on Human Language Technology (NAACL 2003), vol. 1, pp. 252–259 (2004). https://doi.org/10.3115/1073445.1073478

21. Toutanova, K., Manning, C.: Enriching the knowledge sources used in a maximum entropy part-of-speech tagger. In: Proceedings of the Joint SIGDAT Conference on Empirical Methods in Natural Language Processing and Very Large Corpora, pp. 63–70 (2002). https://doi.org/10.3115/1117794.1117802

22. Wang, M.: Bilingual and cross-lingual learning of sequence models with bitext. PhD Thesis, Stanford University (2014)

23. Wang, M., Che, W., Manning, C.: Joint word alignment and bilingual named entity recognition using dual decomposition. In: ACL 2013–Proceedings of the 51st Annual Meeting of the Association for Computational Linguistics, vol. 1, pp. 1073–1082 (2013)

24. Yarowsky, D.: Hierarchical decision lists for word sense disambiguation. Comput. Hum. **34**, 179–186 (2000). https://doi.org/10.1023/A:1002674829964

25. Yarowsky, D., Ngai, G.: Inducing multilingual POS taggers and NP bracketers via robust projection across aligned corpora. In: Proceedings of NAACL (2001). https://doi.org/10.3115/1073336.1073362

A Group Clustering Recommendation Approach Based on Energy Distance

Tu Cam Thi Tran[1]iD, Lan Phuong Phan[2], and Hiep Xuan Huynh[2]([✉])iD

[1] Vinh Long University of Technology Education,
Vinh Long, Vinh Long Province, Vietnam
`tuttc@vlute.edu.vn`
[2] Can Tho University, Can Tho City, Vietnam
`{pplan,hxhiep}@ctu.edu.vn`

Abstract. Recommendation system focus on the individual recommendation and using relationships between users and users or between items and items by distance or similarity measures. In reality, there are many situations when recommending to individuals is not as important as recommending to groups. Researches on the relationship between a group and a group have not yet been interested in the recommendation. This paper mainly focus on applying the energy distance for group recommendation system. The proposed recommendation model is evaluated on the Jester5k and the MovieLens datasets. The experiment result shows the feasibility of applying the potential energy for the group recommendation problems.

Keywords: Group recommendation · Individual recommendation · Energy distance · Energy model · Collaborative filtering

1 Introduction

Most recommendation systems are designed to recommend for the individual users [1,15,16]. When a new user needs to be recommended, the system will find the similarity between that user and all users in the database, then suggest the suitable items to that user. Several studies on recommendation systems are based on the groups of the users [2,4,5] with similar interests. In reality there are many situations when recommending to individuals is not as important as recommending to groups. Most of researches show a linear relationship between individual and individual, a linear relationship between group and group whereas researches on the non-linear relationship between groups of users have not yet been considered.

In this paper, we propose a new collaborative filtering recommendation model using nonlinear relationships instead of linear relationships. This approach: is based on the energy distance between a user and a user to create an individual recommendation model; then, uses the potential energy to determine the

---

Supported by organization x.

relationship between a user group and a user group to perform the group recommendations.

The paper is structured as follows. Section 2 presents background and related work, including: the method of K-groups clustering [11,12], the energy distance, the group recommendation, and evaluation for the recommendation result. Section 3 proposes the model and the algorithm of the group recommendation [9] by energy. Section 4 shows the experiment on the Jester5k and the MovieLens datasets. Section 5 is the conclusion.

2   Background and Related Work

2.1   Energy Distance

Let U_1 and U_2 be independent random vectors in R^d, X and Y are cumulative distribution functions of U_1 and U_2 respectively. The energy distance [5,7,8] can be determined according to the expected distance between two random vectors as the follow.

$$D^2(X,Y) = 2E\,\|U_1 - U_2\| - E\,\|U_1 - U_1'\| - E\,\|U_2 - U_2'\| \geq 0 \qquad (1)$$

where $\|.\|$ represents the Euclidean norm, E is the expected value, and the random variable U_1' represents an independent and identically distributed (iid) copy of U_1; U_1' and U_2' are iid of U_1 and U_2.

Consider the null hypothesis that two random vectors, U_1 and U_2, have the same cumulative distribution functions: $X = Y$. For samples $u_{11}, ..u_{1n}$ from U_1 and $u_{21}, ..u_{2m}$ from U_2, the Energy for testing this null hypothesis [8] is calculated by formula (2).

$$\varepsilon(U_1, U_2) = 2A - B - C \qquad (2)$$

where A, B, and C are simply averages of pairwise distances:

$$A = \frac{1}{nm} \sum_{i=1}^{n} \sum_{j=1}^{m} \|u_{1i} - u_{2j}\| \qquad (3)$$

$$B = \frac{1}{n^2} \sum_{i=1}^{n} \sum_{j=1}^{n} \|u_{1i} - u_{1j}\| \qquad (4)$$

$$C = \frac{1}{m^2} \sum_{i=1}^{m} \sum_{j=1}^{m} \|u_{2i} - u_{2j}\| \qquad (5)$$

It is proved that $\varepsilon(U_1, U_2) = D^2(X, Y)$ is zero if and only if U_1 and U_2 have the same distribution $(X = Y)$. It is also true that $\varepsilon(U_1, U_2)$ is always non-negative. When the null hypothesis of equal distributions is true, the test statistic T [8] is the follow.

$$T = \frac{n \times m}{n + m} \varepsilon_{n,m}(U_1, U_2) \tag{6}$$

The energy distances are a powerful tool for the multivariate analysis. It applies to random vectors with the unlimited sizes. The energy distances can be applied to measure the compatibility between a sample and a hypothetical distribution, or it is also applied to measure the compatibility between two or more samples (the sizes of the samples are arbitrary, not necessarily equal).

2.2   Energy Distance Based K-Group Clustering

K-groups clustering [11,12] by energy distance [13,14] performs on the set of the samples. The K-groups clustering is created by using Gini mean distance. The formula for the k-groups distance [13,14] is:

$$K_j = \frac{n_j}{2} Gini(G_j) = \frac{1}{2n_j} \sum_{i,m=1}^{n_j} |U_{i,j} - U_{m,j}| \tag{7}$$

where the cluster Gini mean distance [2,14] is calculated by (8) the formula.

$$Gini(G_j) = \frac{1}{n_2^j} \sum_{i,m=1}^{n_j} |U_{i,j} - U_{m,j}| \tag{8}$$

In (8), G_j is the cluster j; $U_{i,j}$ and $U_{m,j}$ is the vectors U_i and U_m in cluster j.

2.3   Group Recommendation

The group recommendation systems [10] have been developed to support the recommendation involving a group of the users. The state-of-the-art in group recommendation is presented by [3,9]. Two common used methods of group recommendation [13] are: aggregation recommendation, and aggregation preferences. In aggregation recommendation, the recommendation for each user is proposed, then the recommendation for the group is aggregated from the recommendation of all member of group. In aggregation preferences, the recommendation combines the group member's preferences into a group preferences model. The determination of the group recommendation depends on the aggregation strategy function [13]. Three basic methods of the aggregation functions include: majority-based, consensus-base and borderline methods. The majority-based method represents aggregation mechanisms that focus on the most popular. Some functions are used by this method: Approval Voting; Copeland Rule. The consensus-base method represents aggregation mechanisms that take into account the preferences of all group members. Some functions are used by this method: Additive Utilitarian; Average; Average without Misery. The borderline method represents aggregation mechanisms that take into account only a subset of the user preferences. Some functions are used by this method: Least Misery; Most Respected Person.

2.4 Evaluation

Two metrics which are often used for evaluating error in individual recommendation systems [13,15] are: Mean Absolute Error (MAE) and Root Mean Square Error (RMSE).

Let $p(i,j)$ be the real rating of user i for item j; $\hat{p}(i,j)$ be the predicted rating of user i for item j and K is the set of all user-item pairings (i,j).

Root mean square error between the real ratings and the predicted ratings is defined by formula (9).

$$RMSE = \sqrt{\frac{\sum_{(i,j)\epsilon K}(p(i,j) - \hat{p}(i,j))^2}{|K|}} \tag{9}$$

Mean absolute error between the real ratings and the predicted ratings is presented by formula (10).

$$MAE = \frac{1}{|K|}\sum_{(i,j)\epsilon K}|p(i,j) - \hat{p}(i,j)| \tag{10}$$

To evaluate the group recommendation system, Mean Absolute Error group $MAE(g)$ [5] is used. $MAE(g)$ between the groups is calculated by (11) where g is the number of groups; R_g denotes the set of ratings of group g contained the ratings of user; $p(g,j)$ is the real rating of group g for the item j; $\hat{p}(i,j)$ is the predicted rating of group g for the item j.

$$MAE(g) = \frac{\sum_{p(g,j)\epsilon R_g}|p(g,j) - \hat{p}(g,j)|}{|R_g|} \tag{11}$$

The rating of group g, $p(g,j)$ is calculated by (12).

$$p(g,j) = \frac{\sum_{i\epsilon g}p(i,j)}{|g|} \tag{12}$$

The prediction of group g, $\hat{p}(g,j)$ is calculated by (13).

$$\hat{p}(g,j) = \frac{\sum_{i\epsilon g}\hat{p}(i,j)}{|g|} \tag{13}$$

3 Energy-Based Group Recommendation

3.1 Model

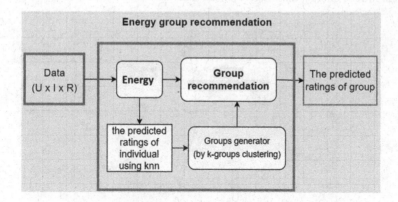

Fig. 1. Energy-based group recommendation model.

Figure 1 presents a general overview of a group based recommendation model. The input of this model is a data matrix. The output is the predicted ratings of group which are used for recommending the Top-N items to group. The model consists of four components.

- Data matrix (U × I × R).
  + U is the set of users (the closed universe), $U = \{u_1, u_2, .., u_n\}$ where $u_k \epsilon U, k = 1..n$ is a finite set of n objects (a nonempty set).
  + I is the set of items $I = \{i_1, i_2, .., i_m\}$ where $i_j \epsilon I, j = 1..m$ is a finite set of m attributes (a nonempty set).
  + R is the rating matrix where each R_{ui} is a value of R:

$$R_{ui} = \begin{cases} r_{ui} & \text{if the user } u \text{ rates the item } i \\ \oslash & \text{if f the user } u \text{ dose not rate the item } i \end{cases}$$

- The component "energy" is a utility tool for calculating energy between users and users, between groups and groups. Energy distance [17] is calculated by formulae (2) and (6).
- The component "predicted ratings of individuals" uses k-nearest neighbor method and the energy distance to predict the missing ratings such that MAE and RMSE errors are minimal.
- The component "groups generator" classifies the users into k groups based k-group clustering using energy distance.
- The component "group recommendation" will be used to predict the missing ratings of group.

Figure 2 illustrates the method to find three nearest neighbors based on the energy distance in the component "predicted ratings of individuals". For example, the user U_a, who is need to recommend, with knn = 3, the users U_2, U_5, U_6 with the lowest energy distance are selected.

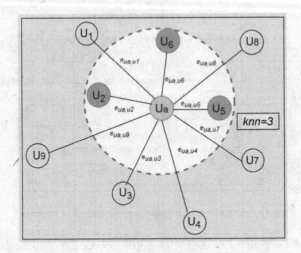

Fig. 2. Finding 3-nearest neighbors using energy distance.

Figure 3 illustrates three user groups that they are classified the energy distance.

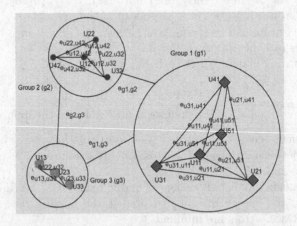

Fig. 3. Classifying users into 3 gropus using energy distance.

3.2 Algorithm

The energy-based group recommendation algorithm is the follow:

Algorithm. Energy-based group recommendation

Input: The Data Matrix (U x I x R);
Output: Top-N items to be recommended to a group;
Begin
Step 1: Calculating the energy between a user and a user in U by fomular (2).
$< Matrix\,[i]\,[j] = Energy\,[u_i \times u_j] >$;
Step 2: Predicting the missing ratings of the R matrix by knn method using the energy calculated by step 1.
Step 3: Partitioning users into groups by k-groups clustering using the energy distance.
$< Group\_List\_N\,[i]\,[j] = Energy\,[Group\_N\,[i]]\,,[Group\_N\,[j]] >$;
Step 4: Predicting the ratings of the user group based on the average method of group by fomular (12).
Step 5: Sorting the predicted ratings of the user group in descending order.
$< Sort(Group\_List\_N\,[i]) >$;
Step 6: Recommending n items with the highest predicted ratings to group.
$< Print(Top - N\,[i] >)$;
End.

---

4 Experiment

4.1 Datasets

Jester5k[1], data set contains a 5000 × 100 rating matrix (5000 users and 100 jokes) with ratings between −10.00 and +10.00. All selected users have rated 36 or more jokes. Jester dataset contains a sample of 5000 users from the anonymous ratings data from the Jester Online Joke Recommendation System collected between April 1999 and May 2003.

MovieLens[2], (MovieLens Dataset (100 k)) contains a sample of 100.000 users were released 4/1998 by GroupLens Reseach (https://grouplens.org), The dataset was collected through the MovieLens web site (movielens.umn.edu) during the seven-month period from September 19th, 1997 through April 22nd, 1998. This dataset includes 100,000 (100 k) from 943 users on 1682 movies with ratings between 1 and 5. Each user has rated at least 20 movies.

4.2 Tool

The propose model is developed by R language and it is integrated with the 'rrecsys' package [6] and 'energy' package [8]. The results of the proposed model will be compared with the two models available in 'rrecsys' package using the MAE, RMSE error.

---

[1] https://rdrr.io/cran/recommenderlab/man/Jester5k.html.
[2] https://rdrr.io/cran/rrecsys/man/ml100k.html.

4.3   Scenario 1: Energy-Based Individual Recommendation

This scenario predicts the missing ratings by using the k-nearest neigborhood method with the energy measure.

To predict the missing ratings of the R matrix, we compared the predicted ratings with the real ratings of each individual on two models: (1) EUB: The model predicts the missing ratings based on knn by using energy measure; and (2) CUB: The model predicts the missing ratings based on knn by using the Cosine measure.

Figure 4 and Fig. 5 show the MAE, RMSE errors of the two models CUB and EUB with knn = 5, 10, 20, 30 on the Jester5k dataset.

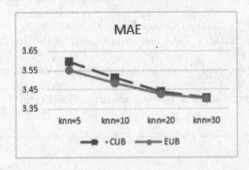

Fig. 4. Evaluating the MAE error for the individual recommendations with Jesterk5k dataset.

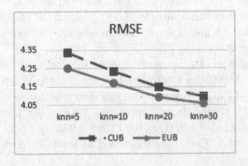

Fig. 5. Evaluating the RMSE error for the individual recommendations with Jesterk5k dataset.

The experimental results show in Fig. 4, and Fig. 5 that the MAE and RMSE errors of the EUB model are always smaller than the MAE and RMSE errors of the CUB model.

Figure 6 and Fig. 7 show the results of the two models on the MovieLens dataset with knn = 5, 10, 20.

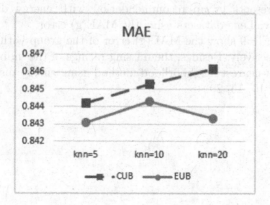

Fig. 6. Evaluating the MAE error for the individual recommendations with MovieLens dataset.

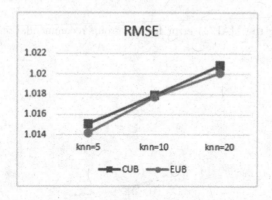

Fig. 7. Evaluating the RMSE error for the individual recommendations with MovieLens dataset.

The experimental results in Fig. 6 and Fig. 7 show that the MAE and RMSE errors of the EUB model are also smaller than the MAE and RMSE errors of the CUB model.

From the experimental results on the Jester5k and MovieLens datasets, the k-nearest neighborhood method using the energy distance gives the lower errors than the k-nearest neighborhood method using the cosine distance.

4.4   Scenario 2: Energy-Based Group Recommendation

This scenario presents group recommendation with energy distance on the Jester5k and MovieLens datasets using the MAE(g) error.

Figure 8 and Fig. 9 show the MAE(g) error of the group with the sizes to be 50, 70, 100, respectively. Besides, the missing ratings of the individuals are predicted based on k-nearest neighborhood method using energy distance (scenario 1) with knn = 5, 10, 15, 20.

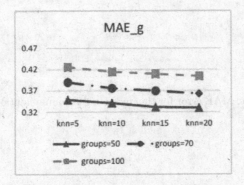

Fig. 8. Evaluating the MAE(g) error for the group recommendations with Jester5k dataset.

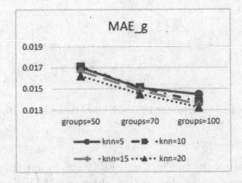

Fig. 9. Evaluating the MAE(g) error for the group recommendations with MovieLens dataset.

Experimental results show that the larger the size of the group, the smaller the error. The experimental results also show that the larger the number of the k neighbors (k = 20), the smaller the error.

5 Conclusion

We propose a new energy-based group recommendation model by using the energy distance to predict the missing ratings of the individuals and to cluster the groups of the users, then predict the missing ratings of the group, and finally recommend the most suitable the items for the user who need recommending based on the predicted ratings of group. The proposed group recommendation model has been evaluated on two common datasets which are Jester5k and MovieLens using the MAE, RMSE errors for individual recommendation, and the MAE(g) error for group recommendation. General, the individual recommendation model based on energy distance gives a smaller error than the compared model based on cosine on both datasets; and the energy-based group recommendation model shows that when the MAE(g) error is used for evaluating the recommendation result, the larger the size of the group and the larger the number of the k nearest neighbors, the smaller the errors. Therefore, the energy-based group recommendation model using the energy distance shows the feasibility of applying the potential energy for the recommendation problems.

References

1. Adomavicius, G., Tuzhilin, A.: Toward the next generation of recommender systems: a survey of the state-of-the-art and possible extensions. IEEE Trans. Knowl. Data Eng. **17**(6), 734–749 (2005)
2. Boratto, L., Carta, S., Satta, M.: Groups identification and individual recommendations in group recommendation algorithms. In: CEUR Workshop Proceedings, pp. 27–34 (2010)
3. Boratto, L., Carta, S.: State-of-the-art in group recommendation and new approaches for automatic identification of groups. In: Soro, A., Vargiu, E., Armano, G., Paddeu, G. (eds.) Information Retrieval and Mining in Distributed Environments. Studies in Computational Intelligence, vol. 324, pp. 1–20. Springer, Heidelberg (2010). https://doi.org/10.1007/978-3-642-16089-9_1
4. Liu, H., et al.: Self-supervised learning for fair recommender systems. Appl. Soft Comput. **125**, 109126 (2022)
5. Felfernig, A., Boratto, L., Stettinger, M., Tkalčič, M.: Evaluating group recommender systems. In: Group Recommender Systems. SECE, pp. 59–71. Springer, Cham (2018). https://doi.org/10.1007/978-3-319-75067-5_3
6. Çoba, L., Zanker, M., Symeonidis, P.: Environment for evaluating recommender systems (2019). https://rdrr.io/cran/rrecsys/. Repository CRAN
7. Gábor, S.J., Maria, R.L., Nail, K.B.: Measuring and testing dependence by correlation of distances. Ann. Stat. **35**(6), 2769–2794 (2007)
8. Rizzo, M., Székely, G.: Energy distance. Wiley Interdiscip. Rev. Comput. Stat. **8**(1), 27–38 (2016)

9. Jameson, A., Smyth, B.: Recommendation to groups. In: Brusilovsky, P., Kobsa, A., Nejdl, W. (eds.) The Adaptive Web. LNCS, vol. 4321, pp. 596–627. Springer, Heidelberg (2007). https://doi.org/10.1007/978-3-540-72079-9_20

10. Ntoutsi, I., Stefanidis, K., Norvag, K., Kriegel, H.-P.: gRecs: a group recommendation system based on user clustering. In: Lee, S., Peng, Z., Zhou, X., Moon, Y.-S., Unland, R., Yoo, J. (eds.) DASFAA 2012. LNCS, vol. 7239, pp. 299–303. Springer, Heidelberg (2012). https://doi.org/10.1007/978-3-642-29035-0_25

11. Li, S.: K-groups: a generalization of K-means by energy distance. Ph.D. thesis, Bowling Green State University (2015)

12. Li, S., Rizzo, M.L.: K-groups: a generalization of K-means clustering. arXiv preprint arXiv:1711.04359 (2017)

13. Felfernig, A., Boratto, L., Stettinger, M., Tkalčič, M.: Algorithms for group recommendation. In: Group Recommender Systems. SECE, pp. 27–58. Springer, Cham (2018). https://doi.org/10.1007/978-3-319-75067-5_2

14. Yang, X., Joukova, A., Ayanso, A., Zihayat, M.: Social influence-based contrast language analysis framework for clinical decision support systems. Decis. Support Syst. **159**, 113813 (2022)

15. Dong, M., Yuan, F., Yao, L., Wang, X., Xu, X., Zhu, L.: A survey for trust-aware recommender systems: a deep learning perspective. Knowl. Based Syst. **249**, 108954 (2022)

16. Chandrashekhar, H., Bhasker, B.: Personalized recommender system using entropy based collaborative filtering technique. J. Electron. Commer. Res. **12**(3), 214 (2011)

17. Tran, T.C.T., Phan, L.P., Huynh, H.X.: Energy-based collaborative filtering recommendation. Int. J. Adv. Comput. Sci. Appl. **13**(7), 557–562 (2022)

Security and Blockchain

An Implementation and Evaluation of Layer 2 for Ethereum with zk-Rollup

An Cong Tran[✉], Vu Vo Thanh, Nghi Cong Tran, and Hai Thanh Nguyen

Can Tho University, Can Tho, Vietnam
{tcan,nthai}@cit.ctu.edu.vn, tcnghi@ctu.edu.vn

Abstract. Ethereum is a very popular blockchain platform. However, due to the limit of the Transaction Per Second (TPS) of this platform, the transaction processing time in Ethereum is very slow, which greatly affects the user experience and wastes time and other fees. Therefore, the scalability of Ethereum becomes an urgent problem to be solved. In this study, we try to improve the scalability problem of Ethereum by building a layer 2 with the zk-Rollup protocol. An evaluation of the implementation is also conducted. Experimental results show that the cost of transactions decreases depending on the batch size, with the gas cost decreasing by more than 85% for a batch size of 50 transactions. Other evaluation results reveal that deposits incur the most cost and increase faster with the batch size.

Keywords: Ethereum · Layer 2 · Blockchain · Zero Knowledge Proof

1 Introduction

Etherum (ETH) is a blockchain platform proposed by Russian-Canadian programmer Vitalik Buterin in 2013 and officially launched it two years later [1]. By integrating the smart contract development feature, Ethereum allows anyone to design, develop, deploy, and interact with smart contracts, and allows developers to create distributed applications (DApps) more easily. This is creating a diverse and rich ecosystem on Ethereum, with many applications built on the Ethereum blockchain.

The birth of Ethereum is a great step forward in blockchain technology with some improvements in privacy, performance, scaling, and capacity [2] However, there are two sides to everything, and the Ethereum blockchain has some weaknesses that make it difficult to reach many users. One of them can be performance and cost. To invoke and execute the smart contract, we need to pay a certain fee. The user has to pay the miner an amount in ETH to execute the transactions and write them to the blockchain. For more than half a second, Ethereum can only meet from 13 transactions per second (TPS) to 17 TPS [3]. Furthermore, users who want their transactions to be executed faster have to pay more. Therefore, this leads to inflation, and users can have to pay more to have their transactions executed, or we have to wait for hours for the transaction to be executed.

© The Author(s), under exclusive license to Springer Nature Switzerland AG 2023
T. N. Dinh and M. Li (Eds.): CSoNet 2022, LNCS 13831, pp. 107–115, 2023.
https://doi.org/10.1007/978-3-031-26303-3_10

To handle the above problems, many solutions have been proposed. The core team of Ethereum develops Ethereum 2.0 can do thousands of transactions per second with very little cost [4]. However, this requirement can take a long time to develop. Therefore, one has to find other alternatives. Instead of doing everything on the blockchain, one would do it on an off-chain layer and somehow determine whether these operations are valid or not. Thereby reducing the computational load on the on-chain, saving, and reducing costs significantly proved that zk-Rollup is one feasible method.

Many studies have been concerned with the speed of blockchain, as stated in [5]. The work in [6] pointed out that blockchain potentially could improve speed compared to traditional methods. The evaluation of possible solutions for enhancing blockchain capabilities was discussed in [7]. In [8], the authors enhanced the transaction Speed using Parallel Proof of Work with manager selection and distribution of work and a reward system. The work in [9] showed some advantages of blockchain with a low processing speed and attempted to improve the performance. NoSQL Caching System examination with a customized SHA-256 hash core was attempted in [10] to increase the throughput of the blockchain. A European Territorial Cooperation Program INTERREG project SmartLog (Smart Logistics and Freight Villages Initiative) detailed in [11] introduced an approach of Blockchain-Based Solution for Supply Chain Traceability which could describe the lead times in transporting goods by 3–8%.

An extended version from layer 2, a separate Ethereum blockchain, was introduced in [2] to improve Ethereum throughout with a gas fee decrease of up to 100x compared to layer 1. With the idea of off-chain data processing, solutions like state channel, side chain, plasma, or layer 2 were born. Each method has its strengths and weaknesses. Plasma[1] is a pioneering technique in Ethereum. It is not easy to build in practice. The Rollup method is another solution with more advancements with two popular variations, Optimistic rollups, and zk-Rollup [12]. zk-Rollup is also widely-used technique to improve the capabilities of Ethereum such as studies in [13].

In this research, we try to improve the scalability problem of Ethereum by building a layer 2 with the zk-Rollup protocol. Also, an experiment is conducted to evaluate our implementation.

2   Methods

An idea of an off-chain layer has been around for a long time in the community. The main idea is that people do those transactions outside instead of doing transactions with complex calculations inside the blockchain network. Blockchain, then it is somehow guaranteed that the transactions are valid. Users can deposit digital assets into these layers, perform calculations, and withdraw these assets as needed.

In this way, the cost of executing a transaction in Ethereum is significantly reduced, but the question is how do we ensure that the way we execute transactions is correct to be clarified. The tool zk-SNARK is almost a perfect tool

---

[1] https://plasma.io, accessed on 01 July 2022.

to do this. In each transaction, we generate a proof of that transaction and then send that proof to the blockchain network. The network has a smart contract that checks if the proof is true. From there, we update the new state to the smart contract by calling such a group of transactions a batch. We also perform data strokes in the batch to save more cost. For more than half, we can use the Merkle tree to save the current state of the accounts. The problem to verify now is to check the update of the Merkle tree. Using a Merkle tree here greatly reduces the cost of checking and updating the state, similar to how blockchain networks use Merkle trees to convert the data in the block to the Merkle root. On the blockchain network, we store the value of the Merkle root hash on the blockchain. With a batch, we can verify the state update, compute the new Merkle root and update the Merkle root stored on the blockchain. This is the basic idea of zk-Rollup and the basic components of a zk-Rollup layer is described in Fig. 1.

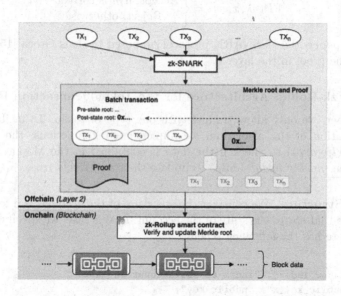

Fig. 1. Basic components of a zk-Rollup layer.

2.1 Zero Knowledge Proof

Zero Knowledge Proof (ZKP) was first proposed in 1985 by Shafi Goldwasser, Silvio Micali, and Charles Rackoff in a paper entitled The Knowledge Complexity of Interactive Proof-Systems, which then established a new research direction in security science [14]. ZKP is a protocol that allows one party (called a Prover) can prove to another party (called a Verifier) that a given statement x is true without conveying any information related to x. In the context of the blockchain, x can be a set of values, and it is proven that x satisfies a given problem. A ZKP protocol must satisfy 3 properties, i.e., completeness, soundness, and zero-knowledge.

2.2   Zero-Knowledge Succinct Non-interactive ARgument of Knowledge (zk-SNARK)

zk-SNARK is a ZKP protocol that requires no communication between the Prover and the Verifier [15]. The basic idea of this protocol can be described as follows. It is supported that the problem $F(\lambda) = x$ must be proved. The zk-SNARK protocol performs the following steps to provide the proof:

- Step 1. A **key generator** (G) takes the secret parameter λ and generates two keys – a proving key (pk) and a verification key (vk): $G(\lambda) = (pk, vk)$.
- Step 2. A **Prover function** (PF) takes pk, x, and a private input w as its arguments and then generates the proof (pr): $pr = PF(pk, x, w)$.
- Step 3. A Verifier function (VF) verifies the proof given vk and x and returns Accept if the proof is correct or Reject if the proof is incorrect:

$$VF(vk, x, pr) = \begin{cases} \text{Accept, if } pr \text{ is correct} \\ \text{Reject, otherwise} \end{cases}$$

In this paper, the zk-SNARK protocol proposed by Jens Groth [15] is used to create the proof in the layer 2.

2.3   The zk-Rollup Architecture for the Token Transaction Problem

In this paper, we only allow trading with one type of token. To facilitate the implementation of the algorithm, we implemented 2 functions: the function $merkle\_verify(data)$ to check if the data is contained in the Merkle tree and the function $merkle\_update(data)$ to update the root Merkle tree.

Account Structure. We consider the structure of the account as a leaf of the Merkle tree that stores the state of the account. In this case, we specify that the account state has the following schema:

```
{
    "layer_index": "index of the account in layer",
    "public_key_x": "public key",
    "public_key_y": "public key",
    "nonce": "number of transactions performed",
    "balance": "number of tokens in the account"
}
```

Where $public\_key\_x, public\_key\_y$ is the public key of the digital signature using EdDSA [16] of the account that owns the $layer\_index$ in the Merkle tree, $layer\_index$ indicates the account's ordinal number, and also indicates the location of the account in the layer's Merkle tree of the layer, balance is the remaining amount of the account, a nonce is the number of transactions that the account has made on layer 2. The hash value of the leaf node is calculated as follows: $leaf\_hash = Hash(public\_key\_x, public\_key\_y, layer\_index, balance, nonce)$.

We also use digital signatures to ensure that all state changes come from the person who has authority over the account, not some bad actor.

Transaction Structure. Following is the structure of transactions in our implementation:

```
{
    "sender": "layer index of sender",
    "receiver": "layer index of receiver",
    "amount": "amount",
    "sig": { A", "R", "s" }
}
```

Where *sender* and *receiver* respectively are the index of the sender account, the *amount* is the number of tokens you want to send, and *sig* is the sender's signature to attest to the action they want to perform. Depending on the type of transaction, we define separate mechanisms to handle them. Here We use EdDSA digital signature algorithm because it is relatively zk-SNARK friendly.

Transaction Proof Mechanism in Layer 2. Suppose we have a transaction that carries the content we specified in the transaction format as above. To generate a proof for a transaction, we do the following steps:

1. Prove that the sender node belongs to the Merkle tree using the function *merkle_verify()*.
2. Prove that the signature matches the sender's signature.
3. Reduce sender's balance.
4. Increase the sender's nonce.
5. Updating the Merkle tree with *merkle_update()*, we also get a new Merkle root. We call this value *intermediate_root*.
6. Prove that the receiver belongs to the Merkle tree using *merkle_verify()*.
7. Increase the receiver's balance.
8. Updating the Merkle tree again with *merkle_update()* with the data of the receiver and the *intermediate_root*, we get a new Merkle root called *final_root*.

When we want to prove a new transaction, we do the same thing, but with the *final_root* we just received when we processed the previous transaction. So the transactions in the patch will overlap, from the first transaction to the last transaction. So finally we have a proof. We send this proof to Ethereum, check it, and do the necessary calculations for the update.

One-to-One Proofing Mechanism for a Deposit or Withdrawal Transaction. We apply this mechanism when a user wants to send the token amount to our layer. According to the convention, a transaction is now performed in layer 2 from index zero to the sender's index account. If the account is already in the layer, we need to execute the order to transfer the token to the layer and create a proof of this transaction. Otherwise, we first need to create a new account for this address on layer 2 and then do the same for the first case. Following the convention, we make a transaction from the index account of the person who wants to withdraw money to index zero.

Generate a Proof and Verify the Transactions for a Batch. In the layer 2 smart contract, we verify transactions and send tokens in or out of accounts when there is an order to deposit or withdraw tokens. In the smart contract, we save the current Merkle root of the Merkle tree in this layer. Then, when we update a batch, we need to check the problem:

$$Verify(merkle\_root, new\_merkle\_root, proof)$$

If this is the case, it means that the bash transaction are correct. Next, we call the smart contract's rollup method, which allows us to verify that a batch is correct, and all is well, updates back to $merkle\_root$. So now the function $rollup()$ has 2 input arguments: the $proof$ and the new_merkle_root.

Instead of storing the account states in the smart contract and then double-correcting them when a new batch is submitted, we do not save the zk-rollup account states to the contract, but instead, put the transaction information in an object with the rollup function number. To ensure the submitted transaction data is correct, we create another problem: namely, checking $sha256(batchdata) = batch\_hash$. That is, the proof must satisfy the problem:

$$\begin{cases} Verify(merkle\_root, new\_merkle\_root, proof) = true \\ Verify(batch\_data, batch\_hash, proof) = true \end{cases}$$

3 Experimental Results

3.1 Environmental Settings

In our implementation, we choose the Ethsnarks [17] toolkit written in C++, Python, Javascript, and Solidity v0.5.0 with OpenZeppelin library version of 2.5.0. Ethsnarks provides tools to create zk-SNARK-based systems more conveniently. In addition, we also use Hardhat, a toolkit to create test nets and write and test smart contracts in Ethereum. The evaluation ran on a Macbook Pro laptop (CPU: 2.9 GHz Dual-core Intel Core i5, RAM: 8 GB, 1867 MHz DDR3, OS: MacOS Catalina).

The application consists of 2 components. The Prover is responsible for generating proof from input transactions. A smart contract is responsible for checking the proof provided by the Prover and updating the new status of the contract. Benchmark for Merkle tree [18] with $2^8 = 256$ nodes.

Gas represents a computational effort that we make to do something on Ethereum. For example, when we create a transaction, that transaction does certain things. The gas value of that transaction represents the number of computing resources the EVM used. As we know that the user wants to perform a transaction, he has to pay the miner a certain amount of money. This fee is calculated as follows: $fee = (GasCost) \times (GasPrice)$.

Gas cost is the amount of gas we spend for that transaction, and gas price is the amount in Wei (1 Wei = 10^{-18} ETH) paid for one gas in that transaction. In Ethereum, the maximum gas value that can be conducted in a block is currently about 15 million gas [19].

Table 1. Gas comparison between the proposed method and traditional method in various batch on transfer transactions: Average gas on each transaction (1), Average gas on each transaction added token transfer gas (2).

Transactions per batch	Total of gas	(1)	(2)	% of improvement
10	298,795	29,879.5	30,697.5	40.73
15	302,680	20,178.7	20,996.7	59.46
20	306,600	15,330.0	16,148.0	68.82
25	310,520	12,420.8	13,238.8	74.44
30	314,429	10,481.0	11,299.0	78.18
35	318,385	9,096.7	9,914.7	80.86
40	322,414	8,060.4	8,878.4	82.86
45	326,358	7,252.4	8,070.4	84.42
50	330,327	6,606.5	7,424.5	85.66
Traditional method			51,789.0	

3.2 Cost Comparison Results

In this section, we present experimental results with the following scenarios. The Merkle tree for this layer has a depth of 8 (contains 28 = 256 accounts). We create 10 accounts for the layer, then make 30 deposit orders from these 10 accounts, then make 30 random token transfer orders within the layer, and finally withdraw all the tokens from the layer to the Ethereum addresses of these accounts. Here we do not charge for creating an account on layer 2.

Table 1 presents a comparison between transfer transactions in the traditional way and transfer transactions using zk-Rollup method. Gas denotes the fee required to perform a transaction on the Ethereum blockchain platform successfully. Without using zk-Rollup, the gas for running a transaction is about 51,789 gas. When using zk-Rollup, if we only perform transactions on the side of gas consumption is only more than 30,697 with a batch size of 10 transactions, and reducing to the number of 7,424 with 50 transactions per batch. The average cost per transaction is 10,482 gas, which is much lower than the cost incurred when making transactions directly on Ethereum.

The cost for verifying with zk-Rollup is more than 290,000. With the current gas block limit of 15 million [20], the number of possible transactions is $(15,000,000 - 290,000)/1,000 = 14,710$ transactions. These calculations are considered under optimal conditions. Although not achieving such perfect results, it still gives very positive results. Below is a table of gas costs for the cases where the number of transactions per batch is 25, 30, 35, 40, and 50.

Table 2 exhibits the cost of 3 types of transactions. As observed, the deposit consumes the most time while the transfer action takes less time than the others. The cost of deposit also increases faster than the 2 other types.

Table 2. Gas comparison of various batch sizes on 3 types of transactions.

Transactions per batch	Deposit	Transfer	Withdraw
10	498,051	298,795	419,390
15	596,987	302,680	479,008
20	684,238	306,600	516,255
25	725,162	310,520	531,321
30	766,049	314,429	535,172
35	806,937	318,385	539,178
40	847,860	322,414	531,909
45	888,783	326,358	546,975
50	929,695	330,327	539,729

4   Conclusion

This study investigates the efficiency of the zk-Rollup in building the layer 2 for Ethereum, which can help to trade tokens on Ethereum with a lower cost. Even though the application is still so rudimentary, it proves the feasibility of the evaluated solution, and thus helps to get a first insight into the scaling problem on public blockchains like Ethereum. However, there are still some limitations that may evolve in the future. The structure of the batches is still basic, and the problem to be proven is still complicated. When the number of batches is large, it takes a long time to generate proofs. It can be used for token transfer, but cannot generate proofs in the general case. There is also no charging mechanism for the layer 2, and also the solution to increase the reliability of layer. Finally, there is a fact that zk-Rollup also supports the transfer of tokens to addresses that do not participate in the layer was not made clear in our layer construction.

References

1. Wood, G.: Ethereum: a secure decentralised generalised transaction ledger. Ethereum Project Yellow Paper **151** (2014)
2. Zhang, W., Anand, T.: Layer 2 and Ethereum 2. In: Blockchain and Ethereum Smart Contract Solution Development, pp. 341–378. Apress (2022). https://doi.org/10.1007/978-1-4842-8164-2_9
3. Nambiampurath, R.: The 5 best Ethereum layer 2 solutions. https://www.makeuseof.com/best-ethereum-layer-2-solutions/
4. Cortes-Goicoechea, M., Franceschini, L., Bautista-Gomez, L.: Resource analysis of Ethereum 2.0 clients. In: 2021 3rd Conference on Blockchain Research and Applications for Innovative Networks and Services (BRAINS), pp. 1–8 (2021)
5. Wilkie, A., Smith, S.S.: Blockchain: speed, efficiency, decreased costs, and technical challenges. In: The Emerald Handbook of Blockchain for Business, pp. 157–170. Emerald Publishing Limited (2021). https://doi.org/10.1108/978-1-83982-198-120211014

6. Kaur, S., Jaswal, N., Singh, H.: Blockchain technology. In: Applications, Challenges, and Opportunities of Blockchain Technology in Banking and Insurance, pp. 204–212. IGI Global (2022). https://doi.org/10.4018/978-1-6684-4133-6.ch012

7. Mechkaroska, D., Dimitrova, V., Popovska-Mitrovikj, A.: Analysis of the possibilities for improvement of blockchain technology. In: 2018 26th Telecommunications Forum (TELFOR), pp. 1–4 (2018)

8. Hazari, S.S., Mahmoud, Q.H.: Improving transaction speed and scalability of blockchain systems via parallel proof of work. Future Internet **12**(8), 125 (2020). https://doi.org/10.3390/fi12080125

9. Kwak, K.H., Kong, J.T., Cho, S.I., Phuong, H.T., Gim, G.Y.: A study on the design of efficient private blockchain. In: Lee, R. (ed.) CSII 2018. SCI, vol. 787, pp. 93–121. Springer, Cham (2019). https://doi.org/10.1007/978-3-319-96806-3_8

10. Sanka, A.I., Cheung, R.C.: Efficient high performance FPGA based NoSQL caching system for blockchain scalability and throughput improvement. In: 2018 26th International Conference on Systems Engineering (ICSEng), pp. 1–8 (2018)

11. Pilvik, R., Kaare, K.K., Koppel, O.: Blockchain-based solution for supply chain traceability: the case of SmartLog project. In: 2021 9th International Conference on Traffic and Logistic Engineering (ICTLE), pp. 57–63 (2021)

12. Gluchowski, A.: ZK rollup: scaling with zero-knowledge proofs. Matter Labs (2019)

13. Gjøsteen, K., Raikwar, M., Wu, S.: PriBank: confidential blockchain scaling using short commit-and-proof NIZK argument. In: Galbraith, S.D. (ed.) CT-RSA 2022. LNCS, vol. 13161, pp. 589–619. Springer, Cham (2022). https://doi.org/10.1007/978-3-030-95312-6_24

14. Goldwasser, S., Micali, S., Rackoff, C.: The knowledge complexity of interactive proof systems. In: 17th Annual ACM Symposium on Theory of Computing, vol. 10 (1985)

15. Groth, J.: On the size of pairing-based non-interactive arguments. In: Fischlin, M., Coron, J.-S. (eds.) EUROCRYPT 2016. LNCS, vol. 9666, pp. 305–326. Springer, Heidelberg (2016). https://doi.org/10.1007/978-3-662-49896-5_11

16. Josefsson, S., Liusvaara, I.: Edwards-curve digital signature algorithm (EdDSA), p. 60. https://datatracker.ietf.org/doc/rfc8032

17. HaRold: EthSnarks. https://github.com/HarryR/ethsnarks

18. Carminati, B.: Merkle trees. In: Liu, L., Özsu, M.T. (eds.) Encyclopedia of Database Systems, pp. 1714–1715. Springer, Heidelberg (2009). https://doi.org/10.1007/978-0-387-39940-9_1492

19. Wood, G.: Ethereum: a secure decentralized generalized transaction ledger; Ethereum yellow paper (2022)

20. Harper, C., Kim, C.: Ethereum gas limit hits 15m as ETH price soars - CoinDesk. https://www.coindesk.com/tech/2021/04/22/ethereum-gas-limit-hits-15m-as-eth-price-soars/

Targeted Attack of the Air Transportation Network Global Component

Issa Moussa Diop[1]([✉]) [iD], Chantal Cherifi[2] [iD], Cherif Diallo[1] [iD],
and Hocine Cherifi[3] [iD]

[1] LACCA Laboratory, Gaston Berger University, PB 234, Saint-Louis, Senegal
{diop.issa-moussa,cherif.diallo}@ugb.edu.sn
[2] DISP Lab, University of Lyon 2, Lyon, France
chantal.bonnercherifi@univ-lyon2.fr
[3] LIB EA 7534, University of Burgundy Franche-Comté, 21078 Dijon, France
hocine.cherifi@u-bourgogne.fr

Abstract. Targeted attacks aim to disintegrate a network by removing nodes or links. Classically, in node attacks, one removes nodes in descending order of a given centrality measure. Although most real-world networks exhibit a mesoscopic structure, strategies proposed in the literature do not exploit this ubiquitous property. We introduce an attack strategy based on the network component structure to overcome these drawbacks. The component structure of a network consists of local components and global components. The local components are the dense parts of the network, and the global components are the nodes and links connecting them. We investigate this attack on the world air transportation network using Degree and Betweenness centrality measures. Results show that the degree-based attack on the global component is more effective than the classical attack on the entire network. In contrast, the classical Betweenness attack slightly outperforms the Betweenness attack on the global component. However, the latter is more efficient.

Keywords: Robustness · World air transportation network · Component structure

1 Introduction

The robustness analysis of networks is very active, especially for critical systems such as transportation, internet, and power-grid systems [1,2]. It investigates how the network resists an attack on the nodes or links of a network. The goal is to design a strategy to dismantle the network by deleting the least number of nodes or links. Generally, one relies on centrality measures to extract essential nodes. Nodes are then removed according to their centrality rank. In this sense, a lot of work has been done. However, most papers do not consider the mesoscopic structure of complex networks in developing attack strategies unless

T. N. Dinh and M. Li (Eds.): CSoNet 2022, LNCS 13831, pp. 116–127, 2023.
https://doi.org/10.1007/978-3-031-26303-3_11

the centrality [3] measure exploits the community structure. We present some related papers on targeted attacks based on the mesoscopic structure. In [4], the authors propose algorithms to dismantle a network based on the community structure. They target the inter-community nodes and the inter-community links. Adding intercommunity links, they also introduce a strategy for improving network resilience. The algorithms are evaluated using twelve networks from different regions (air, bus, rail, road). For node attacks, they compare eight attack strategies (4 attacks based on centrality measures and four based on alternative network dismantling). The paper explores five algorithms for edge attacks. Results show that, for node attacks, their algorithm is an excellent candidate to dismantle a network with a linear execution speed. The attacks based on the Betweenness centrality of the entire network outperform their strategy. Nevertheless, the latter requires more time, especially for large networks. For attacks, their method outperforms all the other methods. The approach to improve the robustness is more effective than its classical alternatives.

The authors of [5], investigate an algorithm Module-Based Attack (MBA). It relies on the nodes or edges interconnecting the communities to fragment a network. Indeed, once the communities are extracted, these nodes or edges are removed in descending order of their Betweenness centrality (no sequential attacks). Ten real networks (4 infrastructural, three biological, and three social networks) are investigated. For attack comparisons, the Betweenness-based attack, Degree-based attack, and Longest Path attack are explored. Results show that the MBA is more efficient in disintegrating a network than the centrality measures. Nevertheless, this efficiency depends on the number of communities and the community size.

This paper introduces a strategy to dismantle a network based on its global component. The global component is a part of the component structure, a new mesoscopic structure introduced in [6]. We investigate the world air transportation network in which the nodes represent the airports, and the links are the flights. In addition, one analyzes the impact of the attacks on world air transportation and the local components.

2   Data and Methods

2.1   Data

The undirected and unweighted network we consider originates from FlightAware [7]. The extracted data covered six days of flight information, from May 17, 2018, to May 22, 2018. Nodes represent airports, and links represent direct flights between airports. Table 1 summarize its basic topological properties. The global air network is sparse and requires twelve hops to join at maximum any two airports. It is disassortative, and its clustering coefficient is small.

2.2   Component Structure

The component structure splits the network into local and global components. The local components capture the dense part of the network. The global

Table 1. Basic topological properties of the world air transportation network, the 7 large local components, and the largest global component. N is the network size. $|E|$ is the number of edges. *diam* is the diameter. l is the average shortest path length. ν is the density. ζ is the transitivity, also called the global clustering coefficient. $k_{nn}(k)$ is the assortativity, also called the Degree correlation coefficient. η is the hub dominance.

Components	N	$|E|$	*diam*	l	ν	ζ	$k_{nn}(k)$	η
World air transportation network	2734	16665	12	3,86	0,004	0,26	−0,05	0.09
North America-Caribbean	657	3828	7	2.88	0.018	0.28	−0.325	0.29
Europe	493	5181	6	2.58	0.042	0.32	−0.2	0.33
East and Southeast Asia	416	2495	7	2.9	0.029	0.34	−0.22	0.32
Africa-Middle East-India	337	1197	6	3.25	0.021	0.28	−0.15	0.23
Oceania	234	464	9	3.5	0.017	0.18	−0.22	0.24
South America	215	527	6	3.15	0.023	0.23	−0.35	0.22
Russia-Central Asia-Transcaucasia	112	427	4	2.23	0.07	0.23	−0.39	0.73
Large global component	513	2194	8	3.28	0.017	0.13	−0.25	0.2

components derive from the nodes and links that connect the local components. One can use a community structure or multiple core-periphery algorithms to uncover the dense groups. Then, one removes the connections between the dense groups to form the local components. Finally, one removes the links within the dense groups. The remained non-isolated nodes and links form the global components. Figure 1A illustrates the process of uncovering the component structure. This example relies on the non-overlapping community detection algorithm to extract the dense parts. Then, by eliminating the inter-community links (black lines), we form the local components. The global components are formed by eliminating the intra-community links (blue, yellow, and pink colors).

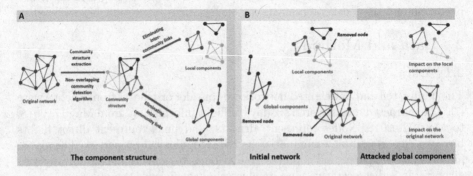

Fig. 1. A)The component structure. B) Attack on the global component. (Color figure online)

2.3   Targeted Attack on the Largest Global Component

We aim to disconnect the network while removing as few airports as possible. Centrality measures play an important role in extracting those vital nodes. A strong attack strategy eliminates the nodes in descending order of a given centrality measure. This paper investigates the Degree and Betweenness centrality measures.

We propose a global component-based attack strategy. Then, we evaluate the impact of this attack on the local components and the whole network. The robustness evaluation process proceeds as follows:

1. Disconnect a node from the largest global component according to an attack strategy.
2. Disconnect the same node from its local component.
3. Disconnect the same node from the world air transportation network.
4. Extract the Largest Connected Component of the world air transportation network.
5. Extract the Largest Connected Component of the local components.

Note that the attack strategy based on the centrality is computed only once before the first attack. Figure 1B displays an example of attack on the global components.

2.4   Performance Measure

The Largest Connected Component (LCC) is widely used to measure the robustness of a network. It is the size of the largest connected component when the network is broken down into multiple components after an attack. The smallest the LCC the most effective the attack with the same budget (number of deleted airports). Let $LCC(i)$ be the size of the LCC when i nodes are removed from a network of size N, we call $\lambda(i) = \frac{LCC(i)*100}{N}$ the fraction of the LCC remaining after attack.

3   Experimental Results

3.1   Component Structure

We use the Louvain community detection algorithm to extract the dense parts of the network. It uncovers 27 communities that correspond to the local components. There are 20 small components located in a single country and seven large components covering different regions of the world [6]. The large components split the world into the following areas: 1) North America-Caribbean, 2) Europe, 3) East and Southeast Asia, 4) Africa-Middle East-Southern Asia, 5) Oceania, 6) South America, and 7) Russia-Central Asia- Transcaucasia.

The global components include nine small global components, and one large global component [6]. The airports of the largest global component are distributed over the world. In the following, we focus on the large local and global components. Table 1 presents their basic topological properties.

3.2 Degree Attacks

The **Degree** of a node is the number of its connections. Figure 2A gives the fraction of the LCC of the large local components and the world air transportation network [8] as a function of the fraction of top Degree nodes removed from the global component. One can see that the South America and Africa-Middle East-Southern Asia components are the most sensitive to attacks. The most resilient components seem to be Europe and East and Southeast Asia. The LLC of the North America-Caribbean, Europe, and East and Southeast Asia components decrease similarly. Indeed, when an airport is removed, the LCC decreases slowly. Removing an airport from the Oceania, South America, and Russia-Central Asia-Transcaucasia components affects their LCC significantly. The LCC of the Africa-Middle East-Southern Asia component and the world air transportation network combines the two schemes already mentioned when an airport is removed. Indeed, alternatively, their LCC decreases slowly and sharply.

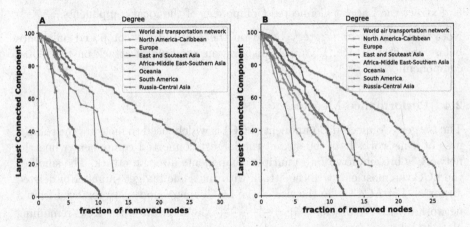

Fig. 2. A) Attack from the largest global component. B) Attack from the world air transportation network.

The North America-Caribbean and the South America components are isolated from the world air transportation network after 5% of the airports are removed. Thus, these two components are more connected than the other components. This detached block represents 26% of the airports in the global network. Traveling between airports in this block is not easy. Indeed, the network is not very dense, and it takes a maximum of 19 hops and an average of 6 hops to connect two airports. The world air transportation network is more sensitive to degree attacks based on the global component than degree attacks based on the world air transportation network. Taking out 10% of the airports is enough to dismantle the overall network for attacks on the global component. While for the attacks on the world air transportation network, removing 12% is necessary to dismantle the network.

By removing more nodes from the global component, the local components become disconnected from the global air transportation network. Table 2 displays information about the local components, once they are isolated. Specifically, we determine how many nodes are required to isolate it from the global air transportation network. Additionally, it reports their giant components' size and basic topological properties.

The North America-Caribbean component is isolated from the network very early. 5% of the overall network's airports must be removed from the top degree global component airports to separate it. Removing the Calgary Airport in Canada provokes the isolation of this component. The LCC of this component contains 69% of the airport of the initial component. These airports cover all the countries of this component. It becomes challenging to travel in this LCC compared to the initial component. Indeed, several hubs are removed by the attacks. Thus, it requires more hops at maximum and, on average, to join any two airports in this LCC. In addition, this LCC is less dense than the initial component.

The South America component is simultaneously isolated from the global air network as the North America-Caribbean component. The attacks directly remove only 3% of the airports of this component. The Comodoro A M Benítez Airport in Chili is the last removed airport before the separation. The LCC of this component keeps 68.16% of the airports of the initial component. These airports are distributed over the countries of this component. On average, it requires the same number of hops to connect two airports in this LCC as in the initial component. Nevertheless, the diameter and the density of the LCC increase. Traveling is no more difficult than in the initial component.

Table 2. Basic topological properties of the five local components isolated by the Degree attack. λ is the Largest Connected Component size. ρ is the fraction of top Strength nodes removed from the world air transportation network to isolate the component. δ is the diameter. l is the average shortest path length. μ is the density.

Components	$\rho(\%)$	$\lambda(\%)$	δ	l	μ
North America-Caribbean	5	69	9	3.44	0.014
Europe	6.87	76.8	14	3.85	0.024
East and Southeast Asia	4.25	76.5	9	3.7	0.023
Africa-Middle East-Southern Asia	6.87	27.4	9	3.85	0.06
Oceania	6.1	27.4	15	5	0.037
South America	5	66.6	10	3.72	0.032
Russia-Central Asia-Transcaucasia	6.32	27.4	7	3.15	0.01

The third isolated component is Oceania when removing 6.1% of the airports. Among these airports, 3.4% belong to this component. The Nadi Airport in Fiji is the last airport connecting this component to the world. Thus, the initial component separates into two components. The LCC includes 35% of the airports of the initial component. These airports cover only the North of Australia and New Zealand. The second largest connected component, which contains 30% of the airport, covers the countries such as Vanuatu, French Polynesia, and the Solomon Islands. Traveling in the LCC becomes more challenging than in the initial component. Indeed, more hops are required to connect at maximum and, on average, two airports in the LCC.

The East and Southeast Asia component is the next region to separate. This isolation occurs when removing 6.25% of the top hubs. In these 6.25%, there are 5.2% of the airports of this component. The Chubu Centrair Airport in Japan is the last connected airport to the world before isolation. The LCC contains 75.4% of the airports of the initial component. The LCC covers all countries of this component. Nevertheless, traveling is more challenging in the LCC than in the initial component. Indeed, the LCC's diameter and the average shortest path length increase while the density decreases.

Removing 6.32% of the top degree airports from the global component, isolates the Russia-Central Asia-Transcaucasia component from the world. Among these airports, there are 14.2% belong to this component. The Kurumoch Airport in Russia is the last connected airport before the separation. The LCC includes 30.35% of the airports of the initial component. These airports are mainly in Russia, except for one airport in Tajikistan and another in Uzbekistan. The diameter almost double; on average, one more jump is required to join two airports in this LCC. Moreover, the diameter decreases profoundly. Thus, it is more challenging to get from one airport to another.

The Africa-Middle East-Southern Asia is the following separated component when 6.87% of the airports are removed from the global component. 12.5% of the airports of this component are included in the 6.87%. The Kigali Airport is the last connected airport to Europe before the isolation. The LCC consists of 15.8% of the airports of the initial component. These airports are in the South of Africa. The diameter increase, whereas the average shortest path is the same. In addition, the LCC is denser than the initial component. One can say that traveling in the LCC is almost the same as in the initial component.

Once the Africa-Middle East-Southern Asia is isolated from the overall network, only the European airports remain. Although its LCC keeps 76.8% of the airport of the initial component, traveling within this LCC becomes more challenging. Indeed, this component is the most attacked. Several hubs are removed. Thus, the diameter value double, the average shortest path increases, and the density decrease.

3.3   Betweenness Attacks

The **Betweenness** of a node is the fraction of the shortest path passing through it. Figure 3A represents the evolution of the LCC of the large component and the world air transportation network when removing the top Betweenness airports from the largest global component. One can see that South America and Oceania components are the most sensitive to the consecutive suppression of top

Betweenness airports. The other components decrease in the same way until 3% of the airports are removed. Beyond this value, the Africa-Middle East-Southern Asia and East and Southeast Asia components are the most resilient. The European component becomes more robust 8% of the airports are removed.

The same tendencies of curves that we see in the attack of the top degrees are present in the Betweenness attacks. The evolution of the LCC of the components is the same except for the Africa-Middle East-Southern Asia component. Indeed, removing an airport slightly impacts this component. Table 3 reports some topological properties of the LCCs once they are isolated.

East and Southeast Asia, the Africa-Middle East-Southern Asia, and a part of the Russia-Central Asia-Transcaucasia component are isolated from the world air transportation network after 6.14% of the airports are removed from the largest global component. Removing the Irkutsk Airport in Russia isolates these three components simultaneously. This LCC includes 21% of the airports of the global air network. Nevertheless, traveling within these three regions is more challenging. Indeed, the diameter and the average shortest path length of this LCC have almost doubled the ones of the global air network, even though one can easily find a stopover to join two airports. The global air transport network is more resistant to Betweenness-based attacks on the global component than the global air network. Nevertheless, the difference is not very big. These two attacks are almost comparable. Indeed, for the attack based on the global component, it is necessary to remove 7.8% of the airports to dismantle the global network, whereas, removing 7.4% on the global network is enough.

The Oceania component is the first isolated part of the global air transportation network. Removing 4.6% of the airports from the global component provokes this separation. Among these airports, only 3% belonging to this component are directly removed. This component becomes unreachable when removing the Brisbane Airport. The LCC contains 59.4% of the airports of the initial component. The LCC covers the West of Australia and the other countries of this component. Traveling within this LCC is very challenging, despite increasing density. Indeed, a maximum of 19 jumps are necessary to join any two airports. On average, one more hop is required to join two airports compared to the initial component.

Removing 4.9% of the airports separates South America from the global air network. 3.25% of the airports of this component belongs to the 4.9%. The Comodoro Arturo M B Airport is the last airport that ensures the connection between this region and the world. After the separation, the LCC contains 61.8% of its airports, which are distributed over this component, except Venezuela, Colombia, and Ecuador. On average, the same number of jumps is required to travel. Nevertheless, long-distance travel requires more hops and the density increase.

East and Southeast Asia is the next region to separate from the world. 6.14% of the airports are sufficient to isolate it. 4.5% of the airports of this component are included in the 6.14%. The Tan Son Nhat Airport in Vietnam is the world's last connected airport. 83.17% of the airports of the initial component, covering all the countries, belong to the LCC. Traveling within this LCC is a bit more

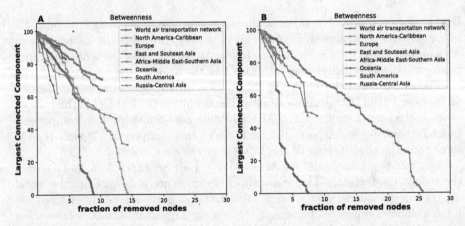

Fig. 3. A) Attack from the largest global component. B) Attack from the world air transportation network.

challenging. Indeed, the diameter increases by more than two jumps, and the average shortest path increase by more than one hop. In addition, the density decreases.

Table 3. Basic topological properties of the five local components isolated by the Betweenness attack. λ is the Largest Connected Component size. ρ is the fraction of top Strength nodes removed from the world air transportation network to isolate the component. δ is the diameter. l is the average shortest path length. μ is the density.

Components	$\rho(\%)$	$\lambda(\%)$	δ	l	μ
North America-Caribbean	6.65	71.38	12	3.57	0.013
Europe	6.65	71	10	3	0.032
East and Southeast Asia	6.14	83.17	9	3.5	0.023
Africa-Middle East-Southern Asia	6.14	63.5	11	5.64	0.02
Oceania	4.9	61.8	18	4.66	0.021
South America	5	66.6	10	3.72	0.033
Russia-Central Asia-Transcaucasia	6.14	30.35	7	3.17	0.09

The Africa-Middle East-Southern Asia component is the next region to split. The Bahrain Airport in Bahrain is the last airport connected to the other region. 10% of the airports of this component is directly eliminated. The LCC keeps 63.5% of the airports of the initial component. These airports cover the region. Nevertheless, taking off from one airport to another requires almost twice as many hops on average and at maximum.

The Russia-Central Asia-Transcaucasia component is separated from the world air transportation network when the Irkutsk Airport in Russia is removed.

Among the removed airports, there 14% of the airports of this component. The LCC contains 30.35% of the airports of the initial component. It covers only Russia, except Tajikistan and Uzbekistan. The diameter has almost doubled compared to the one of the initial component, and the average shortest path increases by one. But, The density increases. Thus, it becomes difficult to travel between any two airports.

Removing 6.65% of the airport of the global air network isolates the European component. Among these airports, there are 10.14% of the European airports. Eliminating the Nice-Côte d'Azur Airport in France separates the Europe component from the North America-Caribbean component. The LCC contains 71% of the airports of the initial component. These airports are located in all the countries of this component. On average, the same number of hops are required to join two airports. Nevertheless, the diameter increases by three jumps, and the density decreases. Thus, traveling within this LCC is a bit challenging compared to the initial component.

Once all these components are separated, the North America-Caribbean component remains in the global air network. Its LCC includes 71.38% of the airport of the initial component. These airports cover all the components except Alaska in the United States. Traveling within the LCC becomes difficult. Indeed, the diameter and the average shortest have almost doubled, while the density decreases.

Table 4. γ, ρ, λ are the order of isolation of local components, the proportion of airports removed from the global air network before isolation, and LCCs of the local components after the separation.

Components	Degree						Betweenness					
	Global component			World air transportation network			Global component			World air transportation network		
	γ	$\rho(\%)$	$\lambda(\%)$	γ	$\rho(\%)$	$\lambda(\%)$	γ	$\rho(\%)$	$\lambda(\%)$	γ	$\rho(\%)$	$\lambda(\%)$
North America-Caribbean	1	5	69	2	9.7	29.8	7	6.65	71.38	4	2.9	79.6
Europe	7	6.87	76.8	7	11.4	27.5	6	6.65	71	7	5	61.8
East and Southeast Asia	4	4.25	76.5	4	11	17.2	3	6.14	83.17	3	2.3	82.4
Africa-Middle East-Southern Asia	6	6.87	27.4	5	11.2	23.5	3	6.14	63.5	5	3.9	47.6
Oceania	3	6.1	27.4	1	9	24.3	1	4.9	61.8	2	2.1	90.2
South America	1	5	66.5	3	9.8	34.4	2	5	66.6	1	1.6	93.5
Russia-Central Asia-Transcaucasia	5	6.32	27.4	6	11.4	18.7	3	6.14	30.35	6	5	47.32

4   Discussion and Conclusion

This paper investigates the vulnerability of air transport to targeted attacks on the global component of the network using two centrality measures (Degree and Betweenness). In addition, we explore their impact on the various local components. The global air transportation network is more vulnerable to attacks on the global component based on Betweenness than Degree. For the Degree, the attacks on the global component cause more damage than the attacks on the world air transportation network. For the Betweenness, removing an airport from the world air transportation network is more efficient than the global component. Nevertheless, the difference in efficiency is not large. Indeed, eliminating 7.5% of the airports from the world air transportation network is sufficient to dismantle it. In contrast, one needs to remove 7.9% of the airports from the global component to fragment the overall network. Consequently, the strategy proposed by this study can be an alternative to the Betweenness-based attacks on the global air transport network. Indeed, the computation of the Betweenness on large networks is less efficient.

Table 4 reports the order of isolation of the local components, the proportion of airports removed from the global air network before isolation, and the LCCs of the local components after separation. It allows us to compare the attacks of the global component and the impacts on the local components and the world's air transportation. The local components progressively separate from the world network as airports are removed from the global component. Nevertheless, the order of isolation of the components differs with the attacks. For the Degree, the North America-Caribbean and the South America components separate first from the overall network. Then Oceania, East and Southeast Asia, Russia-Central Asia-Transcaucasia, Africa-Middle East-Southern Asia, and Europe follow in that order. For Betweenness attacks, Oceania and the South America components are first isolated from the overall network. Then, the East and Southeast Asia, Africa-Middle East-Southern Asia, and Russia-Central Asia-Transcaucasia components are unreachable simultaneously from the overall network. The Europe and North America-Caribbean components are the last isolated components. Thus, Oceania and South America are the most vulnerable components. The European component is the most robust for Degree attacks, while the North America-Caribbean component is the most resilient for Betweenness attacks.

The local components separate faster from the global air network with the attacks on the global component than on the overall network for the Degree attack. Indeed, it is necessary to remove half as many airports on the global component for the first local component to be isolated. The LCCs from the attacks on the global component are larger and connect more countries. However, two categories emerge. At first, one can see that the LCC of North America-Caribbean, Europe, East, and Southeast Asia, and South America, from the attacks on the global component, are twice as large. Traveling within its LCCs is more accessible than the LCCs from the attacks on the overall network. The size LCCs of the remained components are comparable. Nevertheless, the LCC from the attack on the global component of Africa-Middle East-Southern Asia is usable. In the second category, it is the opposite. Indeed, with much fewer airports, three times less, the attack on the world air network isolates a first local component. In addition, the LCCs from this attack tend to include more airports, except the LCCs of Europe, East and Southeast Asia, and Africa-Middle East-Southern Asia components. Except for the Africa-Middle East-Southern Asia component, traveling within the LCCs from the attack on the overall network is less challenging. This paper exclusively explores a targeted attack on the global component of the world air transportation network with two centrality measures (Degree and Betweenness). Future work will consider the impact of various attacks on the local components. Additionally, we plan to investigate other transportation networks.

References

1. Sun, X., Wandelt, S.: Robustness of air transportation as complex networks: systematic review of 15 years of research and outlook into the future. Sustainability **13**(11), 6446 (2021)
2. Wandelt, S., Sun, X., Feng, D., Zanin, M., Havlin, S.: A comparative analysis of approaches to network-dismantling. Sci. Rep. **8**(1), 1–15 (2018)
3. Ghalmane, Z., Cherifi, C., Cherifi, H., El Hassouni, M.: Centrality in complex networks with overlapping community structure. Sci. Rep. **9**(1), 1–29 (2019)
4. Wandelt, S., Shi, X., Sun, X.: Estimation and improvement of transportation network robustness by exploiting communities. Reliab. Eng. Syst. Saf. **206**, 107307 (2021)
5. da Cunha, B.R., González-Avella, J.C., Gonçalves, S.: Fast fragmentation of networks using module-based attacks. PLOS ONE **10**(11), 1–15 (2015)
6. Diop, I.M., Cherifi, C., Diallo, C., Cherifi, H.: Revealing the component structure of the world air transportation network. Appl. Netw. Sci. **6**(1), 1–50 (2021)
7. Flightaware (2018). https://flightaware.com/
8. Diop, I.M., Cherifi, C., Diallo, C., Cherifi, H.: Targeted attacks on the world air transportation network: impact on its regional structure. arXiv preprint arXiv:2209.05617 (2022)

Measuring Cryptocurrency Mining in Public Cloud Services: A Security Perspective

Ayodeji Adeniran and David Mohaisen[✉]

University of Central Florida, Orlando, USA
mohaisen@ucf.edu

Abstract. Cryptocurrencies, arguably the most prominent application of blockchain systems, have been on the rise with wide mainstream acceptance. A central entity in cryptocurrencies is "mining pools", groups of cooperating cryptocurrency miners who agree to share block rewards in proportion to their contributed mining hash power. Despite the many promised benefits of cryptocurrencies, they are equally utilized for malicious activities, e.g., ransomware payments, stealthy command, and control, etc. Thus, understanding the interplay between cryptocurrencies, particularly the mining pools, and other essential infrastructures for profiling and characterization is necessary.

In this paper, we initiate the study of the interplay between mining pools and public clouds by analyzing their communication association through passive domain name system (pDNS) traces. We observe that 24 cloud providers have some association with mining pools as observed from the pDNS query traces, where popular public cloud providers, namely Amazon and Google, have almost 48% of such an association. Moreover, we found that the cloud provider presence and cloud provider-to-mining pool association exhibit a heavy-tailed distribution, emphasizing an intrinsic preferential attachment model with both mining pools and cloud providers. We measure the security risk and exposure of the cloud providers, as that might aid in understanding the intent of the mining. Among the top two cloud providers, we found almost 35% and 30% of their associated endpoints are positively detected to be associated with malicious activities, per the virustotal.com scan. Finally, we found that the mining pools presented in our dataset are predominantly used for mining Metaverse currencies, highlighting a shift in cryptocurrency use and demonstrating the prevalence of mining using public clouds.

1 Introduction

Cryptocurrency has recently been on the rise, with the top three cryptocurrencies amounting to over a trillion USD in value [11,12]. With cryptocurrency gradually gaining acceptance, different cryptocurrencies are still emerging. Many cryptocurrencies directly apply blockchain technology, a distributed ledger over a distributed network of nodes that record transactions. The blockchain

T. N. Dinh and M. Li (Eds.): CSoNet 2022, LNCS 13831, pp. 128–140, 2023.
https://doi.org/10.1007/978-3-031-26303-3_12

distributed system provides a much safer architecture against failure and cryptographic primitives that ensure transactions are safer from being altered. Given the importance and value of those cryptocurrencies, cybercriminals have used them to enable their criminal activities.

Cybercrimes have been evolving over the years, where cybercriminals have been continuously coming up with new ways to violate system security properties, steal information, hijack resources, and demand ransom [7]. The emergence of new attacks has been a continuous race between the attackers and the defenders. The attackers have used several platforms to launch attacks. Attackers were noticed to have changed their strategy to defeat defenses. For instance, with the emergence of blockchain-based technologies, malicious transactions are placed on the blockchain to facilitate malicious activities through the distribution of stealthy command and control channels [3]. Given the significant valuation of cryptocurrencies, cryptojacking, an intentional effort to use others' machines and resources for mining cryptocurrencies, has been on the rise [13]. Notably, the use of cloud resources has been hypothesized to be the main entry point of mining cryptocurrencies [6], although not systematically analyzed.

The focus of this paper is to understand the prevalence of public clouds for mining purposes, possibly mining with malicious intent (e.g., with compromised cloud instances). We hypothesize that utilizing cloud resources for such activities is more consistent with the general compute trends and (from a security standpoint) adversaries? incentives than ever before. For instance, launching cyber-attacks from the private server(s) that can be traced and shut down is no longer popular among attackers because, apart from the ease and flexibility of springing up resources in the cloud, they also make less return on investment compared to setting up private servers. There are several cloud providers, from enterprise-scale to small-scale, the users and adversaries alike can move from one provider to another and set up attack fronts fast. Exploiting blockchain technology in conjunction with cloud resources, the attackers have a vast and cheap cloud to benefit from and a difficult-to-decipher blockchain technology to hide their malicious activities, mainly when those cloud resources are obtained free of charge (i.e., compromised). While security analysis is a byproduct of our analysis, it also highlights the general trend in this space.

Organization. The related work is presented in Sect. 2, followed by our dataset overview in Sect. 3, the main results in Sect. 4, the discussion in Sect. 5, and concluding remarks in Sect. 6.

2 `Related Work

This work is broadly associated with a body of work on crypto mining (irrespective of the tools used for that mining). Tahir et al. [15] studied the abuse of virtual machines in cloud services for mining digital currencies, which is the most related work to ours. Huang et al. [5] were among the first to notice the illegal use of CPU cycles for malware-induced mining. The initial work on web-based crypto mining was presented by Saad et al. [13] and Ruth et al. [10], who measured the prevalence of cryptojacking among websites (i.e., utilizing mining on visitors' machines). To do that, Ruth *et al.* [10] obtained blacklisted URLs using the no coin web extension, mapped them on a large corpus of websites obtained from the Alexa Top 1M list, and identified 1491 suspect websites involved in cryptojacking. In a concurrent work, Saad *et al.* [13] conducted a similar study, but on a larger number of websites; 5703 sites in total. Concurrently, Eskandari *et al.* [4] examined the prevalence of cryptojacking among websites and the use of *Coinhive* as the most popular platform for cryptojacking. All of these studies highlight the issue of cryptojacking through measurements and the emerging use of cryptojacking as an alternative to online ads. Saad *et al.* [13] goes further by conducting code analysis toward detecting cryptojacking codes and their economic impact. Bertino and Nayeem [2] highlighted worms in IoT devices that hijacked them for mining purposes, pointing to the infamous *Linux.Darlloz* worm that hijacked devices running Linux on Intelx86 chip architecture for mining [1]. Krishnan *et al.* [8] studied a series of computer malware, such as *TrojanRansom.Win32.Linkup* and *HKTL_BITCOINMINE*, that turned host machines into mining pools. Sari and Kilik [14] used Open Source Intelligence (OSINT) to study vulnerabilities in mining pools with Mirai botnet as a case study.

3   Dataset and Preprocessing

The dataset used in this study couples an enumeration of mining pools and associations between them and cloud instances that belong to various cloud providers utilizing DNS query data. To this end, the first part of the input data used for this analysis is the mining pools and their associated addresses. The list of the mining pools was sourced by manually listing the domains from the Stelareum mining pool website. This list contains a set of mining pool addresses that are publicly available[1]. We examined the mining pools and copied the corresponding URLs for each. Subsequently, using the popular Digital Envoy IP allocation dataset[2], we enumerated the IP pools of some top public cloud providers, which formed the second input data.

To establish an association between the various cloud providers and the mining pools, we scanned over all the IPs allocated to the public providers on one side and the mining pools addresses on the other side using the passive domain name system (pDNS) dataset used in [9]. The scan utilizes the pDNS dataset to map a relationship between the mining pools and endpoints in the cloud where traffic is sent from those endpoints to the mining pools, and vice versa, at some point in time in the past. The scan output consists of the mining pool's originating IP address and the cloud providers' corresponding IP subnet.

The scan result includes the mining pool domain, pool source IP address, and public IP subnets. The IP subnets are then converted to their respective domains to get the name of the cloud providers. The data contains thousands of response lines (as there could be various pDNS entries associated with the same pair of endpoints). As such, the data was cleaned by rearranging the data

---

[1] https://www.stelareum.io/en/mining/pool.html.
[2] https://www.digitalenvoy.com/.

in descending order and removing those with the least number of responses. To obtain the geographical distribution of the cloud providers in our data, we further augment the data with the country where the cloud provider is located.

4   Main Results

This section presents our main results by measuring and mapping the association landscape between cryptocurrencies and public clouds. Before we dive into our analysis, we review the main dimensions of our analysis.

4.1   Analysis Dimensions

This study is concerned with various dimensions that highlight the interplay between public clouds and cryptocurrencies. Namely, we are concerned with cloud providers (associated with cryptocurrencies) and their geographical affinities, pool size, mining pools, and their distribution, cloud provider-specific distribution, and (potential) illicit activities. We define each of those dimensions as we present the associated results.

4.2   Results and Discussion

❶ **Cloud Providers and Country of Origin.** A large number of public cloud providers make up the cloud ecosystem. While the significant providers are only a few (e.g., Amazon, Google, Azure, and Cloudflare), there are more than thousands of such providers. Understanding the affinity between mining pools and those cloud providers through distribution analysis is essential for two reasons. First, such an analysis will allow us to understand the regional distribution of those providers and associated mining activities. Second, this analysis will further shed light on whether the large providers, in general, are still dominant in their use for mining. Answering this question, possibly positively, would allow us to devise effective policies to counter cryptomining threats. Moreover, insight for this analysis would draw a representative picture of the overall computing ecosystem and associated security characteristics.

Observations. Table 1 shows the cloud providers mapped to the countries of their domain registrations. Interestingly, we find that the distribution of traffic from the mining pools is vastly distributed, covering a large number of providers, and spanning several continents. Moreover, we found that the distribution of the cloud providers' representation with respect to the studied mining pools and their association is quite skewed (heavy-tailed): while there are 24 different cloud providers represented in the dataset, the top 2 (Amazon and Google) have a representation of 48%, while the next 12 providers have 42%.

By the same token, Table 1 highlights the geographical distribution of the different cloud providers, where the distribution is also heavy-tailed over 15 countries, led by the US (57%), followed by Russia (11%), South Korea (6%),

Table 1. Cloud distribution across the country with the frequency of appearance.

Cloud Provider	Country	#	%
Amazon	USA	738	24%
Google	USA	737	24%
KORNET	South Korea	188	6%
Cloudflare	USA	175	6%
Asia Pacific Net	Japan	161	5%
CL-KARELIA	Russia	155	5%
SCL66-rented1	Cyprus	135	4%
MACROREGIONAL	Russia	117	4%
HIPL-SG	USA	69	2%
Corpori	Brazil	58	2%
HOSTERION-SRL	Romania	50	2%
KAZAKTELECOM	Kazakhstan	46	2%
MOTIV-DC-1	Netherlands	46	2%
MOTIV-DC-3	Netherlands	46	2%
DNAP-081217	Finland	44	1%
AOSOZVEZDIE-NET	Russia	38	1%
IPNET-DS-WBS	South Africa	33	1%
RS-KOPERNIKUS	Rep of Serbia	33	1%
PS-1_2177	Kazakhstan	32	1%
BTC-TEMP1	Bulgaria	31	1%
TR-RTNET-981210	Turkey	31	1%
CLOUDFLARENET-EU	USA	28	1%
RU-MOS-SMILE	Russia	28	1%
UK-NTLI-990527	United Kingdom	24	1%
	Total	3043	100%

and Japan (5%). The remaining 11 countries have 21% of the cloud endpoints shared among them collectively. The cloud distribution is obtained from the result of the pDNS scan.

❷ **Mining Pools and Associated Size.** In this study, we measured two major pools in terms of their presence in the cloud. The size of the pool is measured by counting the number of individual (cloud) IP addresses associated with it (i.e., issuing queries). Understanding the pool size would highlight which pool is more popular and central in the cryptocurrency ecosystem, and possibly which cryptocurrency is being mined by the pool utilizing cloud resources.

Observations. We emphasize that we conducted scans of several mining pools in our initial data gathering (i.e., all those present in our initial set), although

Table 2. Mining pools distribution. A heavy-tailed distribution in terms of the number of public cloud associated with the different mining pool (sub)domains.

Pools (subdomain)	#	%
sandpool.org	1,470	41%
etp.sandpool.org	1,205	34%
eu.miningethereum.net	351	10%
miningethereum.net	348	10%
www.miningethereum.net	73	2%
www.sandpool.org	55	2%
ru.etp.sandpool.org	26	1%
dev.sandpool.org	20	1%
Total	3,043	100%

we only got a response from two mining pools domains, sandpool.org and miningetherium.net—which means that the other mining pools did not have any association with public clouds. In our scan, we noticed that the two mining pool domains contain other subdomains which responded to the query from the pDNS. Table 2 shows the mining domain and the subdomain with the corresponding representation in terms of cloud presence. Based on these results, we narrow down the focus of this paper to the two pools and their associations with the public cloud.

Cryptocurrency mining has global acceptance with mining activities being carried out in several parts of the world. Some mining pools are rated to be in the top tier because of the amount of mining traffic recorded and associated with them. These pools are located in different countries. Most of the top-rated pools are located in China. For instance, in the general cryptocurrency ecosystem, pools like F2Pool, AntPool, BTCC, and BW account for more than 60% of all the new bitcoins. While the dataset from the DNS scan recorded most traffic to Amazon and Google cloud providers, a search on the reported top 10 mining pools using censys.io, a search engine that scans the internet for connected devices. Alibaba's cloud network recorded a higher traffic rate for the mining pools based in China, while Amazon had higher traffic for pools located in the US and other countries. The geographical location of the mining pool could be a factor in determining the preferred cloud provider before selecting other available providers.

By the same token, we found that the top two subdomains (by sandpool) represent 75% of the overall cloud associations, with ETP, the second-largest association, representing 34% of the associations (1,205 cloud endpoints). Upon further exploration, we found that this pool is used mostly for mining Metaverse ETP, a cryptocurrency that powers the Metaverse blockchain-as-a-service (BAAS) platform and is located in Europe. Among those two pools, sandpool.org

Table 3. Google vs Amazon Cloud distribution between the two pools. While the general trend of heavy-tailed distribution still applies to the individual cloud shares against the subdomains of the pools, Amazon has a more skewed distribution in contrast to a more evenly distributed share of Google's cloud instances.

Pools	Google	Amazon	Google%	Amazon%
etp.sandpool.org	307	213	42%	36%
sandpool.org	186	357	25%	59%
miningethereum.net	106	9	14%	1%
eu.miningethereum.net	104	5	14%	1%
www.miningethereum.net	19	3	3%	0%
www.sandpool.org	7	9	1%	1%
ru.etp.sandpool.org	5	6	1%	1%
dev.sandpool.org	3	5	0%	1%
Total	737	607	100	100

had 78% (or 2,750) of the cloud associations overall, while miningetherium.net had 22% (or 798) of the cloud associations in total.

❸ **Cloud Providers Distribution vs Pools.** The current cloud ranking by market share places Amazon with the largest share, followed by Microsoft Azure, then Google. In our measurement, we found that Amazon and Google represented almost fifty percent of the total cloud instances connected to the mining pools in our dataset, while the other different cloud providers make up the remaining half. We want to further understand the detailed distribution between the two major cloud providers to the two mining pools to characterize their similarities and differences, as depicted in Table 3.

Observations. From those results, we made two key observations. First, the per-cloud distribution follows a similar trend of heavy-tail as in the generation distribution, although less skewed in the case of Google, where cloud share is distributed more evenly on a larger number of pool domains. Second, while Azure is quite popular in the abstract, and on par with the popularity of Google Cloud, it is absent from this analysis. We still speculate that cloud popularity may have played a factor in the distribution, and the absence of Microsoft Azure cloud in the distribution would possibly point to other factors that may have influenced the selection of Amazon and Google (e.g., strict abuse policies, or the popularity of this cloud in a given country). Some smaller cloud providers also reported traffic from the pool that indicated association to the clouds that could be either randomly or selectively. The information in the dataset is not explicit enough to accurately provide the details, but we hypothesize that factors like cost, security, or restrictions for some category of users, such as in the case of Azure, could be responsible for the choice and the clear trend.

④ Malicious Associations. The popularity of cloud services over the years has made them attractive for both benign and malicious use. Services and applications previously hosted on private servers are now hosted in the cloud and many public and private companies are still migrating their workloads to the cloud. It is no surprise that mining activities are shifting to the cloud, considering the cost and the flexibility offered. Setting up servers in the cloud takes a few minutes at a significantly lower cost. The flexibility of moving from one cloud provider to another could be another factor besides the cost that attracted the miners to shift their activities to using cloud resources.

Cybercriminals operate by masquerading their malicious traffic from detection using different techniques. The servers are shut down once detected, which is a big loss for cybercriminals. Cloud services provide an easy solution for cybercriminals addressing this issue. For instance, to make their activities more discreet, cybercriminals make use of blockchain technology when operating from the cloud. In case of detection, they quickly move to another cloud provider to set up their servers in a few minutes with very minimal disruption to their services and activities. They continue to operate in these cycles at a relatively small cost and manage to keep afloat for a while before being detected.

A central question in our analysis is whether some of those cloud instances used for mining cryptocurrencies are involved in malicious activities. Unfortunately, the dataset we have is limited in many ways, particularly the absence of a payload for the DNS resolution or subsequent application-layer communication, which makes it impossible to draw such a conclusion. However, utilize our knowledge of the endpoint on the cloud to frame the question into a plausibility analysis: among those IP addresses associated with the cloud providers, how many of them are associated with malicious activities?

Observations. In order to address this question, we conducted an additional scan on the IP addresses associated with the cloud instances using virustotal.com, which is one of the three online scanning sites we used in our analysis. The results of the scan are shown in Table 4. Among the 24 cloud providers reported in Table 1, only five cloud providers have positive scan results in virustotal.com, namely Amazon, Google, KORNET, Cloudflare, and CL-KARELIA. Among them, CloudFlare had the largest detection rate, with 44% of the cloud instances reported by VirusTotal as having some security issues (i.e., flagged as a source of malicious activity). The percentage is followed by Amazon (at 34.69%) and Google (29.85%). Among those cloud providers KORNET had the least positive rate, with only around 1% of the instances detected by virustotal.com.

While those results are inconclusive, and cannot be used to argue for intent associated with the mining activities taking place on those cloud providers for the different mining pools, or whether mining is taking place if at all, the fact that a positive detection is associated with a number of those public cloud IPs highlights the potential risk associated with those instances.

Table 4. A distribution analysis of the malicious cloud instances (IPs) in contrast to the total number of IP count associated with the given cloud provider, and the associated percentage. Notice that with the top cloud providers, a significant number of instances are shown to be associated with a malicious activity at some point in time, per virustotal.com scan. Cloud providers not present in this table returned negative scan results.

Cloud Provider	Count	Malicious	Percentage
Amazon	738	256	34.69 %
Google	737	220	29.85 %
KORNET	188	2	1.06 %
CloudFlare	175	77	44 %
CL-KARELIA	161	10	6.21 %

5   Discussion

From our analysis, we notice that more associations are reported on the Google cloud platform than on the Amazon platform. The preference for Google Cloud over Amazon by the miners could be due to various reasons. For instance, computing power and the cost of electricity are among the challenges in cryptomining. Both cloud providers have instances with computer power resources to handle the cryptomining activities, but the cost might be the main differentiating factor because the profit determines the attractiveness of mining. The two top cloud providers are known to have high reliability and availability, and provide a range of offerings that suit the applications associated with the highlighted mining activities; virtual reality. Given the expectations of those applications, the failure rate is low in these clouds and miners could run their mining system uninterrupted.

By the same token, the less popular and smaller cloud providers have fewer resource offerings and may lack the capacity required by cryptominers. Cybercriminals using blockchain for sending malicious traffic might prefer smaller cloud providers because they may be cheaper, prone to exploits, and less secure in general, while they may accommodate such activity to drive traffic on their cloud platform. There is the possibility of using popular clouds as well by exploiting the vulnerabilities in such clouds, especially by hijacking user accounts with weak credentials and hiding their malicious traffic among millions of packets originating from the cloud. Our security analysis above highlights the potential of this hypothesis, as a number of public cloud nodes (identified by their addresses) are shown to be associated with malicious activities in virustotal.com scan.

Security in the cloud is a shared model system whereby both the cloud provider and the customer have their responsibilities shared. While several security measures are recommended, some fail to implement the required minimum security and may have their account hijacked and used for malicious purposes, including mining cryptocurrency. Hijacking user accounts for mining activity is common because of the high computing resources required for mining and this comes with a price: using someone else account in this manner transfers the cost to the account owner while the miners earn rewards for their mining activity. These kinds of account hijacking for mining activities were reported by a number of cloud providers, especially by the top providers. For instance, 86% of the hijacked accounts in Google clouds are used for cryptomining [16]. Our results allude to a similar outcome, as many of the cloud addresses shown in our analysis are associated with malicious activities.

In this study, we noticed that the two main mining pools uncovered in our analysis are quite protective of registration information. Upon digging into the DNS records of their domains, we found that their DNS resolution and hosting are done by Cloudflare. Neither of those main pools nor their associated subdomains is detected by virustotal.com. To contrast this result with other major mining pools, we evaluated the top mining pools (besides those studied thus far). The results are shown in Table 5. We noticed all the top ten mining pools are hosted in the cloud as well, and mostly in Cloudflare. Out of the ten mining pools, nine are hosted on the Cloudflare cloud, highlighting a persistent trend in the utilization of cloud resources for running mining pools, perhaps for their high availability. The public cloud provides some security measures, but that does not necessarily prevent the pools from being used for malicious purposes. We then scanned all the mining pools IP addresses using virustotal.com. While all of them returned negative detection results (although many returned "unrated" result for the scanned addresses), indicating that they were not involved in reported malicious activities. However, two types of detection were reported: passive DNS (pDNS) replication and communication file detection.

Interestingly, none of the domain names associated with those pools have pDNS flag; pDNS is what we use for retrieving the association between clouds and mining pools. The pDNS Replication provides temporary storage for DNS queries and captures the queries on the network and stores them for later retrieval. The stored queries are in historical form, which can be analyzed later by security experts. virustotal.com explains the main idea behind passive DNS as inter-server DNS which captures messages and forwards them to a collection point for analysis and storing of the individual DNS records in a database where they are indexed and queried after the processing. Given the lack of pDNS data for those domains, it is not surprising that we could not see them in our initial association dataset. To this end, our analysis comes with a caveat: the estimated association between those cloud providers and mining pools is a lower bound, and only captures those that are explicit about their association.

The files entry in Table 5 highlights the number of files that have been determined to perform some kind of communication with the IP address of the domain under consideration. These files are not considered malicious in nature, but indicative of an association with other addresses, which highlights our hypothesis that the estimated association is a lower bound.

Table 5. Top 10 mining pools scan. IP addresses are masked for privacy.

Mining Pool	IP Address	Domain	pDNS	Files
Binance.com	13.226.**.**	Amazon	0	2
slushpool.com	104.26.**.**	Cloudflare	0	1
f2pool.com	104.18.**.**	Cloudflare	0	10+
pool.btc.com	104.18.**.**	Cloudflare	1	10+
viabtc.com	104.16.**.**	Cloudflare	0	3
v3.antpool.com	104.18.**.**	Cloudflare	0	4
poolin.com	104.22.**.**	Cloudflare	0	5
bw.com	172.66.**.**	Cloudflare	0	7
bitfury.com	104.26.**.**	Cloudflare	0	10+
v3.antpool.com	104.18.**.**	Cloudflare	0	4

The analysis we have conducted thus far is based on virustotal.com, which is the golden standard for evaluating security through detection against a range of scanners and antivirus vendors. We scan the pools and associated subdomains using various in-house products of threat intelligence. Interestingly, and as shown in Table 6, we found that a number of those pools are reported as involved in malicious activities (in the description, CMC threat intelligence reported that malware is hosted on the listed mining pools). While a detection that is not replicated by the other major vendors in virustotal.com, highlights a divide in the security industry on how mining is perceived.

Table 6. Mining Pool IP address Malicious scan.

Pool	Security Vendor	Number
v3.antpool.com	CMC Threat Intelligence	1
bitfury.com	CMC Threat Intelligence	1
poolin.com	CMC Threat Intelligence	1
v3.antpool.com	CMC Threat Intelligence	1
viabtc.com	CMC Threat Intelligence	1
pool.btc.com	CMC Threat Intelligence	1

6   Conclusion

In this paper, we initiate the systematic study between public clouds and cryptocurrencies, one of the most prominent applications of blockchain systems. Through pDNS traces, we establish the association between two mining pools and cloud providers. Unsurprisingly, we found that the major cloud providers are popular in their association with mining pools, with a heavy-tailed distribution and global presence. Upon examining the security of the associated cloud endpoints associated with mining pools, we found that a significant number of them (above 30% in three cases) are malicious by virustotal.com scan results. By examining the hosting patterns of mining pools, we found that they are also heavily utilizing cloud providers, and the view of those mining pools, from a security standpoint, is divided. While our study is limited by the lack of payload from which one could understand the intent of the different associations between cloud and mining pools, it calls for further actions in this direction by providing preliminary anecdotes through characterization.

Acknowledgement. This research was supported by the Global Research Laboratory (GRL) Program through the National Research Foundation of Korea funded by the Ministry of Science and ICT (NRF-2016K1A1A2912757). Part of this work was additionally supported by CyberFlorida Seed Grant (2021–2022).

References

1. Bansal, S.K.: Linux worm targets internet-enabled home appliances to mine cryptocurrencies (2014). https://thehackernews.com/2014/03/linux-worm-targets-internet-enabled.html
2. Bertino, E., Islam, N.: Botnets and internet of things security. Computer **50**(2), 76–79 (2017)
3. Böck, L., Alexopoulos, N., Saracoglu, E., Mühlhäuser, M., Vasilomanolakis, E.: Assessing the threat of blockchain-based botnets. In: 2019 APWG Symposium on Electronic Crime Research (eCrime), pp. 1–11. IEEE (2019)
4. Eskandari, S., Leoutsarakos, A., Mursch, T., Clark, J.: A first look at browser-based cryptojacking. In: IEEE European Symposium on Security and Privacy Workshops, EuroS&P Workshops, London, United Kingdom, pp. 58–66 (2018). https://doi.org/10.1109/EuroSPW.2018.00014
5. Huang, D.Y., et al.: Botcoin: monetizing stolen cycles. In: The Network and Distributed System Security Symposium (2014)
6. Kharraz, A., et al.: Outguard: detecting in-browser covert cryptocurrency mining in the wild. In: Liu, L., et al. (eds.) The World Wide Web Conference, WWW 2019, San Francisco, CA, USA, 13–17 May 2019, pp. 840–852. ACM (2019). https://doi.org/10.1145/3308558.3313665
7. Kim, H., Park, J., Kwon, H., Jang, K., Choi, S.J., Seo, H.: Detecting block cipher encryption for defense against crypto ransomware on low-end internet of things. In: You, I. (ed.) WISA 2020. LNCS, vol. 12583, pp. 16–30. Springer, Cham (2020). https://doi.org/10.1007/978-3-030-65299-9_2
8. Krishnan, H., Saketh, S., Vaibhav, V.: Cryptocurrency mining transition to cloud. Int. J. Adv. Comput. Sci. Appl. **6** (2015)
9. Perdisci, R., Papastergiou, T., Alrawi, O., Antonakakis, M.: IoTFinder: efficient large-scale identification of IoT devices via passive DNS traffic analysis. In: IEEE European Symposium on Security and Privacy, EuroS&P 2020, pp. 474–489. IEEE (2020). https://doi.org/10.1109/EuroSP48549.2020.00037
10. Rüth, J., Zimmermann, T., Wolsing, K., Hohlfeld, O.: Digging into browser-based crypto mining. In: Proceedings of the Internet Measurement Conference 2018, IMC 2018. ACM, New York (2018). https://doi.org/10.1145/3278532.3278539

11. Saad, M., Chen, S., Mohaisen, D.: SyncAttack: double-spending in bitcoin without mining power. In: ACM Conference on Computer and Communications Security, pp. 1668–1685. ACM (2021). https://doi.org/10.1145/3460120.3484568
12. Saad, M., Choi, J., Nyang, D., Kim, J., Mohaisen, A.: Toward characterizing blockchain-based cryptocurrencies for highly accurate predictions. IEEE Syst. J. **14**(1), 321–332 (2020). https://doi.org/10.1109/JSYST.2019.2927707
13. Saad, M., Khormali, A., Mohaisen, A.: Dine and dash: static, dynamic, and economic analysis of in-browser cryptojacking. In: 2019 APWG Symposium on Electronic Crime Research, eCrime 2019, Pittsburgh, PA, USA, 13–15 November 2019, pp. 1–12. IEEE (2019). https://doi.org/10.1109/eCrime47957.2019.9037576
14. Sari, A., Kilic, S.: Exploiting cryptocurrency miners with OISNT techniques. Trans. Netw. Commun. **5**(6) (2017)
15. Tahir, R., et al.: Mining on someone else's dime: mitigating covert mining operations in clouds and enterprises. In: Dacier, M., Bailey, M., Polychronakis, M., Antonakakis, M. (eds.) RAID 2017. LNCS, vol. 10453, pp. 287–310. Springer, Cham (2017). https://doi.org/10.1007/978-3-319-66332-6_13
16. The Guardian: Cryptocurrency miners using hacked cloud accounts, google warns (2021). shorturl.at/bgwA9

Do Content Management Systems Impact the Security of Free Content Websites?

Mohamed Alqadhi[1], Abdulrahman Alabduljabbar[1], Kyle Thomas[1], Saeed Salem[2], DaeHun Nyang[3], and David Mohaisen[1(✉)]

[1] University of Central Florida, Orlando, USA
mohaisen@ucf.edu
[2] Qatar University, Doha, Qatar
[3] Ewha Womans University, Seoul, South Korea

Abstract. This paper investigates the potential causes of the vulnerabilities of free content websites to address risks and maliciousness. Assembling more than 1,500 websites with free and premium content, we identify their content management system (CMS) and malicious attributes. We use frequency analysis at both the aggregate and per category of content (books, games, movies, music, and software), utilizing the unpatched vulnerabilities, total vulnerabilities, malicious count, and percentiles to uncover trends and affinities of usage and maliciousness of CMS's and their contribution to those websites. Moreover, we find that, despite the significant number of custom code websites, the use of CMS's is pervasive, with varying trends across types and categories. Finally, we find that even a small number of unpatched vulnerabilities in popular CMS's could be a potential cause for significant maliciousness.

Keywords: Web security · Free content websites · CMS · Measurement

1 Introduction

Today, free content websites are an essential part of the Internet, providing ample resources to users in free books, movies, software, and games. The security of free content websites has always been a focal point of debate and studies. The main questions around the study of free content websites have been their security and privacy: what are the costs associated with using those websites? Those costs have been studied by contrasting free content websites with premium websites–websites that provide similar content but charge fees–across multiple dimensions, e.g., vulnerabilities in code, infrastructure utilization, and the richness of the privacy policies [1,5,11,15,17].

For instance, it was found that there is a higher level of maliciousness in free content websites than in premium websites, which makes free content websites unsafe for the users [2]. Digital certificates used by those websites are shown to be problematic [3]. Their privacy policies are shown to be limited in covering

T. N. Dinh and M. Li (Eds.): CSoNet 2022, LNCS 13831, pp. 141–154, 2023.
https://doi.org/10.1007/978-3-031-26303-3_13

essential policy elements [4]. Despite the importance of the literature, it falls short in determining the root cause for the lack of security and privacy in free content websites. The contrast provided in the literature highlights that free content websites are a source of lurking risks and vulnerabilities that could expose users and their data to significant security costs. However, there is a lack of a study that looks into various potential contributors to the vulnerability to better understand a mitigation strategy for those risks.

To address this gap, we revisit the security analysis of free content websites. The critical insight we utilize for our analysis is that the security of any website is best understood by understanding the codebase of its content. In essence, we also hypothesize that many of the vulnerabilities associated with those websites could be caused by a repeated software design pattern in their codebase, as is the case with other web technologies. We find that we can understand the repeated patterns by studying the utilization of third-party content management systems (CMS's), which are heavily used in today's websites.

In this paper, we contribute to the state-of-the-art by analyzing and contrasting the security of free content websites through the lenses of CMS analysis using 1,562 websites. We annotate the websites with their malicious attributes and systematically evaluate the role of CMS as a contributing factor. We find that a significant number of the websites (≈44%) use CMS's, which comes with vulnerabilities and contributes to maliciousness. We find that the use pattern of CMS's is unique across different types of websites and categories. The top-used CMS's have several aspects in common, such as unpatched vulnerabilities, which help explain the maliciousness of websites using them.

The rest of this paper is as follows. In Sect. 2, we review the related work. In Sect. 3, we review our dataset and its annotation. In Sect. 4, we provide an overview of the methods utilized in this paper. In Sect. 5, we provide the results and the discussion. Finally, in Sect. 6, we provide the conclusion and recommendation for future research or work.

2   Related Work

In the following, we sample and review the most related pieces of prior work.

Online Website Analysis. Researchers have held that diverse constituents might be subject to increased risks when using free content websites, given the evolution of online services and web applications. These risks have been examined across various website features including digital certificates, content, and addressing infrastructure. [3]. Another study conducted component and website-level analyses to understand vulnerabilities utilizing two main off-the-shelf tools, VirusTotal and Sucuri [2], linking free content websites to significant threats.

Privacy Practices Reporting. Mindful of the implicit security cost, another work has looked into the interplay between privacy policies and the quality of those websites. Namely, the prior work examined user comprehension of risks linked to service use through privacy policy understanding [4]. The researchers passed several filtered privacy policies into a custom pipeline that annotates the policies against various categories (e.g., first and third-party usage, data retention) [10]. The authors found that the privacy policies of free content websites are vague, lack essential policy elements, or are lax in specifying the responsibilities of the service provider (website owner) against possible compromise and exposure of user data. On the other hand, they found that the privacy policies of the premium content websites are more transparent and elaborate about reporting their practices on data gathering, sharing, and retention [4].

Tracking and Website Structure. Another study has contributed to this field by revealing the tracking mechanisms of corporate ownership [12]. To comprehend the web tracking phenomenon and subsequently craft material policies to regulate it, the authors argued that it is imperative to know corporations' actual degree and reach that may be subject to the increased regulations. The most significant finding in this research was that 78.07% of websites within Alexa's top million instigated third-party HTTP requests to the domain owned by Google. The researchers observed that the overall trend shown by past surveys is that many of the users of websites value privacy and that the present privacy state online denotes an area of material anxiety. Concerning measurement, the same study highlights that the level of tracking on the web is on the rise and does not show indications of abating.

3   Dataset and Data Annotation

Websites. For this study, we compiled a dataset that contains 1,562 websites, with 834 free content websites and 728 premium websites, which have been used in prior work [2–4]. In selecting those websites, we consider their popularity while maintaining a balance per the sub-category of a website. To determine the popularity of a website, we used the results of search engines Bing, DuckDuckGo, and Google as a proxy, where highly ranked websites are considered popular. To balance the dataset, we undertook a manual verification approach to vet each website across the sub-category (see below). Namely, we sorted the websites into five categories based on the content they predominantly serve: software, music, movies, games, or books. The following are the free and premium content websites count per category: books (154 free, 195 premium), games (80 free, 113 premium), movies (331 free, 152 premium), music (83 free, 86 premium), and software (186 free, 182 premium).

Dataset Annotation. For our analysis, we augment the dataset in various ways. We primarily focused on information reflecting the exposure to the risk of users [3]. We determine whether a website is malicious or benign using the VirusTotal API [18]. VirusTotal is a framework that offers cyber threat detection, which helps us analyze, detect, and correlate threats while reducing the required effort through automation. Specifically, the API allowed us to identify malicious IP addresses, domains, or URLs associated with the websites we use for augmentation.

CMS's. Since this work aims to understand the role of software (CMS, in particular) used across websites and its contribution to threat exposure, we follow a two-step approach: (1) website crawling and (2) manual inspection and annotation. First, we crawl each of the websites and inspect its elements to find the source folder for the website. From the source folder, we list the source and content for each website to identify the CMS used to develop this website. This approach requires us to build a database of the different available CMS's to allow annotation automation through regular expression matching. We cross-validate our annotation utilizing existing online tools used for CMS detection. We use CMS-detector and w3techs, two popular tools, to extract the CMS's used for the list of websites. For automation, we build a wrapper that prepares the query with the website, retrieves the response of the CMS used from the corresponding tool, and compares it to the manually identified set in the previous step. Among the CMS's identified, WordPress is the most popular, followed by Drupal, Django, Next.js, Laravel, CodeIgniter, and DataLife. In total, we find 77 unique CMS's used across the different websites, not including websites that rely on a custom-coded CMS.

Vulnerabilities. Our dataset's final augmentation and annotation are the vulnerability count and patching patterns. For each CMS, we crawl the results available in various portals concerning the current version of the CMS to identify the associated vulnerability. Namely, we crawl such information from cvedetails, snyk.io, openbugbounty, and wordfence. Finally, to determine whether a vulnerability is patched or not (thus counting the number of unpatched vulnerabilities), we query cybersecurity-help. [6–9,13,16,19].

4   Analysis Methods

The key motivation behind our analysis is to understand the potential contribution of CMS's to the (in)security of free content websites, which has been established already in the prior work, as highlighted in Sect. 2. To achieve this goal, we pursue two directions. The first is a holistic analysis geared toward understanding the distribution of various features associated with free content and premium websites (combined). The second is a fine-grained analysis that considers the per-category analysis of vulnerabilities. In essence, our study utilizes frequency analysis of various features to understand trends and affinities and a holistic view of vulnerabilities. The features we utilize are as follows:

❶ **CMS.** This feature signifies the industry name of the content management system utilized by the free content website, premium content website, or both.

❷ **Count.** This feature signifies the total number of websites that utilize the given CMS for their operation. In particular, we assume each given website utilizes only one CMS, which has been the case in our analysis.

❸ **Percent.** This feature signifies the normalization of the count feature by the total number of studied websites. We use the percentage to understand a relative order of the CMS's contribution that is easier to interpret.

④ Malicious count (MC). This feature is calculated per CMS. It highlights the total number of websites utilizing the given CMS deemed malicious. For our maliciousness check, we utilize the output of VirusTotal, where a website is deemed malicious if at least one scanner has flagged it as malicious.

⑤ Malicious percentage per CMS (MPCP). This feature signifies the normalized MC by the count feature. It highlights the significance (as a percentage) of the malicious websites for the given CMS. It highlights the actual relative contribution of the CMS to the maliciousness of websites taking into account their relative representation in our dataset.

⑥ Malicious percentage (MP). This feature signifies the MC feature normalized by the total MC value (i.e., the overall number of malicious websites) by capturing a given CMS's relative contribution to the total number of malicious websites. It signifies the contribution irrespective of the representation of that CMS in our dataset. The gap between MPCP and MP signifies whether a given CMS is more secure in the abstract or not.

⑦ Total vulnerabilities (TV). This feature signifies the total number of vulnerabilities associated with the given CMS.

⑧ Unpatched vulnerabilities (UV). This feature signifies the total number of unpatched vulnerabilities associated with the given CMS.

⑨ Correlation analysis. This feature identifies the relationship between the CMS's and the maliciousness of the sites. For that, we use the Pearson correlation coefficient, defined as $\rho_{X,Y} = \frac{\text{cov}(X,Y)}{\sigma_X \sigma_Y}$. Here, X is a random variable associated with free content/premium content type (malicious vs. benign), and Y is a random variable capturing the CMS's associated with the given type.

5 Results and Discussion

5.1 Overall Analysis Results and Discussion

First, we explore the distribution of the various features outlined earlier in a holistic manner, considering the free vs. premium labels of the websites. The results are shown in Table 1, and we make the following observations.

(1) The total number of malicious websites is 525 out of 1,561 websites, corresponding to 33.63% of them. This number is surprisingly large, especially in contrast to general website maliciousness levels, which are estimated at 1%[1].

(2) In terms of vulnerability, the maliciousness of those websites corresponds to 2,760 vulnerabilities in the CMS's the websites employ. Among them, 145 are unpatched at the time of our scanning. While small as a percentage (only 5.25%), we note that some of those unpatched vulnerabilities are associated with the most popular CMS's our dataset. For example, one unpatched vulnerability is associated with WordPress, which is used by more than 24% of websites and is associated with 32% of the total number of malicious websites. This supports our hypothesis on the role of CMS's as an amplification avenue of vulnerabilities and

---

[1] https://ophtek.com/what-are-malicious-websites/.

associated impact, where a single vulnerability could be utilized to contaminate a large number of websites and recruit them into malicious endeavors.

(3) We observe that a majority of the websites (883, or 56.6%) in our dataset use custom code, with 30.5% (or 269) of them being malicious. Custom-coded websites made up 51.2% of the total malicious websites. In contrast, while the websites that used CMS's represented 43.4% of all websites, they had 48.8% of all malicious websites, which corresponds to 37.8% maliciousness among those that utilize CMS's.

(4) Our estimate of the role of CMS's serves only a lower bound, as we do not consider the potential for shared codes among custom websites (i.e., websites that do not use a standard CMS). Those websites might be reusing cross-website codes, which could amplify the vulnerabilities.

(5) We observe a range of percentages of vulnerabilities and maliciousness across the website groups utilizing different CMS's, where that percentage sometimes exceeds 40% (well above the average) even with major CMS's; e.g., WordPress (44.3%), Next.js (53.85%), and Shopify (70%). These results show a significant trend in the maliciousness of the websites based on their platforms.

5.2 Category-Based Analysis Results and Discussion

The main results provided in Sect. 5.1 are profound, although they do not look into the individual categories and how they differ (if at all). To help answer this question, we conduct the same analysis we had in Sect. 5.1, but per category; for books, games, movies, music, and software. Our analysis provides a contrast against the mean (Sect. 5.1) and group (free vs premium).

Table 1. Distribution of the combined free and premium websites across different CMS's. Studied distribution characteristics are for each CMS: the percentage among all websites (percent), the count, malicious count (MC), malicious per CMS websites count (MPCP), the malicious percentage among the total websites (MP), the total number of identified vulnerabilities with the given CMS (TV), and the total number of unpatched vulnerabilities for the given CMS (UV).

CMS	Count	Percent	MC	MPCP	MP	TV	UV
Custom code	883	56.57	269	30.46	17.23	–	–
WordPress	379	24.28	168	44.33	10.76	8	1
Zendesk	26	1.67	11	42.31	0.70	2	2
Drupal	25	1.60	3	12.00	0.19	228	0
Adobe EM	22	1.41	1	4.55	0.06	93	0
Shopify	20	1.28	14	70.00	0.90	0	0
Magento	18	1.15	5	27.78	0.32	210	3
Next.js	13	0.83	7	53.85	0.45	9	0
Laravel	9	0.58	2	22.22	0.13	9	1

(*continued*)

Table 1. (*continued*)

CMS	Count	Percent	MC	MPCP	MP	TV	UV
vBulletin	9	0.58	3	33.33	0.19	0	0
HubSpot CMS	8	0.51	5	62.50	0.32	3	0
Bigcommerce	6	0.38	0	0.00	0.00	20	0
Django framework	6	0.38	1	16.67	0.06	1	0
Salesforce C360	6	0.38	0	0.00	0.00	1	0
Gatsby	5	0.32	2	40.00	0.13	1	0
IPS community	5	0.32	2	40.00	0.13	3	0
Joomla	5	0.32	1	20.00	0.06	83	2
Oracle CX	5	0.32	0	0.00	0.00	25	0
Salesforce cloud	5	0.32	5	100	0.32	56	0
Sitecore CMS	5	0.32	2	40.00	0.13	19	1
Others	101	6.47	24	23.76	1.54	1,989	135
Total	1,561	100	525	–	33.63	2,760	145

Table 2. Distribution of free vs. premium **books content websites** across different CMS's. Studied distribution characteristics are for each CMS; Keys are as in Table 1.

CMS	Count	Percent	MC	MPCP	MP
Free content websites					
Custom code	115	75.16	31	26.96	20.26
WordPress	22	14.38	10	45.45	6.54
Drupal	3	1.96	0	0.00	0.00
Django framework	2	1.31	1	50.00	0.65
vBulletin	2	1.31	1	50.00	0.65
Others	9	5.88	3	33.33	1.96
Total	153	100	46	–	30.07
Premium websites					
Custom code	84	43.08	19	22.62	9.74
WordPress	46	23.59	12	26.09	6.15
Shopify	10	5.13	7	70.00	3.59
Drupal	7	3.59	0	0.00	0.00
Magento	6	3.08	2	33.33	1.03
Others	42	21.53	13	30.95	6.67
Total	195	100	53	–	27.18

General Observations. Before delving into the specific analysis of each category, we make the following broad observations. **(1)** We notice that while the use of CMS's is common among both the free content and premium websites, the usage follows different patterns: whereby the number of CMS's utilized by the free content websites is small, it is prominent in the case of premium websites, with a more significant heavy-tailed distribution (i.e., a significant number of the CMS's have a minimal representation in terms of the websites that utilize them). This is very well-captured in the "others" row in every table, where we combine the CMS's with 1–2 websites. We observe that "others" in the case of premium websites is significantly more than that in the free content websites part of the table. (2) Across the different websites, we observe inconsistent patterns concerning the division between custom code and CMS: where it is significantly greater in the case of free content websites vs premium for books (75% vs 43%), movies (86% vs 51%), music (65% vs 51%), and software (59% vs 30%), the pattern does not hold for games (47% vs 56%).

❶ **Books.** The results in Table 2 show that there are 153 free content websites and 195 premium websites. With 348 websites, 149 use a CMS, and 199 use custom code. Under this category, 46 (30.8%) of free content and 53 (27.2%) of premium websites are malicious. In total, 99 (28.5%) of the books' websites are malicious. This result shows that slightly more free content websites are malicious. In contrast, both types of websites have a malicious percentage that is less than the average (33.6%) per Table 1. Interestingly, the free content websites have a 39.5% chance of being malicious vs. a 30.6% chance for the premium.

It is natural to ask whether the ranking of the CMS's persists in both the free content and the premium websites. While the top CMS is the same in both cases, others in the top 4 for the free content (ordered) are Drupal, Django, vBulletin vs. Shopify, Drupal, and Magento for premium. Shopify is the most malicious CMS (percentage-wise) with 70%. It is used only in the premium books category, in contrast to the top (count-wise) malicious CMS (WordPress) used in both.

Table 3. Distribution of free vs. premium **games content websites** across different CMS's. Studied distribution characteristics are for each CMS; Keys are as in Table 1.

CMS	Count	Percent	MC	MPCP	MP
Free content websites					
Custom code	38	47.50	22	57.89	27.50
WordPress	34	13.75	18	52.94	22.50
DataLife engine	2	2.50	2	100	2.50
vBulletin	2	2.50	0	0.00	0.00
Discuz!	1	1.25	0	0.00	0.00
Others	3	3.75	1	33.33	1.25
Total	80	100	43	–	53.75
Premium websites					
Custom code	64	56.64	15	23.44	13.27
WordPress	22	19.47	8	36.36	7.08
Magento	4	3.54	2	50.00	1.77
Zendesk	4	3.54	1	25.00	0.88
Bigcommerce	2	1.77	0	0.00	0.00
Others	17	15.04	9	52.94	7.96
Total	113	100	35	–	30.97

Table 4. Distribution of free vs. premium **movies content websites** across various CMS's. Studied distribution characteristics are for each CMS; Keys are as in Table 1.

CMS	Count	Percent	MC	MPCP	MP
Free content websites					
Custom code	285	86.10	105	36.84	31.72
Wordpress	34	10.27	17	50.00	5.14
Django framework	2	0.60	0	0.00	0.00
Laravel	2	0.60	1	50.00	0.30
DataLife engine	1	0.30	1	100	0.30
Others	7	2.11	4	57.14	1.21
Total	331	100	128	–	38.67
Premium websites					
Custom code	78	51.32	6	7.69	3.95
WordPress	18	11.84	5	27.78	3.29
Zendesk	11	7.24	5	45.45	3.29
Adobe EM	6	3.95	0	0.00	0.00
Drupal	4	2.63	2	50.00	1.32
Others	35	23.03	5	14.29	3.29
Total	152	100	23	–	15.13

❷ **Games.** Similarly, as shown in Table 3, there are 80 free content and 113 premium websites for games. With 193 websites in total, 91 of them are shown to use CMS, while 102 used custom code. Among the free content games websites, 43 (53.75%) are shown to be malicious in comparison to 35 (31%) of the premium games websites. Put together, the total number of malicious games websites were 78 (roughly 40%). From these results, we make several observations: (1) significantly more free content websites are malicious, (2) both types of websites have a malicious percentage that is close to or significantly higher than the average (33.6%) per Table 1, and (3) the free websites have a 50% chance of being malicious when using a CMS compared to about 40% in the case of the premium websites.

We notice that the top CMS is observed to be the same in both cases. However, others in the top 4 for free content (ordered) are DataLife Engine, vBulletin, Discuz! vs. Magento, Zendesk, and Bigcommerce for premium game websites. DataLife Engine is also shown to be the most malicious CMS (percentage-wise) at 100%, and is only used in the free games category, in contrast to the top (count-wise) malicious CMS (WordPress) used in both categories.

❸ **Movies.** As shown in Table 4, we found that 331 free content websites and 152 premium content websites serve movies. Among the 483 websites, 120 used a CMS, while 363 used a custom code. On the other hand, 128 (38.7%) are shown to be malicious in the free content category vs. 23 (15.1%) for the premium websites category. This result is somewhat expected, given the general association of free movie websites with malicious content distribution. Overall, we found that 151 (30.26%) of the websites in the movies category are malicious. As such, we make the following observations: (1) significantly more free content websites are malicious (23% gap), and (2) both types of websites are slightly less likely to be malicious than the average (33.6%).

We also explore the trend in top CMS's, which are shown to be the same in both categories. Others in the top 4 for the free content (ordered) are Django Framework, Laravel, DataLife Engine vs. Zendesk, Adobe Experience Manager, and Drupal for premium. We notice a draw between WordPress and Laravel for the most malicious CMS (as a percentage) for the free websites with 50%. We also notice that Drupal is the most malicious CMS used only in premium movies with a percentage of 50%. On the other hand, the top (count-wise) malicious CMS is (WordPress) which is used in the free and premium movies websites.

Table 5. Distribution of free vs. premium **music content websites** across different CMS's; Keys are as in Table 1.

CMS	Count	Percent	MC	MPCP	MP
Free content websites					
Custom code	54	65.06	24	44.44	28.92
WordPress	18	21.69	5	27.78	6.02
Drupal	2	2.41	0	0.00	0.00
MediaWiki	2	2.41	0	0.00	0.00
Shopify	2	2.41	1	50.00	1.20
Others	5	6.02	2	40.00	2.41
Total	83	100	32	–	38.55
Premium websites					
Custom code	44	51.16	7	15.91	8.14
WordPress	19	22.09	2	10.53	2.33
Zendesk	4	4.65	2	50.00	2.33
Gatsby	3	3.49	1	33.33	1.16
Oracle CX	2	2.33	0	0.00	0.00
Others	14	16.28	3	21.43	3.49
Total	86	100	15	–	17.44

④ **Music.** As illustrated in Table 5 there are 83 free content and 86 premium content websites, out of which 98 websites use custom code and 71 use a CMS, giving us a total of 169 websites. Overall, 32 (38.6%) of free content were reported as malicious sites compared to 15 (17.44%) of premium content sites. This result conveys the following: (1) Free music websites are significantly more like to be malicious compared to premium music sites. (2) It is noticeable that music websites are slightly less likely to be malicious than the average. Namely, the free websites have twice the chance of being malicious compared to the premium websites (40% vs. 20%).

. Similarly, the most utilized CMS under the music category is WordPress in the free and premium content alike, followed by (ordered) Drupal, MediaWiki, and Shopify in free content, and Zendesk, Gatsby, and Oracle CX Commerce in the premium websites. Where the most malicious CMS (percentage-wise) is Shopify in free music, the results show that Zendesk is the most malicious CMS in premium music with a similar percentage for both CMSs (50%).

⑤ Software. For the last category, results in Table 6 show that 247 websites use CMS's compared to 121 websites that use custom code, for a total of 368 websites; 186 for free content and 182 are premium. Overall, we found 116 (62.4%) of the free content websites are malicious vs. 34 (18.7%) of the premium websites, which adds up to a total of 150 (40.76%) malicious software websites. The results illustrate a significant difference between the free and the premium malicious websites with enormous trends in software-free content. On the other hand, both types have a more malicious percentage than the average (33.63%). Interestingly, the free software websites have more than 60% chance of being malicious against the 20% chance for the premium websites. Unique to this category, the top code base is not the custom code but WordPress, which deviates from the last four categories. Moreover, the other most used CMS's (ordered) are

Table 6. Distribution of free vs. premium **software content websites** across different CMS's. Studied distribution characteristics are for each CMS; Keys are as in Table 1.

CMS	Count	Percent	MC	MPCP	MP
Free content websites					
WordPress	111	59.68	81	72.97	43.55
Custom code	69	37.10	33	47.83	17.74
Contentteller	1	0.54	0	0.00	0.00
IPS community	1	0.54	1	100	0.54
Jimdo	1	0.54	0	0.00	0.00
Others	3	1.61	1	33.33	0.54
Total	186	100	116	–	62.37
Premium websites					
WordPress	55	30.22	10	18.18	5.49
Custom code	52	28.57	7	13.46	3.85
Adobe EM	14	7.69	0	0.00	0.00
Drupal	7	3.85	0	0.00	0.00
Next.js	5	2.75	4	80.00	2.20
Others	49	26.92	13	26.53	7.14
Total	182	100	34	–	18.68

Contentteller, IPS Community Suite, and Jimdo, among free content websites, vs. Adobe Experience Manager, Drupal, and Next.js for the premium content. Nevertheless, Contentteller is shown to be the most malicious CMS with a 100% maliciousness chance and has been used only in the free software category. In contrast to the top (count-wise) malicious CMS (WordPress) used in both free and premium software websites.

5.3   Putting It Together: Discussion

❶ Correlation Heatmap. The correlation heatmap is shown in Fig. 1. Most malicious sites are free content websites based on the correlation heat map. We also find that the free software websites are the most malicious, shown in Table 6 with (62.37%) malicious percent. In contrast, the relation between the maliciousness of a website and the premium category is relatively weak. We also find that the premium category uses more CMS's than the free websites. Ultimately, free content websites using custom code are the most malicious. The second most are the sites using WordPress, which are likely to be malicious for both the free and premium categories. Premium websites using Zendesk and Shopify are the most malicious among the other premium websites.

❷ Further Discussion. Based on the previous results, we infer that CMS websites have a higher malicious percentage than custom code websites. Table 1 shows that 30.46% of the custom code websites are malicious against 38.55% of the websites that use CMS's. We also find that the free content sites have the highest vulnerabilities and maliciousness compared to premium websites in the per-category comparison. We also notice that websites that use specific CMS's, such as WordPress, Shopify, Next.js, Gatsby, and Sitecore, have a high chance of being malicious. It will be hard to generalize these results among the CMS's that only occur once or twice, even if they have a 50% chance of being malicious.

Our assessment of CMS's risk shows they are generally associated with malicious websites. Taking five as a threshold of occurrence with a malicious percent higher than 30%, Shopify is considered high-risk, with 20 websites and 70% of them being malicious. When applying the same criterion to other CMS's, we find that Next.js (53.85%, 13 times), WordPress (44.33%, 379 times), and Zendesk (42.31%, 26 times) are high-risk. We highlight the possibility of reducing the attack surface of websites by not using a high-risk CMS or by fixing the CMS to restrict these vulnerabilities.

We noticed that the CMS with a lower malicious percentage has the highest number of total vulnerabilities but the lowest number of unpatched vulnerabilities. We highlight a pattern to use in practice: those with lower unpatched vulnerabilities are likely CMS's that provide good maintenance and apply the latest security standards. One can recommend reducing the risk of websites using CMS's using the same insights. It has been argued that this could be accomplished by having ongoing monitoring and management of the free content websites [14].

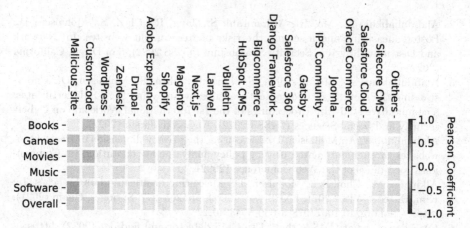

Fig. 1. Correlation Heatmap: Pearson correlation coefficient shows the linear relation between two features. This case shows the most contributor CMS's based on Table 1 and the five content categories (Books, Games, Movies, Music, and Software). It depicts the relation between the maliciousness of the websites and the different categories. The yellow color means that this feature strongly relates to free categories. In contrast, the blue color reflects the strong relation with the premium content. The white color represents the weak relation. (Color figure online)

6   Concluding Remarks and Future Work

Free content websites are an exciting web component. Our study shows various analyses to uncover specific risks associated with those websites in contrast to premium. It highlights the significant challenges with free content websites regarding increased vulnerabilities to maliciousness. Although well-established that free content websites are more likely to be malicious, we tie this likelihood to their utilization of CMS's, in aggregate and at a per-category analysis. Recognizing this problem and the potential role CMS's play in websites security, it is essential to generalize this insight to a more significant number of websites, contrast those trends to other general websites (besides the free content vs. premium), and conduct measurements over time to capture the security.

Acknowledgement. This research was supported by the Global Research Laboratory (GRL) Program through the National Research Foundation of Korea funded by the Ministry of Science and ICT (NRF-2016K1A1A2912757). Part of this work was additionally supported by CyberFlorida Seed Grant (2021–2022).

References

1. Adepoju, S.A., Oyefolahan, I.O., Abdullahi, M.B., Mohammed, A.A., Ibiyo, M.O.: A human-centered usability evaluation of university websites using SNECAAS model. In: Handbook of Research on the Role of Human Factors in IT Project Management, pp. 173–185. IGI Global (2020)

2. Alabduljabbar, A., Ma, R., Alshamrani, S., Jang, R., Chen, S., Mohaisen, D.: Poster: measuring and assessing the risks of free content websites. In: Network and Distributed System Security Symposium (NDSS 2022), San Diego, California (2022)

3. Alabduljabbar, A., Ma, R., Choi, S., Jang, R., Chen, S., Mohaisen, D.: Understanding the security of free content websites by analyzing their SSL certificates: a comparative study. In: Proceedings of the 1st International Workshop on Cybersecurity and Social Sciences (CySSS 2022), Nagasaki, Japan (2022)

4. Alabduljabbar, A., Mohaisen, D.: Measuring the privacy dimension of free content websites through automated privacy policy analysis and annotation. In: Companion Proceedings of the Web Conference (2022)

5. Alkinoon, M., Choi, S.J., Mohaisen, D.: Measuring healthcare data breaches. In: Kim, H. (ed.) WISA 2021. LNCS, vol. 13009, pp. 265–277. Springer, Cham (2021). https://doi.org/10.1007/978-3-030-89432-0_22

6. CMS detect: what CMS is that? Use CMS detector and find out (2022). https://cmsdetect.com/

7. CVE Details: The ultimate security vulnerability datasource (2022). https://www.cvedetails.com/

8. Gall, R.: Wordpress 5.9.2 security update fixes XSS and prototype pollution vulnerabilities (2022). https://www.wordfence.com/

9. Cybersecurity help: vulnerability database (2022). https://www.cybersecurity-help.cz/

10. Jayanthi, S., Sasikala, M.S.: XGraphticsCLUS: web mining hyperlinks and content of terrorism websites for homeland security. Int. J. Adv. Netw. Appl. **2**(6), 941–949 (2011)

11. Al Kinoon, M., Omar, M., Mohaisen, M., Mohaisen, D.: Security breaches in the healthcare domain: a spatiotemporal analysis. In: Mohaisen, D., Jin, R. (eds.) CSoNet 2021. LNCS, vol. 13116, pp. 171–183. Springer, Cham (2021). https://doi.org/10.1007/978-3-030-91434-9_16

12. Libert, T.: Exposing the hidden web: an analysis of third-party HTTP requests on 1 million websites. arXiv preprint arXiv:1511.00619 (2015)

13. openbugbounty: The complete list of bug bounty and security vulnerability disclosure programs launched and operated by open bug bounty community (2022). https://www.openbugbounty.org/

14. Ostroushko, A.: Restricting access to websites as an new procedure of government coercion. Financ. Law Manag. 167–173 (2015)

15. Pan, X., Cao, Y., Liu, S., Zhou, Y., Chen, Y., Zhou, T.: CSPAutoGen: black-box enforcement of content security policy upon real-world websites. In: Proceedings of the 2016 ACM SIGSAC Conference on Computer and Communications Security, pp. 653–665 (2016)

16. Snyk: find and automatically fix vulnerabilities in your code, open source dependencies, containers, and infrastructure as code (2022). https://snyk.io/

17. Verkijika, S.F., De Wet, L.: Quality assessment of e-government websites in Sub-Saharan Africa: a public values perspective. Electron. J. Inf. Syst. Dev. Ctries. **84**(2), e12015 (2018)

18. VirusTotal: Analyze suspicious files and URLs to detect types of malware, automatically (2022). https://www.virustotal.com/

19. W3Techs: W3techs - world wide web technology surveys (2022). https://w3techs.com/sites

Fact-checking, Fake News, and Hate Speech

BNnetXtreme: An Enhanced Methodology for Bangla Fake News Detection Online

Zaman Wahid[1]([⊠])(iD), Abdullah Al Imran[2](iD), and Md Rifatul Islam Rifat[3]

[1] University of Calgary, Calgary, Canada
zaman.wahid@ucalgary.ca
[2] American International University - Bangladesh, Dhaka, Bangladesh
[3] Rajshahi University of Engineering and Technology, Dhaka, Bangladesh

Abstract. In the last couple of years, the government, and the public have shown a lot of interest in fake news on Bangladesh's fast-growing online news sites, as there have been significant events in various cities due to unjustifiable rumors. But the overall progress in study and innovation in the detection of Bangla fake and misleading news is still not adequate in light of the prospects for policymakers in Bangladesh. In this study, an enhanced methodology named BNnetXtreme is proposed for Bangla fake news detection. Applying both embedding based (i.e. word2vec, Glove, fastText) and transformer-based (i.e. BERT) models, we demonstrate that the proposed BNnetXtreme achieves promising performance in detection of Bangla fake news online. After a further comparative analysis, it is also discovered that BNnetXtreme performed superior to BNnet-one of the state-of-the-art architectures for Bangla fake news detection introduced previously. The BNnetXtreme especially BERT Bangla base model performed with an accuracy score of 91% and an AUC score of 98%. Our proposed BNnetXtreme has been successful in improving the performance by an increase of 1.1% in accuracy score, 5.6% in precision, 1.1% in F1 score, and about 9% in AUC score.

Keywords: Bangla · Fake news identification · Text classification · Natural Language Processing · Deep Neural Network

1 Introduction

In Bangladesh, reading news online has become increasingly prevalent, displacing the conventional method of keeping up with current events by reading hard copies of newspapers shipped to one's doorstep every morning. In response to this fact, the number of online news portals has considerably expanded as they adapt to the contemporary method of maximizing profit of which the potential has been underlined exponentially. However, even well-established and renowned newspaper firms in Bangladesh use misleading and false headlines to attract visitors to their websites. Millions of people in Bangladesh shared a news story

T. N. Dinh and M. Li (Eds.): CSoNet 2022, LNCS 13831, pp. 157–166, 2023.
https://doi.org/10.1007/978-3-031-26303-3_14

on a well-known online news portal in 2015, informing them that their favorite actor, Abdur Razzak, had died. Nonetheless, a small number of people may have picked up on the fact that the actor is dead in the latest film he was working on, rather than in real life [5]. People's panic lasted weeks, but the practice of publishing satire and fake news with an impressive methodology of making the news appear true in order to boost website traffic skyrocketed at times. Even though Bangladesh publishes over 1000 newspapers online every day, the inundation of false and satirical news has made finding real news challenging. The fake and satirical news portals are often found to be filled with malicious intentions-having both increasing the number of visitors and stealing the visitor's personal data to sell to third-party organizations.

Natural Language Processing (NLP) has played a key role in understanding written languages and delivering on-demand, informative solutions to respective problem domains, particularly in the detection of fake news, over the years. While the number of studies in fake and satirical news in English are commendable, there has been little progress in Bangla news. In line with the progression required in Bangla news, we contribute to the following in this study:

1. Propose a novel methodology for Bangla fake news detection online
2. Demonstrate a promising and improved detection performance of Bangla fake news by providing a comparative performance analysis among the different classifiers used.

Serving the purpose, we have applied both embedding-based and transformer-based models, especially BERT Bangla base putting an emphasis. Our proposed methodology shows a substantial improvement in detecting fake Bangla news than the formerly introduced architecture BNnet [5]. The remaining parts of the paper are as follows: Related Works, Data Description, Data Preprocessing, Methodology, Result & Analysis, and Conclusion.

2   Related Works

Bangla does not have a lot of literature in the field of NLP since it is a low-resource language. In this section, we go through some of the current NLP applications and developments that are relevant to this research.

A study [5] introduced a Deep Neural Network (DNN) architecture named "BNnet" to identify satire and fake Bangla news widely spread to online news portals and social networking sites in Bangladesh, from quite a range of reputed and fake online news websites. Using web scraping, they have collected diverse Bangla news content from 25 real and 18 fake news portals–43 news portals in total. They found that the proposed BNnet architecture provides the best performance in predicting fake Bangla news–using prominent evaluation metrics–with an accuracy and AUC score of 90%.

Recently, Hossain et al. [2] proposed a Bi-LSTM based model for Bangla fake news detection online over a total of 57000+ samples in a benchmark dataset and reported that deep learning architectures perform better in fake news classification than classical machine learning methods. A deep hybrid model is developed by Adib et al. [4] for detecting Bangla fake news, which uses 1D Convolutional Neural Networks (CNN) for feature extraction and traditional Machine Learning (ML) approaches for classification. Because the neural network takes care of extracting features from the dataset, the proposed approach allows to save time and effort, reporting the performance accuracy is over 90%. George et al. [3] applied Convolution Neural Network and Long Short-Term Memory over a dataset of 50K fake news of Bangla. They reported the performance of their model is 75%. M Z Hossain et al. [6] provided an analysis in developing a benchmark framework with state-of-the-art NLP techniques to classify Bangla fake news by introducing an annotated dataset containing 50K Bangla news. Three neural-based methods: Convolution Neural Network (CNN), Long Short-Term Memory (LSTM), and Bidirectional Encoder Representations from Transformers (BERT) were applied alongside traditional linguistic features. With an F1 score of 0.68, the BERT produces the best performance while LSTM performs the least with an F1 score of 0.53.

M G Hussain et al. [7] conducted an experiment with Support Vector Machine (SVM) and Multinomial Naive Bayes (MNB) classifiers, to detect Bangla fake news. The Term Frequency - Inverse Document Frequency Vectorizer (TF-IDF) and CountVectorizer techniques for feature extraction were utilized. It is reported that SVM with a linear kernel outperforms MNB. However, they only relied on three news portals: ProthomAlo, a popular authentic newspaper in BD; ProthomAlu, BalerKontho, and Moktikontho, all these three online portals are well known for sarcastic and fake news only–not a single article is true. Thus, the class distribution is handled very poorly and left with potential biases.

Based on the news code of the International Press Telecommunications Council (IPTC), Abu Nowshed et al. [8] used TF-IDF approach for feature extraction and Recall-Precision graph to show how the model performs. Several other studies such as [9–14] used doc2vec, Latent Semantic Analysis (LSA), and the Latent Dirichlet Allocation (LDA), DBSCAN methods to investigate content-based recommendation systems and detection of fake Bangla news. Their findings reveal that doc2vec can deliver more contextual suggestions than LSA and LDA, as well as exceed LSA and LDA in terms of accuracy, ranging from 60% to 89%. Primarily the majority of the studies contain typical NLP applications such as newsgroup or subject classification, content-based suggestion, text summarization, and language type classification, as seen in the above reviews. Taking these research scopes into account, we intend to apply several deep learning approaches to design a better methodology for detecting Bangla fake news online rather than following typical classical ML approaches.

3   Preprocessing of Data

A proprietary dataset [5] is used that contains 500 observations with news from 31 different categories such as Politics, Entertainment, Sports, etc. There are

257 observations labeled as real news and 243 observations labeled as fake news in total. For a supervised classification task, the dataset is well balanced and distributed. When data is obtained from different internet sources using web scraping, the malformation and unstructuredness of text data add to the pre-processing challenges [17]. We have completed three pre-processing tasks in this phase: denoising, normalization, and stemming. We wrote a hard-coded program to automate the process of denoising to clean the noises contained in our dataset, such as '\r', '\n', HTML tags: '<br>', '<p>', and extra white-spaces. Later, we used two normalization techniques to normalize our Bangla textual data: i) Punctuation removal; ii) stop word removal. In this study, we used a rule-based Bengali stemmer that proposed in [15,16] to have a language specific stemming.

4   Methodology

Our methodology is split into two parts: one uses embedding-based models and the other uses a transformer-based model. The following Fig. 1 shows the proposed methodology of this experiment.

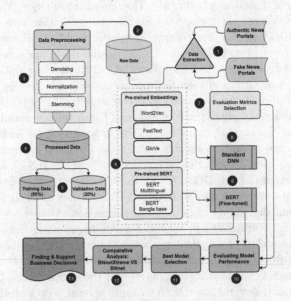

Fig. 1. Proposed methodology for detecting Bangla fake news

4.1   Embedding-Based

We have chosen to apply several pre-trained word embedding models to extract features for this experiment because they often contribute to better results in the different tasks and learn the context and associations of terms from the text

better than count based vectorizers. Therefore, we have utilized the following
three word embedding techniques: Word2vec, FastText, and GloVe. Word2Vec
[20] takes a text corpus as input and produces a collection of vectors of length
300 in the form of feature vectors that describe words contextually in the corpus.
It is a set of model architectures and optimizations that have been shown to perform well on a variety of downstream natural language processing tasks. Next,
the FastText [19] generates word embeddings using subword information. Character n-gram representations are learned, and words are represented as the sum
of the n-gram vectors. By using character-level knowledge, FastText can truly
outperform the competition, particularly when it comes to rare words. Finally,
in GloVe [18], each word is represented as a 300-dimensional GloVe vector and
training is based on aggregated global word-word co-occurrence matrix from a
corpus, rather than local-running windows for words. Since GloVe combines the
benefits of the word2vec skip-gram model, the benefits of matrix factorization
methods can exploit global statistical information. Once the feature extraction
part is complete, we split the data for training and validating the data, following
a standard widely used stratified 80% and 20% splitting ratio. Then train a deep
neural network (DNN) with a four hidden layers and having a dropout layer after
each hidden layer. The number of units in each hidden layer is reduced to 512,
256, 128, and 64 respectively. The architecture of DNN for the embedding based
model is given as follows (Fig. 2): Due to the nature of classification tasks in this
study, we have used the binary-cross-entropy loss function, additionally, keeping
the learning rate is of 0.0001 for Adam optimizer, a stochastic gradient descent
technique, for all the classification algorithms applied. For word2vec, the model
reached the optimized accuracy in its 50th epoch with accuracy: 0.7515. For
fastText and GloVe the optimized number of epochs are 67 and 71 respectively.

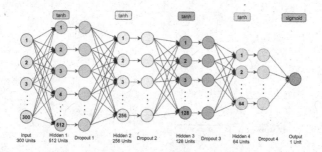

Fig. 2. Proposed DNN architecture for the embedding based model

4.2   Transformer Based: BERT

Ulike traditional models used for NLP-related tasks, BERT's input embeddings
are the combination of the token embeddings, the segmentation embeddings
and the position embeddings. An input sentence is converted into a tokenized
sequence before being transferred to the Token Embeddings layer. In addition,

two special tokens: [CLS] and [SEP] are added to the beginning and end of this tokenized sequence, respectively where, the token [CLS], also known as a classification token, is used to represent a pair of input texts for downstream classification tasks, and the token [SEP] is used to distinguish them. In this study, we used both the BERT multilingual and the BERT Bangla base [1], which are both pre-trained BERT models. First, we initialized the fine-tuned model with the same pre-trained model parameters for the downstream task, Bangla fake news classification, during fine-tuning. Then, using the task-specific labeled data, we fine-tuned all of the parameters from start to finish. For both the Bangla base and multilingual base of BERT, the hyper-parameters were kept the same. The architecture of the BERT Bangla base is as shown in Fig. 3.

5   Experimental Result and Analysis

In this section, we demonstrate and discuss the performance of embedding models. Then we analyze the output of BERT based pre-trained models. Finally we compare the results of this whole experiment with the result presented in [5] paper for detecting Bangla fake news.

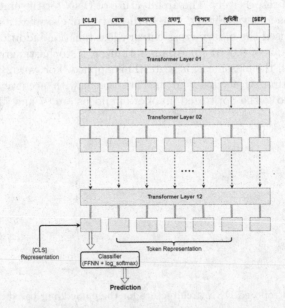

Fig. 3. Architecture of BERT fine-tuning model

5.1   Embedding Result Analysis

For predicting Bangla satire and fake news, we have applied three feature extraction-based embedding models: word2vec, fastText, and GloVe. The results of these algorithms are shown in the following Table 1.

Table 1. Experimental results of embedding algorithms

Classifier	Accuracy	Precision	Recall	F1 Score	AUC
word2vec	**0.77***	0.75	**0.82***	**0.79***	**0.77***
fastText	0.75	0.72	**0.82***	0.77	0.75
GloVe	0.73	**0.77***	0.67	0.72	0.73

Referring to Table 1, it can be seen that both word2vec and fastText perform similarly in terms of Recall score, 0.82, although word2vec performs superior in other measures. GloVe performs worst in most evaluation criteria. On the other hand, word2vec outperforms all the other embedding algorithms applied, with higher scores in Accuracy, Precision, F1 Score, AUC Score of 77%, 75%, 79%, and 77% respectively, which is above 3% better performance than fastText in these four evaluation metrics. It is clear that while embedding algorithms may not perform impressively however the performance still seems reasonable.

5.2 BERT Result Analysis

The following Table 2 demonstrates the results of BERT. The multilingual BERT base outperforms all the other classifiers with a recall score of 0.94, way superior to embedding models. The performance are quite consistent in BERT multilingual model, however, there is a significant raise for both recall and AUC score. The AUC score for a machine learning model shows how balanced the model is as it's not influenced by the size of the test data, AUC is a better metric for binary classifiers. Table 2 demonstrates that both the BERT multilingual and Bangla base performs exceptionally well in terms of AUC Score, 0.95 and 0.98 respectively, making the Bangla base a win.

Table 2. Experimental results of BERT models

Classifier	Accuracy	Precision	Recall	F1 Score	AUC
Multilingual	0.87	0.83	**0.94***	0.88	0.95
Bangla base	**0.91***	**0.94***	0.88	**0.91***	**0.98***

Although the performance of the Bangla base is inferior to BERT multilingual on recall score, the BERT Bangla base yet produces superior results in every other metric. The accuracy is over 4% better and the precision is over 11% better in the BERT Bangla base. Therefore, making it the best classifier between both the embedding and transformer-based algorithms.

5.3   Comparative Result Analysis with the Latest Study

Now we illustrate a comparative analysis of the experimental results between the proposed methodology in this study and BNnet [5]. Table 3 shows the comparison of the results of these both methodologies.

Table 3. Comparative result analysis of BNnet and BNnetXtreme

BNnet

Classifier	Accuracy	Precision	Recall	F1-Score	AUC Score
LR	0.87	0.87	0.88	0.87	0.87
NB	0.8	0.88	0.71	0.78	0.8
DT	0.76	0.78	0.75	0.76	0.76
RF	0.86	0.8	**0.96***	0.88	0.86
KNN	0.88	0.87	0.9	0.89	0.88
SVM	0.86	0.85	0.88	0.87	0.86
DNN	0.9	0.89	0.92	0.9	0.9

BNnetXtreme

Model	Accuracy	Precision	Recall	F1 Score	AUC Score
word2vec + DNN	0.77	0.75	0.82	0.79	0.77
fastText + DNN	0.75	0.72	0.82	0.77	0.75
GloVe + DNN	0.73	0.77	0.67	0.72	0.73
BERT Multilingual	0.87	0.83	0.94	0.88	0.95
BERT Bangla base	**0.91***	**0.94***	0.88	**0.91***	**0.98***

In BNnet [5], it is demonstrated that Decision Tree produces the poorest result in detecting Bangla fake news with both accuracy and AUC score of 0.76. Their proposed DNN architecture yields the best result as high as with 90% accuracy and AUC score. It seems BERT multilingual and the BNnet DNN architecture perform similarly, but not necessarily the same. In terms of recall and AUC score, the multilingual model performs better with scores of 0.94 and 0.95 respectively, while BNnet DNN scores 0.92 and 0.90. On the other hand, BNnet performs better in terms of accuracy, precision, and f1 score: 0.90, 0.89, and 0.90. Since the trade-off between BERT multilingual and BNnet DNN is complicated, it is hard to determine the better one. Now, comparing all the experimental results of BNnet and BNnetXtreme's models, it is evident that BERT Bangla base illustrates a significant improvement in all other evaluation metrics combined except Recall. From Table 3, it can be seen that the BERT Bangla base achieved an improved accuracy by 1.1%, precision by 5.6%, and f1 score by 1.1% in regards to the best performance achieved in BNnet DNN. Furthermore, as the AUC score is the ultimate measure to assess the balance of a model, the BERT Bangla base model in BNnetXtreme produced an excellent score, 0.98, which is 8.9% greater than the BNnet DNN's AUC score of 0.90. Lastly, based on all the analysis, evaluation metrics, and performance, it can be said that the new BNnetXtreme demonstrated a significantly improved performance in detecting Bangla fake news, and the BERT Bangla base classifier is found to be the best.

6 Conclusion

Fake news in the context of Bangladesh is critically vital to consider for advanced research and analysis to provide users with a safe place in online news portals that may potentially be deceptive, fraudulent, or dangerous. In this study, we have introduced a methodology named BNnetXtreme that encompasses the application of both embedding and transformer-based deep learning architectures that streamline and boost the performance of Bangla fake news detection. We demonstrated that transformer based models (BERT) perform superior to the word embedding models. Although the BERT multilingual base model is found to be the best for minimizing the misclassification rate, the BERT Bangla base model wins the race with superior results in all other evaluation metrics, especially an AUC Score of 0.98. Furthermore, with a comparative result analysis, we conclude that our proposed methodology BNnetXtreme is successful in boosting the performance of Bangla fake news detection by an increase of 1.1% in accuracy, 5.6% in precision, 1.1% in f1 score, and most importantly, 9% in AUC score. Future studies can include the incorporation of both character, word, and contextual embeddings for building superior Bangla fake news detection models.

References

1. Sarker, S.: BanglaBERT: Bengali mask language model for Bengali language understanding (2020). https://github.com/sagorbrur/bangla-bert. Accessed 10 Feb 2022
2. Hossain, E., Nadim Kaysar, M., Joy, A.Z.M.J.U., Mizanur Rahman, M., Rahman, W.: A study towards Bangla fake news detection using machine learning and deep learning. In: Shakya, S., Balas, V.E., Kamolphiwong, S., Du, K.L. (eds.) Sentimental Analysis and Deep Learning. Advances in Intelligent Systems and Computing, vol. 1408, pp. 79–95. Springer, Singapore (2022). https://doi.org/10.1007/978-981-16-5157-1_7
3. George, M.Z.H., Hossain, N., Bhuiyan, M.R., Masum, A.K.M., Abujar, S.: Bangla fake news detection based on multichannel combined CNN-LSTM. In: 2021 12th International Conference on Computing Communication and Networking Technologies (ICCCNT), pp. 1–5. IEEE, July 2021
4. Adib, Q.A.R., Mehedi, M.H.K., Sakib, M.S., Patwary, K.K., Hossain, M.S., Rasel, A.A.: A deep hybrid learning approach to detect Bangla fake news. In: 2021 5th International Symposium on Multidisciplinary Studies and Innovative Technologies (ISMSIT), pp. 442–447. IEEE, October 2021
5. Al Imran, A., Wahid, Z., Ahmed, T.: BNnet: a deep neural network for the identification of satire and fake Bangla news. In: Chellappan, S., Choo, K.-K.R., Phan, N.H. (eds.) CSoNet 2020. LNCS, vol. 12575, pp. 464–475. Springer, Cham (2020). https://doi.org/10.1007/978-3-030-66046-8_38

6. Hossain, M.Z., Rahman, M.A., Islam, M.S., Kar, S.: BanFakeNews: a dataset for detecting fake news in Bangla (2020). arXiv preprint arXiv:2004.08789H

7. Hussain, M.G., Hasan, M.R., Rahman, M., Protim, J., Al Hasan, S.: Detection of Bangla fake news using MNB and SVM classifier. In: 2020 International Conference on Computing, Electronics and Communications Engineering (iCCECE), pp. 81–85. IEEE, August 2020

8. Wahid, Z., Satter, A.Z., Al Imran, A., Bhuiyan, T.: Predicting absenteeism at work using tree-based learners. In: Proceedings of the 3rd International Conference on Machine Learning and Soft Computing, pp. 7–11, January 2019

9. Chy, A.N., Seddiqui, M.H., Das, S.: Bangla news classification using naive Bayes classifier. In: 16th International Conference Computer and Information Technology. IEEE (2014)

10. Nath Nandi, R., Arefin Zaman, M.M., Al Muntasir, T., Hosain Sumit, S., Sourov, T., Jamil-Ur Rahman, Md.: Bangla news recommendation using doc2vec. In: 2018 International Conference on Bangla Speech and Language Processing (ICBSLP). IEEE (2018)

11. Dhar, A., Dash, N., Roy, K.: Classification of text documents through distance measurement: an experiment with multi-domain Bangla text documents. In: 2017 3rd International Conference on Advances in Computing, Communication and Automation (ICACCA) (Fall). IEEE (2017)

12. Mouhoub, M., Al Helal, M.: Topic modelling in Bangla language: an LDA approach to optimize topics and news classification. CIS **11**, 77 (2018)

13. Paul, A., et al.: Bangla news summarization. In: Nguyen, N.T., Papadopoulos, G.A., Jędrzejowicz, P., Trawiński, B., Vossen, G. (eds.) ICCCI 2017. LNCS (LNAI), vol. 10449, pp. 479–488. Springer, Cham (2017). https://doi.org/10.1007/978-3-319-67077-5_46

14. Haque, Md.M., Pervin, S., Begum, Z.: Automatic Bengali news documents summarization by introducing sentence frequency and clustering. In: 2015 18th International Conference on Computer and Information Technology (ICCIT). IEEE (2015)

15. Mahmud, Md.R., Afrin, M., Razzaque, Md.A., Miller, E., Iwashige, J.: A rule based Bengali stemmer. In: 2014 International Conference on Advances in Computing, Communications and Informatics (ICACCI). IEEE (2014)

16. Parves, A.B., Al Imran, A., Rahman, Md.R.: Incorporating supervised learning algorithms with NLP techniques to classify Bengali language forms. In: Proceedings of the International Conference on Computing Advancements. ACM (2020). https://doi.org/10.1145/3377049.3377110

17. Pennington, J., Socher, R., Manning, C.D.: Glove: global vectors for word representation. In Proceedings of the 2014 Conference on Empirical Methods in Natural Language Processing (EMNLP), pp. 1532–1543, October 2014

18. Joulin, A., Grave, E., Bojanowski, P., Douze, M., Jégou, H., Mikolov, T.: Fasttext. zip: compressing text classification models (2016). arXiv preprint arXiv:1612.03651

19. Church, K.W.: Word2Vec. Nat. Lang. Eng. **23**(1), 155–162 (2017)

20. Devlin, J., Chang, M.W., Lee, K., Toutanova, K.: BERT: pre-training of deep bidirectional transformers for language understanding (2018). arXiv preprint arXiv:1810.04805

Heuristic Gradient Optimization Approach to Controlling Susceptibility to Manipulation in Online Social Networks

Abiola Osho[1]([✉]) [iD], Shuangqing Wei[2] [iD], and George Amariucai[1] [iD]

[1] Kansas State University, Kansas, USA
{aaarise,amariucai}@ksu.edu
[2] Louisiana State University, Baton Rouge, USA
swei@lsu.edu

Abstract. Manipulation through inferential attacks in online social networks (OSN) can be achieved by learning the user's interests through their network and their interactions with the network. Since some users have a higher propensity for disclosure than others, a one-size-fits-all technique for limiting manipulation proves insufficient. In this work, we propose a model that allows the user to adjust their online persona to limit their susceptibility to manipulation based on their preferred disclosure threshold. Our experiment, using real-world data provides a way to measure manipulation gained from a single tweet. We then proffer solutions that show that manipulation gain derived as a result of participating in OSNs can be minimized and adjusted to meet the user's needs and expectations, giving at least some measure of control to the user.

Keywords: Manipulation · Gradient Optimization · Social Network Analysis

1 Introduction

Online social networks (OSNs) provide a medium where people build social relationships with other people who share similar personal or career content, interests, activities, backgrounds, or real-life connections. While social interaction and influence can be beneficial to users, they can also be detrimental to user privacy and opinion formation. This is because the activities they generate can be used to learn other latent (and sometimes sensitive) information, like their beliefs and orientations. The search to find a balance between privacy preservation, independent thinking and social influence leads us to ask if users can be given control over their privacy. Our research aims to directly address this question by allowing the users to examine their likelihood for manipulation based on their social interactions, giving them insight into their degree of exposure so that they can choose how much protection needs to be implemented based on their privacy needs.

T. N. Dinh and M. Li (Eds.): CSoNet 2022, LNCS 13831, pp. 167–178, 2023.
https://doi.org/10.1007/978-3-031-26303-3_15

We examine manipulation gain in terms of a user's susceptibility to targeted manipulation through inferential attack in a single tweet. We propose a model that first measures a user's probability of engaging with a post in a neutral environment and then measures the degree of deviation of this probability when a profile and posts from that profile are targeting them. By doing this, the user has an idea of how much a particular friend in their network can cause them to change the way they interact. This change can be seen as manipulation gain because it is caused by how much information can be inferred from their activities. We examine a scenario where the sender of a message tries to mislead the receiver by optimizing their (sender) attributes and those of the message to mimic what the receiver will typically show interest in. By doing this, the receiver is gradually manipulated to engage with a post that they would otherwise ignore. The receiver is said to be manipulated if the probability of engagement with the targeted message deviates from what it would have been if the message were produced in the absence of inferred knowledge about that specific user.

Existing privacy protection techniques proffer self-sanitization as a way to address information leakage but the sanitization itself is to be done by the user's friends [13]. With little efforts in limiting manipulation in the wild, it is essential that the user has more control on how accurately spammers, learning models, and third-party vendors make inferences about them. We hypothesize that the profile and posts of the user are representative of their true self and can be used to make observations about them. We adopt Gaussian Process Classification model to learn the user's probability to react to a post shared by their friend, and then use gradient optimization methods to heuristically search for attributes that can be optimized to limit their likelihood to respond to a targeted post.

The contributions of the paper are as follows:

- We present a node-to-node feature analysis model to estimate the probability of reaction to a post based on a set of user and message attributes.
- We fit a stochastic model to estimate a user's susceptibility to manipulation through inferential attack.
- We implement a privacy preservation mechanism controlled by the user based on their propensity for disclosure.
- We provide a metric for estimating manipulation gain based on implemented protection mechanism.
- We draw conclusions regarding the degree of change in disclosed attributes to minimize manipulation.

The paper is organized as follows. The section on *Related Works* reviews the related work on information diffusion in social networks. Section *Model* describes our general approach, classification algorithms and evaluation metrics. Section *Experiment* describes the data collection, the experiment setup and the feature set, while *Results* presents our results. Finally, *Conclusion and Future Work* gives conclusions and insights into possible future works.

2   Related Works

Users post and provide personal information without an understanding of how it might be used or accessed, leading to privacy and information leakage. Privacy controls are provided to limit access to user information but OSN default settings allow unlimited access, unless the controls are enabled by users. To address privacy concerns, users can utilize privacy settings and hide sensitive information, but it has been shown by He et al. [4] and Zheleva et al. [16] that such measures, even though promising, are not sufficient to protect the user's privacy due to the friendship relations, group memberships, or even participating in activities like mentions, tags, shares, and commenting, which can be harvested through 'screen-scraping' [8] or other means. To show that group membership can be increase information leakage, Zheleva et al. [16] proposed eight privacy attacks for sensitive attribute inference using a variety of classifiers and features to show ways in which an adversary can utilize links and groups in predicting private information.

Information leakage can be viewed as the combined probability of sensitive attribute inference from the information available in immediate friends' profiles, Talukder et al. [13] addressed that by presenting a friends rank component to find the amount of match of sensitive attribute values between a user and their friends. To estimate Facebook users' ages, Dey et al. [2] exploited the underlying social network structure to design an iterative algorithm which derives age estimates based on ages of friends and friends of friends, while Li et al. [7] inferred demographic information such as age, gender, education by observing users' exposed location profiles. To address the privacy concerns emanating from neighborhood attacks, Zhou and Pei [17] adopted k-anonymity and l-diversity models from relational data to social network data. In this work we look beyond just privacy and information leakage, and examine how these can be exploited in inferential attacks and proffer solutions to directly address them.

3   Model

3.1   Dataset

To generate the features required for the model, we collect the metadata of Tweet and User JSON (JavaScript Object Notation) objects. We adopted a previously created tool made publicly available on GitHub [10], that crawls the Twitter Streaming API. In collecting Twitter dataset aimed at general conversations, bias can be introduced due to events happening in the real world. For the purpose of this research, we adopted the crawler to collect streams of tweets and associated user profiles instead of specifying usernames, IDs, topics or demographics, with the data collected over multiple 7-day periods.

We remove accounts with no followers and/or activities, and then focus on the set of users with established followership relationship by creating a relationship graph by connecting the users in our dataset. We also limit our data collection to posts in English language. Table 1 shows the distribution of the dataset after

Table 1. Data distribution

Data statistics	Count
Total number of users	12200
Follower-followee relationships	46400
Total number of tweets	81500

cleaning and pre-processing. For each tweet in the dataset, we associate the profiles of the sender (associated account), receiver (the sender's follower), and a binary *engagement label*, if the receiver has generated some reaction to the post.

3.2   Attributes

The proposed framework makes use of 2 categories of features, see Table 2: one for the user and the other for the message. For each user (sender or receiver), we learn 8 directly observable attributes and a social homogeneity feature that is common to both of them. These attributes have been extensively studied in the works of Castillo et al., [1] and Yang et al. [15]. These attributes are important because a user's friends impact the kind and volume of messages that end up in their timeline and the higher the number of followers, the farther the possibility of reach. Twitter posts are very fluid, taking up various forms and content, we learn 6 attributes of the message that describes the content, popularity and sentiment associated with them.

Table 2. Attribute description

		Feature	Description
1	User features	Tweets containing URL	Number of user's tweets containing URLs
2		Presence of user description	Shows user's profile has description (bio)
3		User verified	Shows if user's account is verified
4		Number of followers	Higher follower count means higher reach
5		Number of friends	Average number a user follows
6		Account age	Account age in days
7		Status count	Total number of posts over account's lifetime
8		User favorites count	Number of user's tweets endorsed by others
9		Social homogeneity	Depicts common friends and followers
10	Message features	Presence of hashtags	shows if a tweet contains hashtags
11		Presence of URLs	Shows if a tweet contains URLs
12		Presence of media	Shows if a tweet contains media
13		Tweet favorites count	Favorites count for tweet
14		Retweet count	Retweet count for tweet
15		Sentiment score	Sentiment score for tweet

3.3   Manipulation Gain

In this work, the sender of a message (adversary) learns the behavior of their network by observing activities generated by their followers in terms of topics their followers are interested in, inferring latent and demographic attributes that might not have previously been shared by their followers. The sender can then use this learned information to adapt their profile and posts into one that their followers will find interesting enough to interact with. The receiver (the adversary's follower) is gradually manipulated into engaging with an account that they would generally not interact with because the posts coming from such an account mimic what the receiver will typically show interest in. Engagement is simply the generation of a reaction from the receiver of the post in the form of a reply, retweet, favorite, share, or like. We define manipulation in terms of a user's engagement with a post, caused by inferences made on the user's behavior. A user is said to be manipulated if the probability of engagement with a targeted message deviates from what it would have been if it were a message produced in the absence of any knowledge about the specific user. It should be noted that this deviation can be positive or negative and it is only meant to show that there is a change from the user's regular behavior.

The objective of the sender is to vary the attributes of its messages to maximize the probability of engagement from the receiver of the message. The receiver reacts by varying its published features in accordance with the goal of an adversary/manipulator/sender. The receiver aims to minimize the absolute difference between their probability of interaction with a targeted message and the probability of interaction with a random post on their timeline. At each point of optimization, the sender/receiver takes the observable attributes as the true attributes of the other. Both parties can only modify their attributes before disclosing them and cannot make changes after they are disclosed. The receiver (user B) can only modify their features, while the sender (user A) can optimize over either their own attributes or those of the message (Table 3).

Table 3. Summary of notation

Notation	Description
τ_i	Response from user u_i
θ_i	Attributes of user u_i
θ'_i	Disclosed attributes of user u_i
M_{ij}	Message attributes from user u_i to user u_j
M'_{ij}	Optimized message from user u_i to user u_j
ϵ	Measure of manipulation gain
β	Probability of response to random message
γ	Probability of response to targeted message

The probability that user B (receiver) with attributes θ_B will engage with a random message m_{AB} from user A (sender) with attributes θ_A, given the true attributes of user B is:

$$\beta = P[\tau_B = 1 | \theta_A, M_{AB}, \theta_B]. \tag{1}$$

If user B chooses to modify their disclosed attributes, thus disclosing θ'_B in place of θ_B, then from user A's perspective, the probability that user B will engage with a message m_{AB} is given by:

$$P[\tau_B = 1 | \theta_A, M_{AB}, \theta'_B]. \tag{2}$$

Since user A is only privy to the disclosed attributes, user A optimizes over their own attributes θ_A and over the message attributes M_{AB} to ensure engagement with the post, such that:

$$[\theta'_A(\theta'_B), M'_{AB}(\theta'_B)] = argmax_{\theta_A, M_{AB}} P[\tau_B = 1 | \theta_A, M_{AB}, \theta'_B]. \tag{3}$$

In order to minimize the chances of being manipulated, user B needs to publish an optimal set of attributes $\theta^*(B)$, such that:

$$\theta^*_B = argmin_{\theta'_B} P[\tau_B = 1 | \theta'_A(\theta'_B), M'_{AB}(\theta'_B), \theta_B]. \tag{4}$$

This leads to user A finding the optimal

$$[\theta^*_A(\theta^*_B), M^*_{AB}(\theta^*_B)] = argmax_{\theta_A, M_{AB}} P[\tau_B = 1 | \theta_A, M_{AB}, \theta^*_B]. \tag{5}$$

The challenge with this is that the true probability of engagement is based on user B's true attributes, that is:

$$\gamma = P[\tau_B = 1 | \theta'_A, M'_{AB}, \theta_B]. \tag{6}$$

The model is intended such that user B sets a threshold ϵ on the maximum allowable deviation of γ from β. The user then works to ensure that

$$|\gamma - \beta| \leq \epsilon \tag{7}$$

at every point in time. Manipulation is said to be successful if $|\gamma - \beta| > \epsilon$. The goal of the adversary (user A) is to maximize the LHS of Eq. (7) while user B focuses on minimizing manipulation gain, ϵ arising from inferential attacks by shrinking that value as much as possible.

4   Experiment

4.1   Gaussian Process Classification Model

Gaussian process classification (GPC) [12] is capable of making fine distinctions in the sense that it models $p(y|x)$ as a fixed Bernoulli distribution. In the GPC model, inference is made from the latent function f given the observed data $D =$

$\{(x_i, y_i)|i = 1, ..., n\}$, with $f_i = f(x_i)$, $f = [f_1, f_2, ..., f_n]$, $X = [x_1, x_2, ..., x_n]$, and $y = [y_1, y_2, ..., y_n]$, where X is the collection of inputs and y are the class labels. x_i is a vector representing the sender, receiver and message attributes, while y_i is a binary value depicting if the receiver has responded to the message from the sender. The GPC model requires specifying a kernel that observes the inputs X and class labels y and defines the covariance function of the data. Inference is then made by computing the distribution of the latent variable corresponding to a test case, and subsequently using this distribution over the latent function f to produce a probabilistic prediction.

In GPC problems, the posterior presents to be analytically intractable and inference involves adopting approximation techniques. We adopt a GaussianProcessClassifier [11] where the integral of the function is easily approximated in a binary case such as ours. GPC models have been shown to perform well in probability estimations [5,9]. We adopt the Expected Calibration Error (ECE) as a way to measure our confidence in the model calibration. ECE is the difference in expectation between confidence and accuracy, estimated by simply taking the weighted average over the absolute accuracy/confidence difference [3,14]. Larger ECE values show larger difference between output confidence and actual model accuracy of the prediction - larger miscalibration, while smaller ECE values indicate less miscalibration.

4.2   Gradient Optimization

We are faced with a problem where the receiver looks to minimize their probability of reaction while the sender is looking to maximize this probability. In finding these best values, we explore the gradient method of optimization with the search directions defined by the gradient of the function at the current point, i.e., descent for receiver and ascent for the sender. Due to the intractability of the GPC model for numerical computation, we treat the function as a black-box oracle where at each iteration, we provide the data point and receive the output which is a partial derivative of the attribute vector derived by searching the attribute space to find the values that move us closer to the solution.

We adopt an iterative gradient optimization approach, where at each iteration, we perform a gradient ascent or descent. We set multiple learning rates of 0.01, 0.001, and 0.0001 but discovered that a rate of 0.001 performed best in the model. Since we are performing a greedy search, the first solution is always accepted as optimal. Mixed discrete-continuous variables pose a bound constraint to search in gradient-based optimization approach since discrete variables often derived from categorical (or binary) values have no ordering and offer no meaning to the learning model. This constraint makes computing gradient values for continuous and discrete variables simultaneously challenging as a 0.001 step to a binary value does not carry the same as it does for a continuous variable. To address this constraint, we adopt a search method that does not update both continuous and discrete variables simultaneously [6]. Instead, the search algorithm performs a search in the continuous space, and then searches in the space of discrete variables to find the optimal gradient. In our case, our discrete variables are binary, so a flip in the discrete sub-vector space was needed for computation.

5   Results

5.1   GPC Model and Calibration

To ensure the inferences drawn from the model align with what is expected, we perform a goodness-of-fit test using the ECE metric to confirm that the model fits the sets of observations as it should. The concern here is that the margin of error observed in the GPC model would impact the optimization task. As mentioned previously that a small ECE values indicate less miscalibration, leading to more confidence in the gradient search results. We use additional calibration techniques - Platt scaling and isotonic scaling [3,14], in hopes to find a better calibrated model. In Table 4, we report the ECE values observed from using the model as it is, as against when we use additional calibration techniques. We see that the model without additional calibration techniques performed best, and this is based on the fact that off-the-shelf GPC models already have some calibration implemented in them. We accept this as favorable and adopt the GPC model as-is to be the baseline for further experimentation.

Table 4. Expected Calibration Error (ECE) reported for GPC model using various calibration techniques.

Calibration technique	Expected Calibration Error (ECE)
Model as-is	**0.0161**
Platt scaling	0.0350
Isotonic scaling	0.0218

5.2   Gradient Optimization and Manipulation

By giving users the power to change certain attributes about their online personae, we cause a deviation to the accuracy of assumptions drawn about them. Accomplishing this means controlling the users' susceptibility to manipulation as messages targeted at them would not accurately model their interest. These changes can take the form of noise where the user introduce random behavior, or it could be in the form of making adjustments to their profile, for example creating posts on new topics. The reported ECE value is significant in telling us how good our GPC model is but since we do not have a confidence interval associated with the result, it is safe to assume that in the worst case scenario, the our reported probability of engagement values are ± 0.0161. In a neutral environment with no targeting, we see an average probability of reaction of 0.488 across our data (blue lines in graphs).

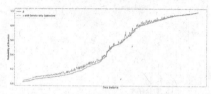
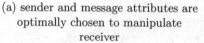

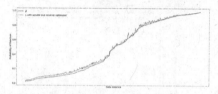

(a) sender and message attributes are optimally chosen to manipulate receiver

(b) receiver optimize their attributes while being targeted

Fig. 1. Probability of response post. (Color figure online)

In Fig. 1a, each point along the x-axis represents a data instance representing the attributes of user A, user B, and the message M_{AB}. The data instances are first ordered with increasing probability of reaction in a neutral environment and this order is maintained throughout experimentation. We see the effect of targeted manipulation on the receiver by showing how their probability of engaging with an optimized profile and targeted message differs from what their probability of reaction would be if no targeting is being done. Since the goal on the part of the sender (adversary) is to increase the probability of reaction, it is not surprising that the observed change in receivers with probability >0.8 is much lower than others. By optimizing their attributes and that of the message, the sender is able to increase the average probability of reaction to 0.510 and a variance of 0.117. With an average deviation of 0.022, see green lines in Fig. 2, the receiver is able to measure the possible changes to their engagement and opinion formation based on exposure to a single tweet. By changing certain attributes about themselves, the receiver optimizes their disclosed attributes to confuse manipulators and models that would have otherwise enhanced their profile or messages to target the user.

With the receiver defending themselves from possible attacks by optimizing their attributes, Fig. 1b, we observe a reduction in the probability of reaction with an average of 0.505 and a variance of 0.105, an average deviation of 0.017 (red lines in Fig. 2). Even though we still observe a deviation from the original probability values (blue line in Fig. 1b), we recall that the receiver gets to set a deviation value as their acceptable threshold. The difference in these probability values, as described in Eq. (7), can be seen in Fig. 2. From this result, the receiver is able to observe the effect of the privacy preservation mechanism and compare the manipulation gain value with their set threshold, ϵ. This observation can serve as a guide for the receiver when making decisions on how much protection they intend to put on their profiles. The average manipulation gain of 0.022 from a single tweet can be reduced to 0.016 when the receiver decided to protect themselves using the protection mechanism. Even though we do not see drastic changes in the probability values given that targeting is done over a single tweet, we are able to show that it is possible for manipulation to occur and to what extent. However, this results would become more interesting when there are several messages targeting the same user.

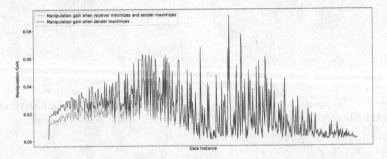

Fig. 2. Estimated manipulation gain when receiver does nothing but sender and message attributes are optimized compared with manipulation gain when receiver optimizes their attributes while being targeted. (Color figure online)

5.3   Attribute Disclosure

Once the gradient search converges and we find the optimal points, note that these optimal points are not necessarily global, the receiver can then view the needed changes to be made to their account. Being a greedy approach it is expected that these optimum values might just be local and having a meta-heuristic function like simulated annealing can aid in ensuring we always find the global optimum. A change in disclosed attribute will involve post creation or deletion for status count, following or unfollowing other accounts, etc.

Foremost, we looked to compare the differences between receivers with lower and higher manipulation gain. This difference is performed by comparing receivers of the same sender as this gives better understanding into their behavior. After looking at the different set of followers, one key finding is that receivers with higher ratio of retweets to original posts tend to experience higher manipulation gain. Additionally, from the social homogeneity score, we observed that users (both sender and receiver) sharing more common friends tend to give off more about their interests, thereby leading to higher risk of effective targeting and ultimately, manipulation gain. Finally, senders creating controversial and alarming posts will generate more interaction from their followers, and through this they can learn the stance or opinion of their followers.

Table 5. Observed changes in disclosed attributes post-optimization.

Features	Sender	Receiver
Number of followers	Increase 20%	Increase 50%
Number of friends	Increase 5%	Reduce 15%
Status count	Increase 5%	Reduce 5%
User favorites count	Increase 30%	Reduce 20%
Social homogeneity	Increase 8%	Reduce 8%

In Table 5, we report the observed change in the users' (both sender and receiver) disclosed attributes post-optimization. We see that the changes involved in the sender's attributes require more interaction with the network: increased posts, number of follower & friends, with an older account. This would be supported by the need to be more visible to generate trust and influence in the network for manipulation purposes. The receiver on the other hand needs to reduce their exposure through the friend's count but maintain an increase in their influence even though there needs to be a reduction in how often they engage with the network leading to a steady decrease in the volume of posts. In reality these changes are gradual as some of them are reliant on others. It is unsurprising that the sender needs to interact more while the receiver needs to reduce engagement but one thing to note here is that this optimization model gives insight into the degree of change that will make an impact. The receiver looking to make minimal changes for privacy preservation will need not cut off interaction completely, but rather make the needed changes based on their need and loss threshold.

6   Conclusion and Future Works

In this work, we set out to build a model that limits a user's susceptibility to targeted manipulation based on inferential attack. The model is designed such that it utilizes the user's probability of engaging with a post as a way to measure their sensitivity to manipulation and provides the user with the ability to make small changes to themselves in order to confuse a manipulator about who they are or what their interests are. The proposed model showed that even though there will be costs to participating in OSNs, as little bits about the user might still be exploited, these costs can be minimized depending on threshold set by the user as their maximum manipulation gain.

One aspect not considered in this work is how a user's influence in the network is affected by the restrictions and noise introduced to their attributes. It will be interesting to examine how this constraint will affect the user's susceptibility to manipulation when there is something to lose as a result of introducing such protection mechanisms.

Acknowledgements. This publication is based upon work supported by the National Science Foundation under Grants OIA-2148878 and CMMI-1952206, and by the NPRP grant #12C-33905-SP-165 from the Qatar National Research Fund (a member of Qatar Foundation). The findings achieved herein are solely the responsibility of the authors.

References

1. Castillo, C., Mendoza, M., Poblete, B.: Information credibility on twitter. In: Proceedings of the 20th International Conference on World Wide Web, pp. 675–684. ACM (2011)

2. Dey, R., Tang, C., Ross, K., Saxena, N.: Estimating age privacy leakage in online social networks. In: 2012 Proceedings IEEE INFOCOM, pp. 2836–2840 (2012). https://doi.org/10.1109/INFCOM.2012.6195711
3. Guo, C., Pleiss, G., Sun, Y., Weinberger, K.Q.: On calibration of modern neural networks. In: International Conference on Machine Learning, pp. 1321–1330. PMLR (2017)
4. He, J., Chu, W.W., Liu, Z.V.: Inferring privacy information from social networks. In: Mehrotra, S., Zeng, D.D., Chen, H., Thuraisingham, B., Wang, F.-Y. (eds.) ISI 2006. LNCS, vol. 3975, pp. 154–165. Springer, Heidelberg (2006). https://doi.org/10.1007/11760146_14
5. Hensman, J., Matthews, A., Ghahramani, Z.: Scalable variational Gaussian process classification. In: Lebanon, G., Vishwanathan, S.V.N. (eds.) Proceedings of the Eighteenth International Conference on Artificial Intelligence and Statistics. Proceedings of Machine Learning Research, San Diego, California, USA, 09–12 May 2015, vol. 38, pp. 351–360. PMLR (2015). https://proceedings.mlr.press/v38/hensman15.html
6. Lewis, R.M., Torczon, V.: Pattern search algorithms for bound constrained minimization. SIAM J. Optim. 9(4), 1082–1099 (1999)
7. Li, H., Zhu, H., Du, S., Liang, X., Shen, X.: Privacy leakage of location sharing in mobile social networks: attacks and defense. IEEE Trans. Dependable Secure Comput. 15(4), 646–660 (2018). https://doi.org/10.1109/TDSC.2016.2604383
8. Mislove, A., Marcon, M., Gummadi, K.P., Druschel, P., Bhattacharjee, B.: Measurement and analysis of online social networks. In: Proceedings of the 7th ACM SIGCOMM Conference on Internet Measurement, pp. 29–42 (2007)
9. Nickisch, H., Rasmussen, C.E.: Approximations for binary Gaussian process classification. J. Mach. Learn. Res. 9(Oct), 2035–2078 (2008)
10. Osho, A., Goodman, C., Amariucai, G.: MIDMod-OSN: A microscopic-level information diffusion model for online social networks. In: Computational Data and Social Networks. pp. 437–450. Springer International Publishing, Cham (2020)
11. Pedregosa, F., et al.: Scikit-learn: machine learning in Python. J. Mach. Learn. Res. 12, 2825–2830 (2011)
12. Rasmussen, C.E., Williams, C.K.I.: Gaussian Processes for Machine Learning. Adaptive Computation and Machine Learning, The MIT Press, Cambridge (2005)
13. Talukder, N., Ouzzani, M., Elmagarmid, A.K., Elmeleegy, H., Yakout, M.: Privometer: privacy protection in social networks. In: 2010 IEEE 26th International Conference on Data Engineering Workshops (ICDEW 2010), pp. 266–269 (2010). https://doi.org/10.1109/ICDEW.2010.5452715
14. Wenger, J., Kjellström, H., Triebel, R.: Non-parametric calibration for classification. In: International Conference on Artificial Intelligence and Statistics, pp. 178–190. PMLR (2020)
15. Yang, F., Liu, Y., Yu, X., Yang, M.: Automatic detection of rumor on sina weibo. In: Proceedings of the ACM SIGKDD Workshop on Mining Data Semantics. p. 13. ACM (2012)
16. Zheleva, E., Getoor, L.: To join or not to join: the illusion of privacy in social networks with mixed public and private user profiles. In: Proceedings of the 18th International Conference on World Wide Web, pp. 531–540 (2009)
17. Zhou, B., Pei, J.: The k-anonymity and l-diversity approaches for privacy preservation in social networks against neighborhood attacks. Knowl. Inf. Syst. 28(1), 47–77 (2011)

Identifying Targeted and Generalized Offensive Speech from Anti-asian Social Media Conversations

Payal Shah and Swapna S. Gokhale[✉]

University of Connecticut, Storrs, CT 06269, USA
`swapna.gokhale@uconn.edu`

Abstract. During the Covid-19 pandemic Asian-Americans have been targets of prejudice and negative stereotyping. There has also been volumes of counter speech condemning this jaundiced attitude. Ironically, however, the dialogue on both sides is filled with offensive and abusive language. While abusive language directed at Asians encourages violence and hate crimes against this ethnic group, the use of derogatory language to insult alternative points of view showcases utter lack of respect and exploits people's fears to stir up social tensions. It is thus important to identify and demote both types of offensive content from anti-Asian social media conversations. The goal of this paper is to present a machine learning framework that can achieve the dual objective of detecting targeted anti-Asian bigotry as well as generalized offensive content. Tweets were collected using the hashtag #*chinavirus*. Each tweet was annotated in two ways; either it condemned or condoned anti-Asian bias, and whether it was offensive or non-offensive. A rich set of features both from the text and accompanying numerical data were extracted. These features were used to train conventional machine learning and deep learning models. Our results show that the Random Forest classifier can detect both generalized and targeted offensive content with around 0.88 accuracy and F1-score. Our results are promising from two perspectives. First, our approach outperforms contemporary efforts on detecting online abuse against Asian-Americans. Second, our unified approach detects both offensive speech targeted specifically at Asian-Americans and also identifies its generalized form which has the potential to mobilize a large number of people in socially challenging situations.

Keywords: Targeted hate · Generalized hate · Asian-Americans

1 Introduction and Motivation

The use of hateful and offensive speech on social media platforms has been on the rise in recent years. Broadly speaking, offensive speech can either be targeted or general. Targeted offensive speech is intended to be derogatory, humiliating, or insulting to the members of a specific group on the basis of attributes such as race, religion, ethnic origin, sexual orientation, disability or gender [18]. Generalized offensive speech may be defined as any strong, impolite, rude or hurtful

© The Author(s), under exclusive license to Springer Nature Switzerland AG 2023
T. N. Dinh and M. Li (Eds.): CSoNet 2022, LNCS 13831, pp. 179–187, 2023.
https://doi.org/10.1007/978-3-031-26303-3_16

language using profanity, that can show debasement of something or someone or show intense emotion, possibly because of an opinion that the person may carry [18].

The trend in using offensive speech was exacerbated during the Covid-19 pandemic [2]. People sought solace in finding someone to blame for the extreme disruption and fear, and became extremely intolerant of others who did not share their opinions and attitudes about everything related to the pandemic. Offensive and hateful rhetoric was also directed towards Asian-Americans, because the origin of the coronavirus was identified to be in Wuhan, China. It did not help that the then President Trump often used the term "China Virus" which encouraged his supporters to echo and act his derogatory choice of words.

When the social and political circumstances are chaotic and turbulent, both generalized and targeted forms of offensive speech online encourages violent actions and behaviors in the offline, physical spaces. Generalized offensive speech can lead to loss of trust in the government, spread of extreme and radical views, and eventually plots to overthrow elected governments [10]. On the other hand, targeted offensive speech can encourage hate crimes and violent acts against the specific ethic group; in fact hate crimes against Asian-Americans increased to unprecedented levels during the pandemic [28]. Moreover, such speech can have a serious impact on the mental health and self-esteem of those who are targeted. Social media companies are not doing nearly enough to identify both targeted and generalized offensive speech that is disseminated via their platforms [26]. It then becomes necessary to automatically identify both types of offensive speech from volumes of social media content, so that it can be subject to appropriate misinformation policies (including demotion, turning off likes and retweets) [6] to prevent the damage it can cause.

The present paper proposes a unified approach that can detect both generalized and targeted offensive speech from social media dialogue tagged using #chinavirus. Tweets were labeled according to whether they condemned or condoned offensive speech against Asians (targeted), and whether they were generally offensive or non offensive (generalized). A suite of textual and non textual features were extracted from the tweet data, and were used to train conventional and deep learning models. These models could identify both targeted and generalized offensive speech with accuracy and F1-score of over 0.88, outperforming most of the contemporary approaches to detect offensive speech. The detection of generalized speech was found to be slightly more challenging than targeted offensive speech. Carefully optimized conventional machine learning models outperformed deep learning and pre-trained transformer models in separating both types of offensive speech. The paper thus opens opportunities to build systems that can detect offensive speech, regardless of its specific type, from social media.

The paper is organized as follows: Sect. 2 summarizes the collection and annotation of tweets. Section 3 describes the unified framework. Section 4 discusses the results. Section 5 compares and contrasts related research. Section 6 concludes and offers directions for future research.

2   Tweet Collection and Annotation

President Trump's endorsement of the term "China Virus" became visible on the national stage during his press secretary's conference in August 2020, a day before the Republican National Convention [5]. Shortly after, there was a noticeable increase in hate crimes and assaults against Asian-Americans in cities such as New York and San Francisco, with a large concentration of this demographic group. Given the tight integration of social media platforms into our society, it was not surprising that following this high-profile event, social media were rife with conversations that supported the term "China Virus" yet also tried to mitigate this hateful dialogue through a significant volume of counter speech [16]. Tweets were collected using #chinavirus a few days after this incident using the rtweet API [17]. The entire sample consisted of about 18,000 tweets. Of these, about 3000 tweets were selected for annotation.

After consulting the news and other sources, we developed a comprehensive coding guide to annotate the tweets along two orthogonal dimensions. Along the first dimension, each tweet was labeled as either condoning ('R') or condemning ('N') the use of "China Virus". Since the tweets that condone the term target Asian-Americans, we refer to this annotation as identifying targeted hate. The coding guide for targeted hate outlined the narratives promoted from each side to justify their philosophy and example tweets:

- **Place of origin:** These individuals believed that because the virus originated in China it was fair game to refer to it as such. They defended their choice that they were following the naming conventions of the previous pandemics, namely, the Spanish flu and the Zika virus. An alternate term "Wuhan virus" was proposed as a more specific nomenclature. For example: *It CAME from CHINA, so it is the CHINA VIRUS. Plain and fucking simple, you fake news media outlet.*
- **Role of Chinese government:** These individuals claimed that the term exposes China's role in the pandemic and it could possibly be xenophobic yet not racist. Similar viewpoints explain that "Chinese virus" rather than "China virus" may be considered racist. Additionally, they hold the Chinese government (CCP) responsible for lying and attempting to cover up the outbreak of the virus and that it did not extend the accusation to ordinary citizens of the country. For example: *I've never called it "China Virus" myself, "CCP Virus" yes. However, replying to the fundamental truth of the tweet I replied to. I agree with OP. China's lies led to death. Condemning China's Government does NO*
- **Conspiracy theories:** These individuals float conspiracy theories supporting the term; popular ones include that the virus was a bioweapon that targeted the western hemisphere, that it was a scam by American democrats to destroy the economy and ruin President Trump's chances of re-election, that Covid-19 is no worse than the flu, and many alternative remedies. For example: *False- best job record in History. It's down right now due to the China virus (idiot). It will and is going back up as we speak.*

– *Intertwining with politics:* These tweets praise President Trump for blocking travel from China, and criticize democrats including Joe Biden, Kamala Harris, Gov. Andrew Cuomo for their conduct during the pandemic. For example: *It was the China virus! Trump shut down flight from China in January when drunk Nancy and dementia Joe were saying he was overreacting. If they were in charge there would be alot more death. Let's not forget*

Tweets condemning the use of "China Virus" argued:

– **Deflection from pandemic response:** These individuals claimed that President Trump was attempting to shift attention away from his administration's poor response to the virus to reduce public outrage. An alternate term "Trump Virus" to counter "China virus" and hold President Trump responsible for the rising toll of the pandemic was popularized. For example: *TRUMP VIRUS! China has it under control!*
– **Promoting racism & xenophobia:** These individuals called out the racist undertones behind the term, and claimed that it was an attempt to incite Americans against a specific ethnic group. President Trump was accused of promoting xenophobia and his press secretary came under fire for using the term in an official White House briefing. For example: *If you call it the China virus, you're racist China is a country bro*
– **Encouraging discrimination & violence:** Those who condemned the use of the term agreed that the phrase only encouraged violence and discrimination towards Asian-Americans and also led to extreme incidents of hate that ended in injuries and fatalities. For example: *she is garbage. Hate crimes against Asian-Americans being on the rise is not preventing her from calling it "the china virus".*
– **Intertwining with politics:** Most of these blame the press secretary and other officials in the Trump administration for emulating their boss and propagating the use of the term, rather than standing up against and questioning its use. For example: *@PressSec @realDonaldTrump I hate that you call it the China virus. What has happened to you in this job? So sad.*

We found that the dialogue from both sides relied on unsavory and unpleasant words for emphasis, insults and attack. This hate was not directed at any protected group, but rather at those with an opposing or an alternative point of view. Therefore, we refer to this as "generalized hate" to distinguish it from "targeted hate" towards Asian-Americans. Relying upon our earlier work [10], we broadly label a tweet as offensive if it insults or bullies someone because of their mental capabilities, political leanings and physical appearances. For example, *Democrats patting the china virus back. Lmfao losers* is a condoning offensive tweet. On the *@PressSec @realDonaldTrump Everything about this shows such a low level of evolution. China Virus Be a grown up. Contribute to truth and respect. Embody what good Americans believe and aspire to. Don't be so base.* is an example of a condemning offensive tweet. Along this dimension, each tweet was labeled as either offensive ('O') or non offensive ('NO').

We note that the annotation into offensive and non offensive classes had to be undertaken with great care as the terms racist, racism and xenophobic were sprinkled all through the tweets. Mere presence of these terms does not deem whether a tweet is offensive or non offensive. Rather these terms had to be considered within context to make the ultimate determination. For example, there is a distinction between using the word "racist" to call an individual racist in an offensive manner, and using the word "racist" to call out racism propagated by the use of the term China virus.

Two annotators labeled each tweet as condemn vs. condone and offensive vs. non offensive based on the coding guides. Only tweets where both annotators matched for both labels were included in the final corpus. The entire corpus consisted of 2019 tweets, split into 974 (48.24%) condone, and 1045 (51.76%) condemn tweets. This split suggests that nearly half of the users are comfortable with the anti-Asian prejudice. The corpus was split into 616 (30.59%) offensive and 1493 (69.49%) non offensive tweets. Thus, about one-third of the users even use foul and offensive language to emphasize their point. If the use of derogatory and insulting language can be viewed as an indicator of how passionately one feels about a particular issue, then it might be concerning that about one-third of the users are sufficiently passionate that they'd even consider disrespecting and mocking the other point of view.

Also of interest was the split of condemning and condoning tweets between offensive and non offensive categories. We found that of the tweets that condone, 19.20% (80.8%) are offensive (non offensive). On the other hand, of the tweets that condemn 41.05% (58.95) are non offensive. Thus, those who condemn Asian hate may be more passionate perhaps out of concern that America should be viewed as a beacon of hope and inclusion, rather than as a hotbed of xenophobia, racism and hate.

3   Identification of Offensive Speech

In this section, we describe aspects of our classification framework to identify both targeted and generalized offensive speech.

- **Textual Features:** We mapped the tokens (words) remaining in the text of the tweets after pre-processing to statistical and semantic features. Statistical features are given by Term Frequency-Inverse Document Frequency, computed as a product of the number of times a word appears in a tweet and the inverse document frequency of the word throughout the set of tweets. Semantic features were computed using Word2Vec embeddings which maps closely related words to multi-dimensional vectors [29].
- **Meta-data features:** Tweet-level features include the number of likes and retweets received by a tweet. User-level features include the numbers of friends, followers, and status updates, whether the account is verified, and how many other users have listed this user. Continuous features were transformed into quartile sets, and binary features using one hot encoding.

- **Training and Testing:** We split the corpus 70%-30% into training and test sets for both problems. Stratified sampling is used to preserve the ratio of the majority and minority classes in both the training and test partitions.
- **Data Balancing:** The corpus is adequately balanced with respect to condone vs. condemn tweets (974 vs. 1045), but imbalanced with respect to offensive vs. non offensive classes (616 vs. 1403). We used a combination of oversampling of the minority class (offensive tweets), and undersampling of the majority class (non offensive tweets) to handle class imbalance. The testing set was left unbalanced.
- **ML Models:** Conventional models include Naive Bayes, Logistic Regression, Support Vector Machines and Random Forest. We also implemented two deep learning models – Long Short-term Memory Recurrent Neural Network (LSTM-RNN) and BERT, a pre-trained transformer-based model trained on Wikipedia and wordsBooksCorpus [9].
- **Performance Metrics:** We designated the condone and offensive classes as positive, and condone and non offensive classes as negative. Based on this designation, we define accuracy, precision, recall and F1-score [29].

4   Results and Discussion

Table 1 lists the performance of the classifiers. Random Forest offers the best accuracy and F1-score for both problems, outperforming the deep learning classifiers. Moreover, across all the classifiers, the following trends can be observed. Precision and recall are nearly similar and balanced, with recall being about a percentage lower than precision. The performance of all the models is lower for generalized compared to targeted offensive speech detection. This disparity is the highest for the Naive Bayes classifier which is also the model that performs the worse. Our results also show that an ensemble classifier with additional features depicting the engagement and activity of authors outperforms the pre-trained transformer (BERT) and LSTM-RNN models. Finally, the Random Forest also outperforms the Logistic Regression classifier which has been shown to have competitive performance in the detection of offensive speech [10].

Table 1. Performance metrics

Targeted speech				Generalized speech			
Model	Accuracy	Precision	Recall	F1-Score	Accuracy	Precision	Recall	F1-Score
LR	0.86	0.86	0.86	0.87	0.83	0.83	0.83	0.83
NB	0.77	0.78	0.77	0.76	0.63	0.70	0.64	0.65
SVM	0.89	0.89	0.88	0.89	0.87	0.85	0.87	0.88
RF	0.91	0.91	0.90	0.90	0.89	0.88	0.89	0.88
LSTM-RNN	0.85	0.79	0.78	0.84	0.75	0.77	0.78	0.77
BERT	0.88	0.89	0.86	0.87	0.84	0.85	0.84	0.83

5   Related Research

There has been a consistent rise in the volume of hateful and offensive speech over social media platforms. Alongside, many research efforts have been undertaken to identify such abusive content from these online conversations.

With respect to detecting generalized offensive speech, Crowdflower and Davidson data sets classify the tweets into three groups, neutral, offensive and hateful [8,12,14,25]. In another data set, tweets are labeled into racism, sexism, and neutral [24]. Data from other social media platforms such as white supremacy fora, wikipedia comments, Facebook and Fox news comments [1,13,20,21,27] have also been analyzed. Many of these research works report accuracy scores in the range of 65% to 80% [19,25].

Detection of hateful and offensive speech targeted towards Asian-Americans during the Covid-19 pandemic from online conversations has gained steam. Cotik *et al.* [7] use a BERT model for Spanish language and report a F1-score of 0.75. Fan *et al.* [11] build a decision tree classifier to classify hate speech annotated by the lexicon-based approach based on emotions. Alshalan *et al.* [3] use a CNN model to detect hate speech targeted against Asians in Arab tweets. Vidgen *et. al.* [23] classify discussions of East Asian prejudice on Twitter into four classes, and achieve a F1-score of 0.83. He *et. al.* [16] classify tweets based on their hatefulness towards Asians into hate, counterspeech or neutral and achieve a F1-score of 0.832, this was further improved to 0.85 by Toliyat *et. al.* [22]. An *et. al.* [4] predict anti-Asian hateful users with a F1-score of around 0.67.

The above efforts either focus on detecting generalized or targeted offensive speech, but not both. In contrast, our work offers a unified approach to identifying both types so that their negative consequences can be mitigated simultaneously. Moreover, we outperform the conventional approaches on both problems.

6   Conclusions and Future Research

This paper presents a unified framework to identify targeted and generalized offensive speech from social media conversations on anti-Asian prejudice. Targeted speech was aimed at blaming Asian-Americans for the Covid-19 pandemic, while generalized offensive speech humiliated and insulted the alternate point of view. Both forms are distinct and have the potential to stir up violence and civil unrest during chaotic and turbulent situations such as those brought about by the Covid-19 pandemic. Our approach can detect both these types, with better accuracy and F1-score compared to the contemporary approaches. Our future research comprises of identifying harassment and bullying from the masking dialogue.

References

1. Abderroauf, C., Oussalah, M.: On online hate speech detection: effects of negated data construction. In Proceedings of the International Conference on Big Data, pp. 5595–5602 (2019)

2. Ahmed, A.: A tsunami of hate: the Covid-19 hate speech pandemic. https://www. humanrightspulse.com/mastercontentblog/a-tsunami-of-the-covid-19-hate-speech-pandemic (2020)
3. Alshalan, R., Al-Khalifa, H., Alsaeed, D., Al-Baity, H., Alshalan, S.: Detection of hate speech in COVID-19 related tweets in the Arab region: deep learning and topic modeling approach. J. Med. Internet Res. **22**(12), e22609 (2020)
4. An, J., Kwak, H., Lee, C.S., Jun, B., Ahn, Y.: Predicting anti-Asian hateful users on Twitter during COVID-19. In: EMNLP, Findings of the Association for Computational Linguistics (2021)
5. Brady, J.S.: Remarks by President Trump in press briefing **19** (2020)
6. Twitter Help Center. Covid-19 misleading information policy. https://help.twitter. com/en/rules-and-policies/medical-misinformation-policy (2021). Accessed 31 Jan 2022
7. Cotik, V.: A study of hate speech in social media during the COVID-19 outbreak. In: ACL 2020 Workshop NLP-COVID (2020)
8. Davidson, T., Warmsley, D., Macy, M., Weber, I.: Automated hate speech detection and the problem of offensive language. In: Proceedings of the International AAAI conference on Web and Social Media (2017)
9. Devlin, J., Chang, M.W., Lee, K., Toutanova, K.: BERT: pre-training of deep bidirectional transformers for language understanding. arXiv preprint. arXiv:1810.04805 (2018)
10. Fahim, M., Gokhale, S.: Detecting offensive content on Twitter during proud boys protests. In: International Conference on Machine Learning and Applications, pp. 1582–1587 (2021)
11. Fan, L., Yu, H., Yin, Z.: Stigmatization in social media: documenting and analyzing hate speech for COVID-19 on Twitter. In: Proceedings of the Association for Information Science and Technology (2020)
12. Data for Everyone Figure Eight. Classification of political social media. https:// www.figure-eight.com/data-for-everyone/ (2015). Accessed 21 Jan 2020
13. Gao, L., Huang, R.: Detecting online hate speech using context aware models. In: Proceedings of the International Conference on Recent Advances in Natural Language Processing, Varna, Bulgaria, pp. 260–266 (2017)
14. Gaydhani, A., Doma, V., Kendre, S., Bhagwat, L.: Detecting hate speech and offensive language on Twitter using machine learning: an N-gram and TF-IDF approach. In: Proceedings of the IEEE International Advanced Computing Conference (2018)
15. Hajibagheri, A., Sukhthankar, G.: Political polarization over global warming: analyzing Twitter data on climate change (poster). In: ASE International Conference on Social Computing, Palo Alto, CA (2014)
16. He, B., Ziems, C., Soni, S., Ramakrishnan, N., Yang, D., Kumar, S.: Racism is a virus: Anti-Asian hate and counterspeech in social media during the COVID-19 crisis. In: Proceedings of the 2021 IEEE/ACM International Conference on Advances in Social Network Analysis and Mining, pp. 90–94 (2021)
17. Kearney, M.W.: Collecting and analyzing Twitter data. https://cran.r-project.org/ web/packages/rtweet/rtweet.pdf (2020)
18. Nghiem, H., Morstatter, F.: Stop Asian Hate!: refining detection of Anti-Asian hate speech during the COVID-19 pandemic. arXiv (2021)
19. Nugroho, K., Noersasongko, E., Fanani, A.Z., Basuki, R.S.: Improving Random Forest method to detect hatespeech and offensive word. In: Proceedings of International Conference on Information and Communications Technology, pp. 514–518 (2018)

20. Salminen, J., Hopf, M., Chowdhury, S.A., Jung, S., Almerekhi, H., Jansen, B.J.: Developing an online hate classifier for multiple social media platforms. Human-centric Comput. Inf. Sci. **10**(1), 1–34 (2020). https://doi.org/10.1186/s13673-019-0205-6

21. Senarath, Y., Purohit, H.: Evaluating semantic feature representations to efficiently detect hate intent on social media. In: Proceedings of the 14th International Conference on Semantic Computing, pp. 199–202 (2020)

22. Toliyat, A., Levitan, S.I., Peng, Z., Etemadpour, R.: Asian hate speech detection on Twitter during COVID-19. Front. Artif. Intell. (2022)

23. Vidgen, B., Margetts, H., Broniatowski, D.A., Hale, S.A.: Detecting East Asian prejudice on social media

24. Waseem, Z., Hovy, D.: Hateful symbols or hateful people? Predictive features for hate speech detection on Twitter. In: Proceedings of the NAACL-HLT, pp. 88–93 (2016)

25. Watanabe, H., Bouazizi, M., Ohtsuki, T.: Hate speech on Twitter: a pragmatic approach to collect hateful and offensive expressions and perform hate speech detection. IEEE Access **6**, 13825–13835 (2018)

26. Wong, Q.: Twitter, Facebook and others are failing to stop anti-Asian hate. https://www.cnet.com/news/politics/twitter-facebook-and-others-are-failing-to-stop-anti-asian-hate/ (2021)

27. Wulczyn, E., Thain, N., Dixon, L.: Ex Machina: personal attacks seen at scale. In: Proceedings of the International World Wide Web Conference Committee (2017)

28. Yam, K.: Anti-Asian hate crimes increased 339 percent nationwide last year, report says. https://www.nbcnews.com/news/asian-america/anti-asian-hate-crimes-increased-339-percent-nationwide-last-year-repo-rcna14282 (2022)

29. Zafarani, R., Abbasi, M.A., Liu, H.: Social Media Mining: An Introduction. Cambridge University Press, New York (2014)

US News and Social Media Framing Around Vaping

Keyu Chen[1], Marzieh Babaeianjelodar[1], Yiwen Shi[1], Rohan Aanegola[1],
Lam Yin Cheung[2], Preslav Ivanov Nakov[3], Shweta Yadav[4], Angus Bancroft[5],
Ashiqur R. KhudaBukhsh[6], Munmun De Choudhury[7], Frederick L. Altice[1],
and Navin Kumar[1(✉)]

[1] Yale University, New Haven, USA
navin183@gmail.com
[2] FTI Consulting, Inc, Washington, USA
[3] Qatar Computing Research Institute, Ar Rayyan, Qatar
[4] University of Illinois Chicago, Chicago, USA
[5] The University of Edinburgh, Edinburgh, UK
[6] Rochester Institute of Technology, Rochester, USA
[7] Georgia Institute of Technology, Atlanta, USA

Abstract. In this paper, we investigate how vaping is framed differently
(2008–2021) between US news and social media. We analyze 15,711 news
articles and 1,231,379 Facebook posts about vaping to study the differences in framing between media varieties. We use word embeddings to
provide two-dimensional visualizations of the semantic changes around
vaping for news and for social media. We detail that news media framing
of vaping shifted over time in line with emergent regulatory trends, such
as; flavored vaping bans, with little discussion around vaping as a smoking cessation tool. We found that social media discussions were far more
varied, with transitions toward vaping both as a public health harm and
as a smoking cessation tool. Our cloze test, dynamic topic models, and
question answering showed similar patterns, where social media, but not
news media, characterizes vaping as combustible cigarette substitute. We
use n-grams and LDA topic models to detail that social media data first
centered on vaping as a smoking cessation tool, and in 2019 moved toward
narratives around vaping regulation, similar to news media frames. Overall, social media tracks the evolution of vaping as a social practice, while
news media reflects more risk based concerns. A strength of our work is
how the different techniques we have applied validate each other.

Keywords: Vaping · Social media · Harm reduction

1 Introduction

The recent introduction of alternative forms of nicotine products into the
marketplace (e.g., e-cigarettes/vapes, heated tobacco products, and smokeless
tobacco) has led to a more complex informational environment [12]. The scientific

T. N. Dinh and M. Li (Eds.): CSoNet 2022, LNCS 13831, pp. 188–199, 2023.
https://doi.org/10.1007/978-3-031-26303-3_17

consensus is that vape aerosol contains fewer numbers and lower levels of toxicants than smoke from combustible tobacco cigarettes [16]. Among youth in the USA, the adolescent nicotine vaping use increased from 2017 to 2019 but then started declining in 2020, which includes a decline in daily vaping as well [14]. Among adults, a Cochrane review found that nicotine vapes probably do help people to stop smoking for at least six months, and working better than nicotine replacement therapy and nicotine-free e-cigarettes [7]. Given that vaping is represented in the public health environment both as a smoking cessation tool and as a harm to youth health, it is highly controversial and polarizing. Early divisions among public health experts led to polarized coverage in the media, confused messages to the public, and inconsistent policymaking between jurisdictions [18]. For many authorities in the United States, the potential health harms of vaping and youth-vaping concerns are overriding considerations of vaping as a smoking cessation tool [20]. The coverage of controversial topics, such as vaping, is often different in social media versus in news media. For example, social media can be more hyperbolic compared to news media [9], or can be used primarily to share information rather than to report news events. News media can shape public perceptions and policy about vaping [19], particularly in the context of major vaping events, such as the August 2019 outbreak of vaping product use-associated lung injury (EVALI). In late 2019, the Centers for Disease Control and Prevention (CDC) began to investigate a steep rise in hospitalizations linked to vaping product use. The condition came to be called EVALI. Researchers identified vitamin E acetate, a chemical added to some THC-containing vaping products, as the main cause of the illness. However, news reports did not always differentiate between THC devices and standard nicotine-based vapes [9], perhaps disproportionately characterizing vaping harms. Such vaping-related news may have triggered national and state-level policy responses, and may have influenced public perceptions (including misperceptions) regarding the harms of vaping [10]. Social media such as Facebook, Twitter, and YouTube have recently become important platforms for health surveillance and social intelligence, providing new insights on vaping to help inform future research, regulations, surveillance, and enforcement efforts. For example, vape tricks, e.g., blowing large vapor clouds or shapes like rings, are popular content on vaping-related social media [11], which can provide insight into risky tobacco use practices. Comparisons between how social media and news media frame vaping are key to managing vaping-related health outcomes. By identifying the differences in such media framing, we can design targeted policy tools specific to each type of medium. For example, messages around curbing youth vaping may be effective on social media, given the proliferation of youth-related vaping content. However, ads on news media designed to mitigate youth vaping may be ineffective as these are not platforms frequented by youth and are perhaps more suited for promoting vaping as a cessation tool for adults who already smoke and wish to quit smoking. Thus, capturing the differences in how vaping is framed on social media vs news media are key to designing targeted policy tools to simultaneously reduce youth vaping and to improve smoking cessation rates, thereby minimizing vape harms and maximizing benefits.

Most research around news media and vaping centered on responses to the 2020 outbreak of vaping-related lung injury (EVALI) [8]. Research regarding vaping on social media generally falls under two domains: machine learning to assist smoking cessation, and content analysis of vaping on social media. While past work provides detail around how vaping is framed on news and social media, there is limited analysis pre-EVALI, and minimal comparisons between news and social media. Thus, we propose a study comparing US social media to US news media framing of vaping. We use Media Cloud to obtain news media articles, and CrowdTangle for Facebook posts, cognizant that these do not provide a representative view of public discourse. Our main research question (RQ) is thus as follows: What are the broad differences between news media (Media Cloud data) and Facebook regarding vaping? Findings suggest that social media tracks the evolution of vaping as a social practice, while news media reflects more risk based concerns.

2   Methods

Data. To capture news media around vaping, we used the open-source media analysis platform Media Cloud (mediacloud.org) to analyze 15,711 media articles between 2008 and 2021 from 271 US media sources. For locating articles related to vaping, we used queries based on a related systematic review: electronic-cigarette, electronic cigarette, electronic cig, e-cig, ecig, e cig, e-cigarette, ecigarette, e cigarette, e cigar, e-juice, ejuice, ejuice, e-liquid, eliquid, e liquid, e-smoke, esmoke, e smoke, vape, vaper, vaping, vape-juice, vape-liquid, vapor, vaporizer, boxmod, cloud chaser, cloudchaser, smoke assist, ehookah, e-hookah, e hookah, smoke & pod, e-tank, electronic nicotine delivery system. We analyzed the articles from 24811 URLs (9100 URLs were broken links) from Media cloud. For social media, we used the CrowdTangle service (crowdtangle.com) to obtain 1,231,379 Facebook posts from 2009 to 2021, with the same set of keywords. The CrowdTangle database includes: 7M+ Facebook pages, groups, and verified profiles. This includes all public Facebook pages with more than 50 K likes, all public Facebook groups with 95 k+ members, all US-based public groups with 2 k+ members, and all verified profiles. The Facebook data we have provided is not solely US-based and contains posts worldwide. Two reviewers independently examined 100 articles to confirm the salience with our research question. The reviewers then discussed their findings and highlighted items deemed relevant across both lists, determining that 95% were relevant. As Facebook posts regarding vaping have a large proportion of advertisements by vaping companies, we removed ads from the CrowdTangle dataset to better identify how the public frames vaping, instead of viewpoints promoted by ads. A total of 91,3726 (74.2%) of the posts were advertisements, resulting in 31,7653 posts that were not ads. More information on how we detected ads is provided below.

Advertisement Classifier. Given the prominence of vaping ads on Facebook, we built a vaping advertisement classifier for our CrowdTangle data. We use a BART large model for NLI-based Zero Shot Text Classification [13] trained on the MultiNLI (MNLI) dataset. BART is a denoising autoencoder for pretraining sequence-to-sequence models. BART is trained by corrupting text with an arbitrary noising function, and learning a model to reconstruct the original text. The classifier was used to classify individual Facebook posts in the CrowdTangle data (0 = not vaping ad, 1 = vaping ad). We extracted a random sample of 2000 posts from the classified data. We then selected three content experts who had published at least ten peer-reviewed articles in the last three years around vaping. Two of these content experts manually classified these 2000 posts (0 = not a vaping ad, 1 = vaping ad). Within these posts, the distribution was as follows, 0:563, 1:1437, yielding >90% agreement with the machine-assigned labels. Examples of ads: *The Smok Alien Kit is now available in blue! Vaping on Blue Raspberry by Emoji Liquids, Get 20% Off 3 Packs of Mistic's Flavored E Cig Cartridges.*

Topic Modelling. To quantitatively analyze the shift in vaping-related topics across media platforms over time, we explore topic distributions with latent Dirichlet allocation (LDA) [2]. LDA is a topic modeling algorithm, which assumes a document is generated from a mixture of topics. Topics represented by a distribution of words and the topic distributions of the documents are learned after LDA application. We generated topic models for each year of the data, separately for CrowdTangle and for Media Cloud. We treated 2009–2012 as a single year due to limited data. To determine the optimal number of topics for each year, we picked the candidate topic numbers with the highest coherence value. The coherence value is based on the normalized point-wise mutual information (PMI) score which measures the statistical independence of two words observed in a close proximity. To further analyze the evolution of vaping-related topics over time and platforms, we used dynamic topic modeling (DTM). While topic modeling analyzes documents to learn meaningful patterns of words, for documents collected in sequence, DTMs capture how these patterns vary over time [1]. DTMs use state space models on the natural parameters of the multinomial distributions that represent the topics [1]. We use BERTopic [5] to visualize the DTMs over time and across platforms.

Semantic Change. Understanding how words central to our research question, such as *vaping*, change their meanings over time is key to comprehending the evolving framing of vaping. Previous work [6] demonstrated how the word *gay* shifted in meaning from *cheerful* to referring to homosexuality from the 1900s-1990s. In a similar way, in this paper we use word embeddings as a diachronic (historical) tool to quantify the semantic change around the word *vaping* over time, thus providing insight into how framing around vaping shifted over time during the period (2009–2021) as well as between media sources. We use an updated version of Gensim word2vec models[1] based on HistWords[2] to provide insight on how *vaping* is framed from 2009–2021, for both CrowdTangle and Media Cloud data. For brevity, we removed words in the fringes of the final output, irrelevant to our RQ, such as *can*, *check*, *need*, and *let*.

---

[1] gist.github.com/zhicongchen.
[2] github.com/williamleif/histwords.

Question Answering. Question answering can help us to understand how news and social media *answer* the same questions about vaping, perhaps revealing differences in vaping frames. For example, news media may be more likely to present vaping harms compared to social media. We used BERT [4] fine-tuned on the SQuAD v1.1 dataset for answer extraction. The model was applied separately on news and social media. Questions were developed based on input from content experts, indicated above. Each content expert first developed a list of ten questions separately. The three experts then discussed their lists to result in a final list of four questions that were broadly similar across all three original lists, and final questions are as follows: Can vaping help you quit smoking regular cigarettes? What are the health consequences of vaping? Why are teens vaping? What is the biggest concern with vaping? We highlighted one question at a time and fed it to the model. While we would have preferred to use more than four questions for our question answering analysis, only four questions were agreed upon by the content experts. This is largely due to disagreement among content experts as to what questions should be included, largely resulting from the controversial nature of vaping, and that academics are in disagreement about the harms and merits of vaping. The model extracts answers for the question leveraging on context information in each article or post. To stay within the admitted input size of the model, we clipped the length of the text (title + body text) to 512 tokens. Each question provided one answer per article or post. We randomly sampled 500, 1000, 1500, and 2000 answers per question. We found that a random sample of 1000 answers provided the greatest range and quality of answers, assessed by two reviewers (80% agreement). We thus randomly sampled 1000 answers per question and content experts then selected the top 5, 10 and 20 most representative answers per question, for both news and social media. We found that selecting the top 5 most representative answers provided the least repetition.

Cloze Tests. Following recent work [15], we used BERT [4] and cloze tests to further understand the differences between Media Cloud and CrowdTangle data. Cloze tests represent a fill-in-the-blank task given a sentence with a missing word [17]. For example, *winter* is a likely completion for the missing word in the following cloze task: In the [MASK], it snows a lot. We developed several cloze tests with input from content experts as described earlier. Each content expert first developed a list of ten cloze tests separately. Then the three experts discussed their lists to come up with a final list of four cloze tests: i) The main issue with vaping is [MASK]; ii) The worst thing about vaping is [MASK]; iii) Teens like vaping because it's very [MASK]; iv) Vaping is [MASK] for smoking. We applied BERT on our Media Cloud and CrowdTangle data to identify the differences in the top five results for each cloze test across the media platforms.

3   Results

Topic Model. Our LDA results for CrowdTangle data indicate that in 2009–2016, some of the most frequent words in the most prominent topics are *happy*

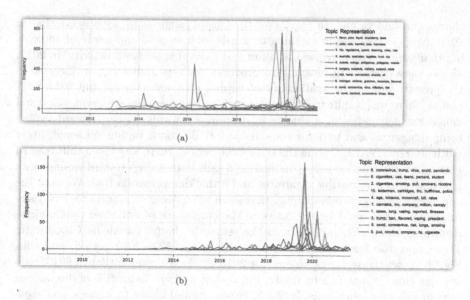

Fig. 1. Dynamic topic model results for CrowdTangle 1b and Media Cloud 1a data from 2009 to 2021

and *wonderful*. From 2017–2021, the most common words in the most prominent topics were *ban*, *lung*, and *teen*. These words are likely in reference to the proposed flavored vaping ban, EVALI, and the youth vaping use. We note the transition within CrowdTangle data from vaping as a fun activity, to a health harm that affects multiple communities. For the Media Cloud data, we note that *ban*, *stop*, and *health* are common across the most prominent topics for topic models over all years. While the CrowdTangle data appears to frame vaping in a range of ways, news media appears persistent at characterizing vaping as a public health issue. Figure 1b represents dynamic topic model results for the Media Cloud data. The five most frequent words for each topic are presented in the Figure legend. Topic proportions are minimal from 2009–2019. From 2019 onward, we observe increases in topics around the COVID-19 pandemic (Topic 7) and EVALI (Topic 1). Topic proportions seem to plateau towards the end of our data. Figure 1a denotes results for CrowdTangle. We noted occasional spikes for topic proportions post-2019, around vaping as a safer alternative to conventional cigarettes (Topic 7), and COVID-19 (Topic 10). Topic proportions appear to plateau toward the conclusion of the analysis period. Broadly, we noted the prevalence of COVID-19-related topics across news and social media [9]. Most importantly, Facebook data contained discussion around vaping as a combustible cigarette alternative, but news media centered instead on vaping harms and regulation.

Semantic Change. Figure 2a and 2b visualizes the semantic change over time for the word *vaping*, from 2009 to 2021. This analysis demonstrates how the meanings of words shift over time, which helps the understanding of semantic changes around vaping. In Fig. 2a for the CrowdTangle data, we observe that in 2010, *vaping* was close in meaning to words such as *good* and *weekend*, indicating that vaping was perceived as a fun and enjoyable weekend activity. In 2011, *vaping* moved closer in meaning to *products*, perhaps indicating an increase in the growth and in the availability of vaping products. During the 2017–2021 period, there was a shift toward words such as *teens, young, epidemic, risks,* and *dangers* which indicates a change in meanings towards vaping, toward vaping being dangerous and perhaps responsible for the youth vaping epidemic, likely motivated by EVALI. During the same time period, there was also a shift toward *quit, alternative,* and *smoking*, indicating a shift in meanings toward vaping as an alternative to combustible cigarettes and a smoking cessation tool. We note the opposing meanings around vaping that seem to co-occur (cessation tool vs. risky product), which could be indicative of the emergence of polarized public views regarding vaping. Figure 2b, shows the semantic change for Media Cloud data. There were too few vaping-related articles in the 2008–2012 period, and thus Fig. 2b shows the semantic change for 2013–2021. We can see that in 2013, *vaping* was close in meaning to *smoke, cities, bar*, perhaps indicative of discussions around indoor vaping bans. In 2015, *vaping* moved closer to *devices* and *regulation*, reflecting the shift in discussion to vaping regulation. From 2016–2021, there was a shift towards *teens* and *youth*, indicative of youth vaping epidemic discussions, and there was also a simultaneous shift towards *juul, flavors, ban*, likely around the planned ban on flavored vaping products. The news media framing of vaping has shifted over time in line with emergent regulatory trends; such as; flavored vaping bans, with little discussion around vaping as a smoking cessation tool. Social media discussions were far more varied, with transitions toward vaping as both a public health risk and a smoking cessation tool.

Table 1. The top four candidate words ranked by BERT probability for the cloze test *"Vaping is [MASK] for smoking"* for Media Cloud and CrowdTangle data.

Media Cloud (probability)	CrowdTangle (probability)
bad (0.148)	better (0.582)
safe (0.105)	bad (0.036)
not (0.093)	not (0.016)
dangerous (0.067)	substitute (0.014)

Question Answering. We present answers to *Can vaping help you quit smoking regular cigarettes?* and the top five most representative answers across news and social media in Table 2. We only found useful results for the indicated question. Other questions did not provide useful results and thus are not presented. For example, the question, *Why are teens vaping?* provided answers such as *bullying* and *health scare*. There was a clear difference in vaping frames for Media Cloud vs CrowdTangle. News media tends to view vaping as a danger, with responses

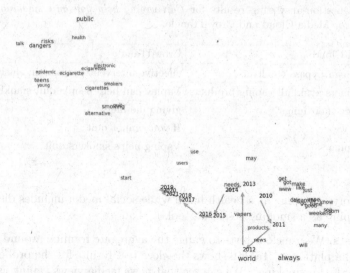

(a) Semantic change for *vaping*, on the CrowdTangle data

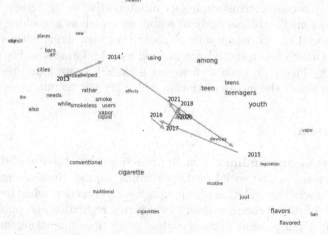

(b) Semantic change for *vaping*, on the Media Cloud data

Fig. 2. Two-dimensional visualizations of the semantic change for the word *vaping*, from 2010 to 2021, for CrowdTangle and for Media Cloud data.

such as *damage their lungs* and *consumers avoid all vaping products*. Social media displays more balanced answers, such as, *effective interventions remain elusive, vaping can help people quit smoking*, and *Vaping helps smokers quit*. News media

Table 2. Question-answering results for *Can vaping help you quit smoking regular cigarettes?* for Media Cloud and CrowdTangle.

Media Cloud	CrowdTangle
if you can bypass the liver	effective interventions remain elusive
consumers avoid all vaping products	Vaping can help people quit smoking
damage their lungs	giving me life
don't vape	If you cannot quit
no vaping	Vaping helps smokers quit

tends to portray vaping as a health harm while social media includes discussion around vaping as a smoking cessation tool.

Cloze Tests. We use cloze tests to gauge the aggregate framing around vaping, across media platforms. Table 2 shows the cloze test results for the probe: "Vaping is [MASK] for smoking". We can see that news media views vaping as *bad* or *dangerous*, but also *safe*. In CrowdTangle, we note a range of vaping frames, with vaping being a *substitute* and *better* for smoking. While both news and social media frame vaping as both dangerous and safe, reflecting diverse viewpoints, only social media characterizes vaping as an alternative to cigarettes. We speculate that news media are less likely to report on vaping as a smoking cessation tool, unlike social media, which may promote that view along with a range of other views. Other cloze tests did not provide useful information. For example, for the probe: "The main issue with vaping is [MASK]", we found results such as *the* and *va* across the media platforms, which are not useful.

4   Discussion

Implications of our findings. Our RQ was to explore the broad differences between news media and social media regarding vaping. Results suggest that social media tracks the evolution of vaping as a social practice, while news media reflects more risk based concerns driven by evolving regulation and public health narratives [3]. Vaping seems to have evolved away from mimicking smoking to being a related but separate practice with its own rituals and subculture so to some extent we can see how social media discussions reflect that evolution, particularly with regard to vaping as a product category. A strength of our work is how the different techniques we have applied validate each other. For example, the semantic change, n-gram and LDA topic model results all detail similar shifts from vaping as a smoking cessation tool to a regulatory issue. Our work is central to understanding how vaping frames shift across media platforms, thus allowing stakeholders to intervene around vaping frames, and to design communication campaigns that improve the way society sees vaping, possibly aiding smoking cessation, and reducing youth vaping.

Limitations. Our findings relied on the validity of data collected with our search terms. We used Media Cloud and CrowdTangle to search for all articles relevant to vaping, and our data contained text aligned with how vaping is framed. We are thus confident in the comprehensiveness of our data. We note that our data sources, Media Cloud and CrowdTangle are not representative enough to be considered public discourse. Thus, we indicate that our data may not be generalizable to how vaping is framed in the US. Similarly, we collected data from a single source for social and news media, and will include more sources in future work. The changes in trend are based on frequency count, and we will use other measures in planned research. A more detailed study on how and why trends change will be conducted in future work. We were not able to obtain statistics about how many times an article was read or shared, or to control for news outlets that are more widely read compared to smaller regional news outlets. We were also not able to distinguish between bias-free publications and opinion/commentary articles. Findings may also not apply to other related issues that are also heavily politicized (e.g., abortion) or other contexts (e.g., vaping frames in Europe). We also note the limitations of BERT, such as its inability to learn in few-shot settings. Such model limitations hampered our ability to analyze how subgroups of vapers are framed, such as LGBT+ vapers, around which there were relatively few news articles.

5   Conclusion and Future Work

We found that news media frames of vaping transitioned over time per emergent regulatory trends, such as; flavored vaping bans; with limited discussion around vaping as a smoking cessation tool. Framing of vaping in social media vaping was more varied; with transitions toward vaping both as a public health harm and a smoking cessation tool. Given the lack of generalizability of our data sources, future work will focus on working towards achieving such representativity. For example, we can include other social media sources, such as Instagram and TikTok. Future work can develop automated platforms to detect inaccurate framing around vaping and promote evidence-based information. Certain vulnerable demographics, such as youth, may be at a greater risk of encountering inaccurate vaping frames. As interventions tailored to specific demographics can be highly efficacious, future work can develop targeted interventions to mitigate inaccurate vaping frames among vulnerable demographics, perhaps enhancing intervention effectiveness and scope. Given the popularity of vaping in low-to-middle-income countries (LMICs) such as Indonesia or China, we suggest that future work explore vaping frames in LMICs. Such work is critical as limited resources, lack of equitable research and funding partnerships, industry strategies to counter tobacco control, and weak tobacco control policies have led to high smoking rates in LMICs. Greater research in this area may mitigate the existing tobacco mortality burden and youth vaping in LMICs. Similarly, First Nation Canadians, Aboriginal Australians, and Indigenous Americans have higher rates of smoking, and we suggest research exploring vaping frames in these demographics. Within tobacco control, there have also been calls to focus on women, sexual and gender minorities, and people of the global majority, and we suggest broadening research to explore vaping frames in these communities.

Acknowledgments. We thank the editor and reviewers for their comments. This study was pre-registered on the Open Science Framework (OSF.IO/kuxjp). This study was funded with a grant from the Foundation for a Smoke-Free World, a US nonprofit 501(c)(3) private foundation with a mission to end smoking in this generation. The Foundation accepts charitable gifts from PMI Global Services Inc. (PMI); under the Foundation's Bylaws and Pledge Agreement with PMI, the Foundation is independent from PMI and the tobacco industry. The contents, selection, and presentation of facts, as well as any opinions expressed herein are the sole responsibility of the authors and under no circumstances shall be regarded as reflecting the positions of the Foundation for a Smoke-Free World, Inc. The contents, selection, and presentation of facts, as well as any opinions expressed herein are the sole responsibility of the authors and under no circumstances shall be regarded as reflecting the positions of FTI Consulting.

Conflict of Interest. Navin Kumar and Keyu Chen declare financial support through a grant from the Foundation for a Smoke-Free World, a US nonprofit 501(c)(3) private foundation with a mission to end smoking in this generation. The Foundation accepts charitable gifts from PMI Global Services Inc. (PMI); under the Foundation's Bylaws and Pledge Agreement with PMI, the Foundation is independent from PMI and the tobacco industry. There are no financial relationships with any other organizations that might have an interest in the submitted work in the previous three years; and no other relationships or activities that could appear to have influenced the submitted work.

References

1. Blei, D.M., Lafferty, J.D.: Dynamic topic models. In: Proceedings of the 23rd International Conference on Machine Learning, pp. 113–120 (2006)
2. Blei, D.M., Ng, A.Y., Jordan, M.I.: Latent dirichlet allocation. J. Mach. Learn. Res. **3**, 993–1022 (2003)
3. Chen, K., et al.: Partisan us news media representations of syrian refugees. arXiv preprint. arXiv:2206.09024 (2022)
4. Devlin, J., Chang, M.W., Lee, K., Toutanova, K.: Bert: pre-training of deep bidirectional transformers for language understanding. arXiv preprint. arXiv:1810.04805 (2018)
5. Grootendorst, M.: Bertopic: leveraging bert and c-TF-IDF to create easily interpretable topics. (2020). https://doi.org/10.5281/zenodo.4381785
6. Hamilton, W.L., Leskovec, J., Jurafsky, D.: Diachronic word embeddings reveal statistical laws of semantic change. arXiv preprint. arXiv:1605.09096 (2016)
7. Hartmann-Boyce, J., et al.: Electronic cigarettes for smoking cessation. Cochrane Database Syst. Rev. (9) (2021)
8. Janmohamed, K., Sakai, S.N., Soale, A.N., Forastiere, L., Kumar, N.: News events and their relationship with US vape sales: an interrupted time series analysis. BMC Public Health **22**, 479 (2022). https://doi.org/10.1186/s12889-022-12858-x
9. Janmohamed, K., et al.: Intersection of the web-based vaping narrative with COVID-19: Topic modeling study. J. Med. Internet Res. **22**(10), e21743 (2020)

10. Jeong, M., Singh, B., Wackowski, O.A., Mukherjee, R., Steinberg, M.B., Delnevo, C.D.: Content analysis of e-cigarette news articles amidst the 2019 vaping associated lung injury (EVALI) outbreak in the US. Nicotine Tob. Res. **24**, 799–803 (2021)

11. Kong, G., LaVallee, H., Rams, A., Ramamurthi, D., Krishnan-Sarin, S.: Promotion of vape tricks on YouTube: content analysis. J. Med. Internet Res. **21**(6), e12709 (2019)

12. Kumar, N., et al.: Interventions to mitigate vaping misinformation: protocol for a scoping review (2021)

13. Lewis, M., et al.: Bart: denoising sequence-to-sequence pre-training for natural language generation, translation, and comprehension. arXiv preprint. arXiv:1910.13461 (2019)

14. Miech, R., Leventhal, A., Johnston, L., O'Malley, P.M., Patrick, M.E., Barrington-Trimis, J.: Trends in use and perceptions of nicotine vaping among us youth from 2017 to 2020. JAMA Pediatr. **175**(2), 185–190 (2021)

15. Palakodety, S., KhudaBukhsh, A.R., Carbonell, J.G.: Mining insights from large-scale corpora using fine-tuned language models. In: ECAI 2020–24th European Conference on Artificial Intelligence. Frontiers in Artificial Intelligence and Applications, vol. 325, pp. 1890–1897. IOS Press (2020)

16. Stratton, K., Kwan, L.Y., Eaton, D.L., et al.: Public health consequences of e-cigarettes: consensus study report (2018)

17. Stubbs, J.B., Tucker, G.R.: The cloze test as a measure of English proficiency. Mod. Lang. J. **58**(5/6), 239–241 (1974)

18. Wackowski, O.A., O'Connor, R.J., Diaz, D., Rashid, M., Lewis, M.J., Greene, K.: '95% less harmful'? exploring reactions to quantitative modified risk claims for snus and e-cigarettes. Tob. Control **31**(6), 730–736 (2021)

19. Wackowski, O.A., Sontag, J.M., Hammond, D.: Youth and young adult exposure to and perceptions of news media coverage about e-cigarettes in the united states, canada and england. Prev. Med. **121**, 7–10 (2019)

20. Wodak, A., Mendelsohn, C.P.: The Australian approach to tobacco harm reduction is even more misguided than the us approach. Am. J. Public Health **110**(6), 783–784 (2020)

Network Analysis

Social Network Analysis
of the Caste-Based Reservation System
in India

Akrati Saxena[1(✉)], Nivedita Sethiya[2], Jaspal Singh Saini[3], Yayati Gupta[4],
and S. R. S. Iyengar[2]

[1] Eindhoven University of Technology, Eindhoven, The Netherlands
a.saxena@tue.nl
[2] Indian Institute of Technology Ropar, Rupnagar, India
[3] Oregon State University, Corvallis, USA
[4] Mahindra Ecole Centrale, Hyderabad, India

Abstract. Being as old as human civilization, discrimination based on various grounds such as race, creed, gender, and caste has existed for a long time. To undo the impact of this long-enduring historical discrimination, governments worldwide have adopted various forms of affirmative action, such as positive discrimination, employment equity, and quota system. In India, people are considered to belong to Backward Class (BC) or Forward Class (FC), and the Indian government designed an affirmative action, locally known as the "Reservation" policy, to reduce the discrimination between both groups. Through this affirmative action, the government provides support to people from the backward class (BC). Although being one of the most controversial and frequently debated issues, the reservation system in India lacks rigorous scientific study and analysis. In this paper, we model the dynamics of the reservation system based on the cultural divide among the Indian population using social network analysis. The mathematical model, using the Erdös-Rényi network, shows that the addition of weak ties between the two groups leads to a logarithmic reduction in the social distance. Our experimental simulations establish the claim for the different clans of frequently studied social network models as well as real-world networks. We further show that a small number of links created by the reservation process are adequate for a society to live in harmony.

1 Introduction

"You do not take a man who for years has been hobbled by chains, liberate him, bring him to the starting line of a race, saying, 'you are free to compete with all the others,' and still justly believe you have been completely fair."
- Lyndon B. Johnson (1965)

Imagine a person chained and locked in a dark room all through his life. After a very long time, he is uncuffed and brought to the outside world. He is asked to compete with outsiders to get an education in schools and jobs in companies. Does he stand any chance with his long years of slavery and discrimination? Is he liberated in real terms?

Slave culture in Africa [1], mistreatment of minority communities in Europe [2], denial of political rights to African Americans in the US [3], religious discrimination against Rohingyas in Myanmar [4] are only a few examples of discrimination experienced by disadvantaged groups of people worldwide. While there are policies to counter the cause and effects of discrimination, most of the discriminated people still remain poor and lack employment and higher education. To address this problem, countries all over the world have adopted different policies, called "Affirmative Action". Affirmative action, though having multifaceted implementations, follows one basic idea, i.e., to give some kind of preference to discriminated groups of people to promote their development. It can be in the form of monetary help, special kind of assistance or positive discrimination in the form of a quota system. Often, affirmative action is perceived as giving unfavorable preference to some members of the society, leading to debates, riots, and court cases [5].

Locally known as the "Reservation" system in India, affirmative action is particularly prominent in the country. The reason behind this is the diversity prevalent in India, with over 3,000 castes and 25,000 sub-castes, each possessing a few hundred to a few million members. Originated thousands of years ago, the concept of caste and resulting discrimination still prevails in the Indian sub-continent [6]. To balance prejudicial social stratification, affirmative steps were undertaken to uplift the backward classes, and the reservation system was introduced wherein a certain number of seats were reserved for the members of socially and economically backward classes at the places of higher education and government jobs. Unsurprisingly, the system soon experienced a lot of backlash from the socially forward community, who claimed it to be unmeritorious and politically inclined towards a few. The critical issue of reservation has become a reason for frequent riots, and protests in the country [7].

While the debate continues, we would like to highlight that the current statistics show a clear tip in favoring the socially forward community, with a majority of the country's shared resources, such as education, wealth, and land-holdings, being mainly accessible to the forward section of the society. This undesirable disparity has been observed by many recent studies [8,9]. Borooah et al. [10] have tried to study the impact of the absence of a reservation system in the country. The study took into account a social group within the country that was of the same social, educational, and economic status as the Scheduled Castes (SC) and Scheduled Tribes (ST) in the pre-independence era. However, the condition of this group was observed to be at a much more elevated state, proving the efficiency of the reservation system. Many similar comparative studies claim that this system is a beacon of hope to wipe out the social disparity in India [11–13].

Most of the studies in the context of the reservation system are survey-based and lack theoretical backing. This paper studies the existing reservation system from a pure network-theoretic perspective. We employ network science concepts, including homophily, weak ties, social capital, opinion formation, and influence propagation, to model the dynamics of the reservation system. As a first step towards modeling this phenomenon, we consider only two social communities, the *socially forward and uplifted* (FC) and the *socially backward and downtrodden* (BC). We establish why and how the reservation system maintains a good balance between the two.

Our motivation comes from the existence of a tangible strength associated with every weak tie, as proposed by Granovetter in his famously cited theory of the *Strength of Weak Ties* [14]. He observed the importance of weak ties to get new opportunities. This is a very prominent network phenomenon that has been ignored in past studies of the reservation system, which we have chosen as the key element of our study. We assume that whenever a reservation is given, it motivates a weak link between BC and FC. The proposed basic mathematical model aids us in finding the number of links between the two communities as a measure of stability for a large-scale social structure. We study the dynamics of the cumulative social capital of the backward classes at discrete time steps. The major contributions of the paper are listed below.

1. We theoretically show that the social distance between BC and FC communities decreases logarithmically as we keep adding the connections between them by providing the reservation facility. In other words, the cumulative social capital of the BC increases logarithmically with an increase in the number of weak ties.
2. The theoretical results are complemented with the experimental analysis of the reservation phenomenon on different kinds of synthetic and real-world complex networks.

The proposed model is a novel but simple and natural model, where it is assumed that the state of the world changes deterministically over time as new network connections are added through time steps. Our study finds it sufficient to insert a minimal number of links between the two clusters in order to foster harmonic relations between the two conflicting groups. As a long-term aim of the reservation system, we see benefits reasonably distributed evenly among members from the forward as well as the backward communities.

The rest of the paper is organized as follows. Section 2 introduces preliminaries and definitions. Section 3 describes the proposed mathematical model. Section 4 discusses experimental simulations. Finally, the paper is concluded in Sect. 5 while highlighting possible future directions.

2 Preliminaries and Definitions

Let $G(V, E)$ represents the undirected social network under consideration, and $G_1(V_1, E_1)$ and $G_2(V_2, E_2)$ represent the induced subgraphs of G, where V_1 is the set of all BC nodes and V_2 is the set of all FC nodes. Therefore, we assume that, $V_1 \cup V_2 = V$ and $V_1 \cap V_2 = \emptyset$. Let n_1 and n_2 be the shorthand notations for the number of individuals in the BC and the FC, respectively, i.e., $|V_1| = n_1$ and $|V_2| = n_2$.

We define the edge set B to be the set of edges $\{u, v\}$, where $u \in V_1$ and $v \in V_2$. Hence, $(B = E - (E_1 \cup E_2))$. Henceforth, we address these edges as *bridges*. The distance $d(u, v)$ between two nodes u and v represents the length of the shortest path between u and v. For each $u \in V_1$, i.e., a person belonging to BC, we define d_u^* as,

$$d_u^* = \min\{k | v \in V_2 \ \& \ d(u, v) = k\}$$

Therefore, d_u^* is the minimum distance of node u from a node in V_2. We refer to this parameter as the *social distance* of node u from the FC.

A path $\langle v_1, v_2, v_3, ..., v_k \rangle$ is called an *entry path* if $v_k \in V_2$ and $v_i \in V_1$ $\forall 1 \leq i \leq k-1$. Therefore, if $d_u^* = l$, then l, the length of the shortest entry path starting from node u, is the social distance of node u from the FC.

3 Caste Reservation System: A Network-Theoretic Approach

The homophily observed in the social structure under consideration is *selection* based [15], i.e., the common characteristics that bound people together are immutable. In this case, the characteristic is the caste of an individual. The reservation system, in its essence, chooses a BC individual and brings him/her in contact with a group of closely knit FC individuals. For example, a BC student getting a seat in a university through reservation implicitly creates friendship ties with a group of closely-knit FC students. The addition of such bridges has two-fold benefits, termed by us as the *forward breeze* effect and the *backward breeze* effect. The *forward breeze* represents the change in the mindset of the FC after coming in contact with the BC. The *backward breeze* represents the increased motivation felt by the BC to be uplifted because of being influenced by the FC members close to them. Our study concentrates on the backward breeze effect.

Our aim is to calculate the gain in the social capital of the BC as a function of the bridges added to the network. There exists no universal definition or technique for measuring social capital [16]. This can be attributed to the inherent subjectivity in the concept of social capital. However, in an exhaustive survey [17], the author differentiates between the social capital of different types:

1. Social capital of an individual with respect to her position in the social network.
2. Social capital of a group with respect to the underlying relationships within the group.
3. Social capital of a group with respect to the network topological connections to other groups.

In the caste reservation scenario, the cumulative social capital to be calculated falls under category 3. The social capital of category 3 was first studied in [18], where the author has shown that the teams with strong outside connections generally performed better as compared to the groups having weaker outside connections. Everett and Borgatti proposed a network measure termed *group centrality* to quantify social capital of type 3 [19]. We adopt a modified version of this definition, which fits well in the reservation scenario. We define the cumulative social capital of BC as the linear sum of the social capital of all individuals present in BC. Further, for every individual u in BC, we assume its social distance (d_u^*) to be a direct measure of its social capital. The lower the social distance of an individual u, the higher is its social capital and vice versa.

As stated earlier, a person from BC getting a reservation implies her/him getting an opportunity to form ties with a set of closely knit FC individuals. In our analysis, we assume that only one tie is created per reservation, i.e., all weak ties originating from a BC are equivalent to one bridge. This is a safe assumption since we are measuring social capital as a function of distance. This distance will rarely change for a pair of nodes in the network when we remove multiple copies of similar functioning edges. However, this assumption does not hold well in all aspects. For example, the presence of multiple weak ties amplifies the strength of the bridge across, in a sense that, even if one link breaks in the future, it does not influence the network topology or social capital significantly.

Next, our aim is to analyze the reduction in d_u^* ($u \in V_1$) as a function of the number of random bridges added to the system. For this, we consider G_1 and G_2 to be *Erdös-Rényi* random graphs [20] with parameters (n_1, p_1) and (n_2, p_2). For every $u \in V_1$ and $v \in V_2$, the edge (u, v) is present with probability b, which we term as the *bridging probability*. We define this new probability space of graphs as *Coupled Erdös-Rényi Graphs* where both communities are represented using two *Erdös-Rényi* random graphs, and reservation links are represented as bridges (weak ties) placed between them.

Generally, real-world social networks depict scale-free degree distribution [21], i.e., the probability that a node u has degree k is, $P(degree(u) = k) \propto k^{-\gamma}$, where $2 < \gamma < 3$. Therefore, we performed empirical analysis on both random and scale-free networks. In scale-free networks, we consider both communities to be scale-free graphs. We observed that the social distance i.e., d_u^* falls at nearly the same rate, independent of whether we consider communities to be scale-free graphs or random graphs for the same number of nodes and edges. Next, we discuss the proposed mathematical analysis.

3.1   Social Distance Analysis on Coupled Erdös-Rényi Graphs

The two classes, BC and FC, are represented by two *Erdös-Rényi* random graphs $G_1(n_1; p_1)$ and $G_2(n_2, p_2)$ respectively. $V(G_1)$ and $V(G_2)$ are the vertex set of graphs G_1 and G_2, respectively, and $V(G_1) = \{1, 2, 3, \ldots, n_1\}$ and $V(G_2) = \{n_1 + 1, n_1 + 2, \ldots, n_1 + n_2\}$. All the results proved in this paper are for asymptotically large graphs G_1 and G_2 i.e. $n_1 \to \infty$ and $n_2 \to \infty$. Every possible edge across the two graphs ($n_1 n_2$ in total) is added with the bridging probability b.

Let u represent an arbitrary node in BC. Our analysis aims at calculating d_u^*, i.e., the social distance of node u from the FC. We begin by developing a few preliminary results.

Lemma 1. $\dfrac{n!}{(n-l)!} \sim n^l$ as $l^2/n \sim 0,$ $\left(f(n) \sim g(n) \text{ if } \lim\limits_{n\to\infty} f(n)/g(n) = 1\right)$

Proof.

$$\frac{n!}{(n-l)!} \sim \sqrt{2\pi n}\left(\frac{n}{e}\right)^n \frac{1}{\sqrt{2\pi(n-l)}}\left(\frac{e}{(n-l)}\right)^{n-l} \quad \text{(using Stirling's approx.)}$$

$$\sim \left(\frac{n}{e}\right)^l \left(1 - \frac{l}{n}\right)^{-(n-l+\frac{1}{2})}$$

$$\sim \left(\frac{n}{e}\right)^l (e)^{l(n-l+\frac{1}{2})/n}$$

$$\sim n^l$$

□

Let M_l represents the total number of possible entry paths of length l with u as one of its endpoints. The next lemma provides an approximation for constant M_l as a function of l.

Lemma 2. $M_l \sim n_2(n_1)^{l-1}$ as $l^2/n_1 \sim 0$.

Proof. To construct an entry path of length l with u as one of its endpoint, we need a vertex from V_2 and a sequence of $l-1$ vertices from $V_1 - u$ i.e. one node has to be selected from n_2 nodes and $l-1$ nodes have to be selected from $n_1 - 1$ nodes. These $l-1$ selected nodes can be permuted in $(l-1)!$ ways.

$$\implies M_l = \binom{n_2}{1}\binom{n_1-1}{l-1}(l-1)!$$

$$= n_2 \frac{(n_1-1)!}{(n_1-l)!}$$

$$\sim n_2(n_1)^{l-1} \quad \text{(from Lemma 1)}$$

□

Further, let X_l be a random variable that represents the number of entry paths of length l with u as one of its endpoints. Next, we calculate the average number of entry paths of length l originating from a BC node u.

Lemma 3. $E[X_l] \sim (n_2 b)(n_1 p_1)^{l-1}$ as $l^2/n_1 \sim 0$

Proof.

$$X_l = \sum_{i=1}^{M_l} Y_i$$

$$\text{where } Y_i = \begin{cases} 1 \text{ if } j^{th} \text{ entry path is present} \\ 0 \text{ otherwise} \end{cases}$$

$$\implies E[X_l] = \sum_{i=1}^{M_l} E[Y_i] \quad \text{(using linearity of expectation)}$$

There exists an entry path $P = \langle v_{\alpha_1}, v_{\alpha_2}, v_{\alpha_3}, \ldots, v_{\alpha_{l+1}} \rangle$ of length l if $(l-1)$ edges $(\{v_{\alpha_1}, v_{\alpha_2}\}, \{v_{\alpha_2}, v_{\alpha_3}\}, \ldots, \{v_{\alpha_{l-1}}, v_{\alpha_l}\})$ are present in G_1 and the bridge $\{v_{\alpha_l}, v_{\alpha_{l+1}}\}$ is also present. Therefore, the probability that the entry path P exists is $(p_1)^{l-1}b$.

$$\implies E[X_l] = M_l(p_1)^{l-1}b \qquad \text{(from Lemma 2)}$$
$$\sim (n_2 b)(n_1 p_1)^{l-1}$$

$\square$

- We are interested only in the case where $b < 1/n_2$, since for $b \geq 1/n_2$, the expected number of bridges per node in BC will be greater than or equal to 1, which is unrealistic in the caste reservation scenario. Henceforth, throughout the analysis, b is assumed to be less than $1/n_2$.

Theorem 1. *For a random graph $G_{n,p}$, $p = \log(n)/n$ is the threshold probability for the property of connectedness.*

A detailed proof is available in [22].

Since we assumed that the two considered graphs G_1 and G_2 are connected, the above theorem provides a lower bound on p_1 and p_2, and hence on the density of graphs G_1 and G_2. Therefore, $n_1 p_1 > log(n_1)$ and $n_2 p_2 > log(n_2)$.

Next, we analyze X_l, the number of entry paths from node u as a function of l. For small values of l, the number of entry paths X_l is negligible ($<< 1$). Our aim is to find the smallest distance d such that there exists at least one entry path of length d from node u. We prove that distance d is equal to $\log_{(n_1 p_1)}(1/n_2 b) + 1$. Henceforth, we represent the quantity $\log_{(n_1 p_1)}(1/n_2 b)$ as d_0 for the sake of clarity.

Theorem 2. *The probability that there exists an entry path of length less than or equal to d_0 with u as its endpoint is almost equal to zero i.e. $P(X_i = 0) \sim 1$ for $1 \leq i \leq d_0$.*

Proof.

$$P(X_i \geq a) \leq E[X_i]/a \qquad \text{(using Markov's inequality)}$$
$$\implies P(X_i \geq 1) \leq E[X_i]$$
$$\sim (n_2 b)(n_1 p_1)^{i-1} \qquad \text{(from Lemma 3)}$$
$$\leq (n_2 b)(n_1 p_1)^{d_o - 1} \qquad \text{(Since } i < d_0)$$
$$= \frac{1}{n_1 p_1}$$
$$< \frac{1}{log(n_1)} \qquad \text{(from Theorem 1)}$$
$$\implies P(X_i \geq 1) \to 0$$
$$\implies P(X_i = 0) \to 1$$

$\square$

Further, we prove that almost always (i.e. with probability close to one), there exists at least one entry path of length $d_0 + 1$ from u, proving our claim that $d_u^* = d_0 + 1$.

Lemma 4. *For any random variable X, $P(X = 0) \leq \dfrac{\sigma_X^2}{\mu_X^2}$, where σ_X and μ_X represent the variance and mean of the random variable X respectively.*

Proof.

$$P(|X - \mu_X| \geq a) \leq \frac{\sigma_X^2}{a^2} \qquad \text{(Chebyshev's inequality)}$$

$$\implies P(|X - \mu_X| \geq \mu_X) \leq \frac{\sigma_X^2}{\mu_X^2}$$

$$P(X = 0) \leq P(|X - \mu_X| \geq \mu_X)$$

$$\leq \frac{\sigma_X^2}{\mu_X^2}$$

$\square$

Lemma 5. *The standard deviation of the random variable X_{d_0+1} approaches zero i.e. $\sigma_{X_{(d_0+1)}} \sim 0$*

Proof.

$$X_l = \sum_{i=1}^{M_l} Y_i$$

$$\implies X_l^2 = \sum_{i=1}^{M_l} \sum_{j=1}^{M_l} Y_i Y_j = \sum_{k=0}^{l} Z_k$$

where Z_k accounts for all $Y_i Y_j$'s, where the ith and j^{th} entry paths have precisely k edges in common. Let $|Z_k|$ represent the number of terms in Z_k's summation.

$$|Z_0| \geq \binom{n_1 - 1}{l - 1}(l - 1)!\binom{n_2}{1}\binom{n_1 - l}{l - 1}(l - 1)!\binom{n_2 - 1}{1}$$

If none of the vertices in the two entry paths are common, then certainly none of its edges are common either. This gives us the above inequality.

$$\binom{n_1 - 1}{l - 1}(l - 1)!\binom{n_2}{1}\binom{n_1 - l}{l - 1}(l - 1)!\binom{n_2 - 1}{1} = \frac{(n_1 - 1)!}{(n_1 - 2l + 1)!}n_2(n_2 - 1)$$

$$\sim \frac{n_2(n_2 - 1)}{n_1(n_1 - 2l + 1)}n_1^{2l}$$

$$\sim n_2^2 n_1^{2l-2}$$

The total number of terms in the summation of X_l^2 are M_l^2 i.e. approximately $n_2^2 n_1^{2l-2}$. Therefore, most of the summation terms of X_l^2 fall into the basket of Z_0.

$$E^2[X_{d_0+1}] \sim 1 \qquad \text{(from Lemma 3)}$$
$$\sigma_{X_l}^2 = E[X_l^2] - E^2[X_d] \qquad \text{(by definition)}$$
$$\implies \sigma_{X_{(d_0+1)}}^2 = E[X_{d_0+1}^2] - E^2[X_{d_0+1}]$$
$$\implies \sigma_{X_{(d_0+1)}}^2 \sim 0$$

$\square$

Theorem 3. *Almost always, there exists an entry path of length equal to $d_0 + 1$ with u as its endpoint i.e. $X_{d_0+1} \geq 1$.*

Proof.

$$P(X_{(d_0+1)} = 0) \leq \frac{\sigma_{X_{(d_0+1)}}^2}{\mu_{X_{(d_0+1)}}^2} \qquad \text{(using Lemma 4)}$$
$$\implies P(X_{(d_0+1)} = 0) \sim 0 \qquad \text{(using Lemma 3 and 5)}$$
$$\implies P(X_{(d_0+1)} \geq 1) \sim 1$$

$\square$

Therefore, almost always, $d_u^* = \log_{n_1 p_1}(1/(n_2 b)) + 1$. Since this formula is independent of u, almost all nodes in the BC have social distance $(d_0 + 1)$. Let x represents the expected number of bridges added to the network.

$$\implies x = n_1 n_2 b$$
$$\implies d_i^* = \log_{n_1 p_1}(n_1/x) + 1$$

$$d_i^* = \frac{\log(n_1) - \log(x)}{\log(n_1 p_1)} + 1 \qquad (1)$$

The above theorem proves that the social distance of any arbitrary node i reduces logarithmically as a function of the number of bridges in the system. Therefore, only the first few bridges are highly effective in reducing the distance between these two communities. The bridges added later don't significantly change the social capital of an individual in BC.

4   Simulation Results

In this section, we experimentally observe the decrease in the social distance d_u^* of an arbitrary node u in the backward class on the addition of subsequent weak ties between the BC and the FC. In our experiments, we model the BC and

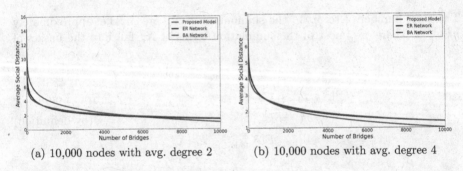

(a) 10,000 nodes with avg. degree 2   (b) 10,000 nodes with avg. degree 4

Fig. 1. Average social distance versus the number of bridges on synthetic networks.

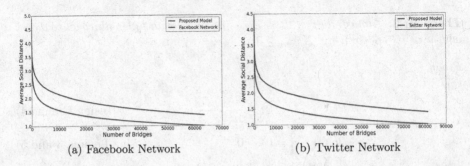

(a) Facebook Network   (b) Twitter Network

Fig. 2. Average social distance versus the number of bridges on real-world networks.

the FC using two popular network generative models, Erdős-Rényi (ER) and Barabási–Albert (BA) [21], and real-world networks. For our experiments, we generate two ER and BA networks having 10,000 nodes and the average degrees of two and four, respectively. We wanted to simulate the reservation dynamics on real-world social networks, but it is challenging to obtain this data. Attributed to a greater similarity between the socialization in the online and offline networks [23,24], we use two online social networking datasets as the underlying graphs for our experiments (i) Facebook (63,392 nodes and 816,831 edges) [25], and (ii) Twitter (81,306 nodes and 1,342,296 edges) [26].

For each of the above-specified networks, the simulation is conducted in the following manner. We create 2 copies of the network at hand, say G_1 and G_2. Initially, the social graph G is a disconnected graph having 2 connected components G_1 and G_2 representing the BC and the FC, respectively. As time proceeds, we add random links, called bridges, between these two components. The simulation is stopped when n_1 number of bridges are added in the system, where $n_1 = |V(G_1)|$, i.e., the size of the BC.

The results for synthetic networks and real-world networks have been reported in Fig. 1 and Fig. 2, respectively. In all the plots, x-axis represents the number of bridges, and y-axis represents the average social distance. The average social distance is computed by averaging the social distance over all nodes of BC.

The experiment is repeated 10 times to compute the average social distance for a given number of bridges. For all kinds of graphs, we observe the convergence of the average social distance on the addition of the n_1 number of bridges. While theoretically, the distance should converge to 1, it converges to a real number between 1 and 2 in our experiments. This is because, in the theoretical model, each BC node is directly connected to an FC node, while in the empirical model, one node of BC can be connected with multiple nodes of FC as the bridges are placed uniformly at random.

However, all plots for random and scale-free networks show that the average social distance decreases logarithmically with the addition of weak ties. Hence it is established that the impact of placing more bridges decreases with time, and a small number of bridges is sufficient to maintain harmonic distance between both communities.

5   Conclusion

The current work is an attempt to present the dynamics of the reservation system using social network analysis. In this model, reservation is considered to instantiate the formation of weak ties between the otherwise sparsely connected communities. As new weak ties are introduced in the system, the social distance between these two communities, FC and BC, decreases. This decrease is observed to be logarithmic, establishing the fact that a minimal number of links lead to a rapid increment in the cumulative benefit to society. To the best of our knowledge, this is a first step towards dissecting the working dynamics of affirmative action with the help of social network analysis. Our results show that the reservation system had a considerable impact on the country's overall development by bridging the gap between the conflicting social groups. The present work highlights that some amount of affirmative action is important to achieve equality in society, but what percentage of it will achieve the best results in a feasible amount of time is an open question. One can pursue this question to arrive at a specific percentage of reservation that should be offered, or an amount of affirmative action, in general. We further plan to extend the proposed model using more realistic parameters and model different affirmative actions.

References

1. Stuckey, S.: Slave Culture: Nationalist Theory and the Foundations of Black America. Oxford University Press, Oxford (2013)
2. Frumkin, H., Walker, E.D., Friedman-Jiménez, G.: Minority workers and communities. Occup. Med. (Philadelphia, Pa.) **14**(3), 495–517 (1998)
3. Omi, M., Winant, H.: Racial Formation in the United States. Routledge, Milton Park (2014)
4. Zawacki, B.: Defining myanmar's rohingya problem. Hum. Rts. Brief **20**, 18 (2012)
5. Ball, H.: The Bakke Case: Race, Education, and Affirmative Action. Landmark Law Cases and American Society, ERIC (2000)

6. Dirks, N.B.: Castes of Mind: Colonialism and the Making of Modern India. Princeton University Press, Princeton (2011)
7. Jaffrelot, C.: The impact of affirmative action in India: more political than socioeconomic. India Rev. **5**(2), 173–189 (2006)
8. Shiva Kumar, A.K., Rustagi, P., et al.: Elementary education in India: Progress, setbacks, and challenges. Technical report (2016)
9. Sedwal, M., Kamat, S.: Education and social equity with a special focus on scheduled castes and scheduled tribes in elementary education (2008)
10. Borooah, V.K., Dubey, A., Iyer, S.: The effectiveness of jobs reservation: caste, religion and economic status in India. Dev. Change **38**(3), 423–445 (2007)
11. Weisskopf, T.E.: Impact of reservation on admissions to higher education in India. Econ. Polit. Wkly **39**, 4339–4349 (2004)
12. Chanana, K.: Accessing higher education: the dilemma of schooling women, minorities, scheduled castes and scheduled tribes in contemporary india. High. Educ. **26**(1), 69–92 (1993). https://doi.org/10.1007/BF01575107
13. Kirpal, V., Gupta, M.: Equality Through Reservations. Rawat Publications, Jaipur (1999)
14. Granovetter, M.: The strength of weak ties. Am. J. Soc. **78**(6), 1360–1380 (1973)
15. McPherson, M., Smith-Lovin, L., Cook, J.M.: Birds of a feather: homophily in social networks. Ann. Rev. Soc. **27**, 415–444 (2001)
16. Aghion, P., Durlauf, S.: Preface to the handbook of economic growth (2005)
17. Borgatti, S.P., Jones, C., Everett, M.G.: Network measures of social capital. Connections **21**(2), 27–36 (1998)
18. Ancona, D.G.: Outward bound: strategic for team survival in an organization. Acad. Manag. J. **33**(2), 334–365 (1990)
19. Everett, M.G., Borgatti, S.P.: The centrality of groups and classes. J. Math. Sociol. **23**(3), 181–201 (1999)
20. Erdös, P., Rényi, A.: On random graphs. Publ. Math. (Debrecen) **6**, 290–297 (1959)
21. Barabasi, A.-L., Albert, R.: Emergence of scaling in random networks. Science **286**(5439), 509–512 (1999)
22. Béla Bollobás.: Modern Graph Theory, vol. 184. Springer Science & Business Media, Berlin (2013)
23. Ocker, R.J., Yaverbaum, G.J.: Asynchronous computer-mediated communication versus face-to-face collaboration: results on student learning, quality and satisfaction. Group Decis. Negot. **8**(5), 427–440 (1999). https://doi.org/10.1023/A:1008621827601
24. van Ingen, E., Wright, K.B.: Predictors of mobilizing online coping versus offline coping resources after negative life events. Comput. Hum. Behav. **59**, 431–439 (2016)
25. Viswanath, B., Mislove, A., Cha, M., Gummadi, K.P.: On the evolution of user interaction in facebook. In: Proceedings of the 2nd ACM Workshop on Online Social Networks, WOSN'09, pp. 37–42, New York, NY, USA (2009)
26. Leskovec, J., Mcauley, J.: Learning to discover social circles in ego networks. In: Advances in Neural Information Processing Systems, pp. 539–547 (2012)

Structure, Stability, Persistence and Entropy of Stock Networks During Financial Crises

Nawee Jaroonchokanan[1], Teerasit Termsaithong[2,3], and Sujin Suwanna[1(✉)]

[1] Department of Physics, Faculty of Science, Mahidol University,
Bangkok 10400, Thailand
sujin.suw@mahidol.ac.th

[2] Learning Institute, King Mongkut's University of Technology Thonburi (KMUTT),
Bangkok 10140, Thailand

[3] Theoretical and Computational Physics (TCP) group, Center of Excellence
in Theoretical and Computational Science (TaCS-CoE), King Mongkut's University
of Technology, Bangkok 10140, Thailand

Abstract. We investigate the network structures of stocks in SET100, NASDAQ100, and FTSE100 from 2006 to 2022, using the correlation distance and the time-space average of correlations as a threshold for connectivity of two stocks. Structure, stability, multifractality, and entropy of the networks are investigated to compare their behaviors before and after financial crises. The results show that during high volatility periods, such as the global financial crisis in 2008 and the COVID pandemic in 2020, the network characteristic path length decreases, while the clustering coefficient increases, suggesting that the network has shrunk in size, and stocks become tightly linked, similar to trends of price and return behaviors observed in many stocks during financial crises. Furthermore, the minimal level of network entropy implies that the market network stability decreases, and each sector has lost its ability to perform independently. We also find that the persistence of the network structure and the network entropy in SET increase during a period of high volatility as evident by a significant increase of the Holder exponent, while results from NASDAQ and FTSE do not exhibit such pronounced behavior, possibly due to having higher market fluctuation. Network features of SET and FTSE show recovery of same values after the 2008 crisis faster than NASDAQ, and in less than 100 trading days; however, they exhibit slower recovery, except for the network entropy, from the COVID-19 pandemic.

Keywords: Structural stock network · Network entropy ·
Multifractality

1 Introduction

Complex systems, whether in physical, biological, economical, or social environments, comprise interacting agents whose collective behaviors give rise to rich

Supported by organization CSoNet2022.

phenomena [4, 13]. A stock market is an economical complex system in which traders play important roles in influencing the market activity and performance. When a stock market experiences a financial stress or crisis, traders can abruptly change investment strategies, e.g., panic selling, which impact stock prices and other financial quantities. A financial stress has effects on not only traders, but also the stocks and markets themselves. Generally, it is difficult to model and predict precisely agent behaviors in a complex system because many factors can affect interaction, hence decision, of agents. However, some collective behaviors can shed insights on the underlying structure and characteristics of the market.

A complex network can be represented by a graph consisting of nodes (stocks) and links (interactions/correlations), respectively. Mathematical quantities in graph topology are usually conducive and effective to quantify the structure of the network, which can probe how the network behaves as it transits through a potentially disruptive event. To that end, the connectivity between two nodes in a network can be defined via the Pearson correlation coefficient, the Fisher information distance, or dynamic time warping [8–10, 12, 14]; and the network structure can be analyzed via, for examples, the characteristic path lengths and clustering coefficients.

Many real-world networks typically fall into scale-free networks where the degree distribution in the network obeys the power law behavior $P(k) \sim ck^{-\alpha}$, with $P(k)$ being a probability of a node having k links. The variation of degrees, also termed heterogeneity, signifies the complexity of a network [6], where higher heterogeneity corresponds to a longer tail, and lower heterogeneity to constant degree, e.g. delta peak, in the degree distribution. The heterogeneity decreases as α increases, suggesting that the network becomes less stable to preserve its heterogeneity and complexity. This can be viewed from the network entropy in the sense that low entropy indicates a highly unstable network [15, 17].

From another perspective, the scaling behavior of a financial quantity can reveal how a market changes its structure and properties. For example, the power spectral density (PSD) $S(f) \sim f^{-\alpha}$ is associated with the Hurst exponent H_e (or correspondingly the Holder exponent h) through $\alpha = 2H_e + 1$, where $\alpha = 2\,(H_e = 0.5)$ signifies a Brownian process. The market behavior is identified as persistent if $H_e > 0.5$, meaning that the market behavior in the next time period tends to follow that from the previous one. Otherwise, it is antipersistent ($H_e < 0.5$), showing an opposite trend to that of the previous period.

In this article, we aim to model and investigate the collective behaviors of stocks in the Stock Exchange of Thailand (SET), National Association of Securities Dealers Automatic Quotation System (NASDAQ), and the Financial Times Stock Exchange (FTSE). The network structures of stocks, including their stability, persistence and entropy, are examined when the market evolves through a financial crisis or high volatility period according to Ref. [1]. In particular, we study whether the market exhibit similar behaviors before and after a crisis, e.g., whether the recovery signals resemble that of the regression.

2   Theoretical Background

We investigate the network structure and its multifractality to examine the collective behaviors of stocks. To that end, we first detail how such a network is constructed.

Correlation Distance: A distance between two stocks i and j can be measured via the Pearson correlation coefficient ρ_{ij} of the stock returns. Let $r_i(t) = \log p_i(t + \Delta t) - \log p_i(t)$ denote the return of the stock i at time t, and $R_i(t) = [r_i(t) - \langle r_i(t) \rangle]/\sigma_i$ its normalized return. Here, $\langle \cdot \rangle$ denotes the ensemble average and σ the standard deviation. Then, $\rho_{ij} = \langle R_i R_j \rangle$, and the distance between the two stocks is defined as $d_{ij} = \sqrt{2(1 - \rho_{ij})}$ [9,10]. Clearly, the distance between two strongly positively correlated stocks is short, and is zero from itself.

Network Features: We briefly introduce three network features used to examine the network structure. Firstly, the characteristic path length from nodes i to j is defined as $\ell = \frac{2}{N(N-1)} \sum_{i<j} \ell_{ij}$, where ℓ_{ij} is the shortest path length between i and j, and N is the total number of nodes [11]. This quantity is used to characterize the size of the network. Secondly, the clustering coefficient $C_i = 2m_i/[n_i(n_i - 1)]$ at node i is a measure of connectivity to its neighbour nodes [11], where n_i is the number of neighbours, and m_i is the number of links that connect to the neighbours. The clustering coefficient of the entire network is defined as $C = \frac{1}{N} \sum_{i=1}^{N} C_i$. Thirdly, the network entropy represents complexity and stability of a network; here we employ the Shannon entropy $H = -\sum_{k=1}^{N-1} P(k) \log P(k)$, where $P(k)$ is the probability of a node having degree k [17].

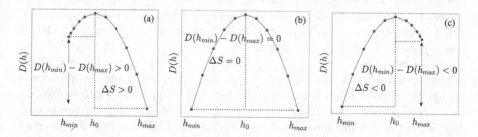

Fig. 1. Multifractal spectra: (a) the signal structure is sensitive to local low fluctuation, and the probability of the highest fluctuation rate surpasses that of the lowest fluctuation rate; (b) the signal structure exhibits left-right reflection symmetry, with equal probability of the minimum and maximum fluctuation rates; (c) the signal structure is sensitive to local high fluctuation, and the probability of the highest fluctuation rate is lower than that of the lowest fluctuation rate [5].

Multifractal Analysis: Many signals in nature are described by multiple values of the Holder exponent h (multifractal) instead of a single value (unifractal). The level of multifractality is defined by the width of a multifractal spectrum

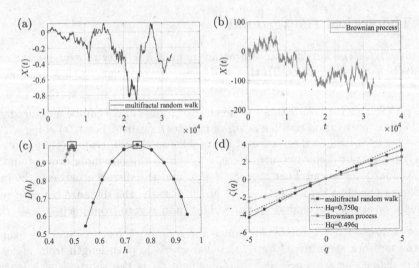

Fig. 2. Multifractal properties compared between a multifractal random walk (blue) and a Brownian process (red), reproduced from Ref. [2]: (a) multifractal random walk, (b) Brownian process, (c) multifractal spectrum, and (d) scaling exponent. (Color figure online)

which indicates a variation of the Holder exponent, as shown in Fig. 1. From many approaches to construct a multifractal spectrum, we have employed the wavelet leader estimate method, in which the power law of the observed data is approximated using the multi-resolution structure function $T(a, t)$, which in turn can yield the structure function $S(q, a)$ by the power law of the time average of the qth power of $T(a, t)$ [7,16]. That is, $S(q, a) = \frac{1}{n_a} \sum_{k=1}^{n_a} |T(a, k)|^q \simeq c_q a^{\zeta(q)}$, where a denotes a scale in the specified range: $a \in [a_m, a_M]$, with $a_M/a_m \gg 1$. $T(a, t)$ is given by wavelet, increment, or box-aggregated coefficients. The Hurst (correspondingly Holder) exponent can be identified through the scaling exponent $\zeta(q)$ in a power series $\zeta(q) = \sum_{p=1}^{\infty} c_p(q^p/p!)$, where c_p are log cumulants, representing the linearity of a multifractal spectrum. For a unifractal process, $\zeta(q)$ yields a linear relation with q such that $\zeta(q) = c_1 q = H_e q$. In other words, we can approximate the Hurst exponent H_e through the slope of the scaling exponent $\zeta(q)$. For instance, Fig. 2 shows comparison of multifractal properties of a multifractal random walk and a Brownian process. It can be seen that the multifractal spectrum of a Brownian process has a narrow width of the Holder exponent range, indicating unifractal with the theoretical value of the Hurst exponent equal 0.5. On the other hand, a multifractal random walk process exhibits a wide multifractal spectrum width containing multiple values of the Holder exponent. The Holder exponent that contributes the highest $D(h)$ represents the slope of the scaling exponent. It is clear that a Brownian process is unifractal as shown from a linear relation of the scaling exponent with q (red dash line in Fig. 2(d)), while the multifractal random walk shows non-linearity with some curvature (blue dash line). Hence, the level of persistence and antipersistence can be inferred and approximated from the Holder exponent $h = c_1$.

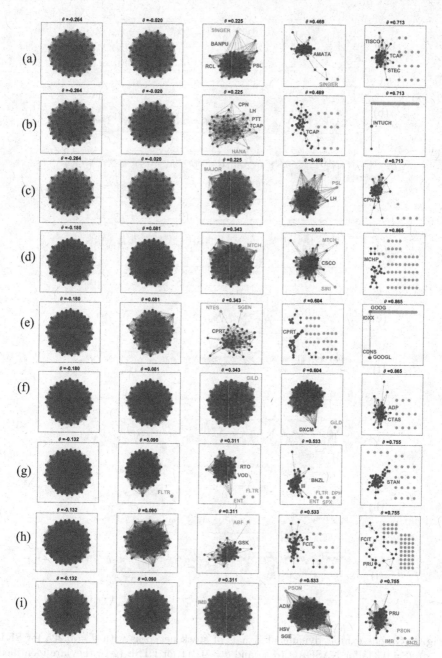

Fig. 3. Network structures represent stocks clustering in SET (a–c), NASDAQ (d–f), and FTSE (g–i) when they approach periods of high and low volatility: (a), (d), and (g) subprime mortgage crisis (2008), (b), (e) and (h) periods of lowest average of squared return (2017), and lastly, (c), (f) and (i) COVID pandemic (2020). Red nodes have the highest number of degrees, while yellow nodes the smallest number of degrees. (Color figure online)

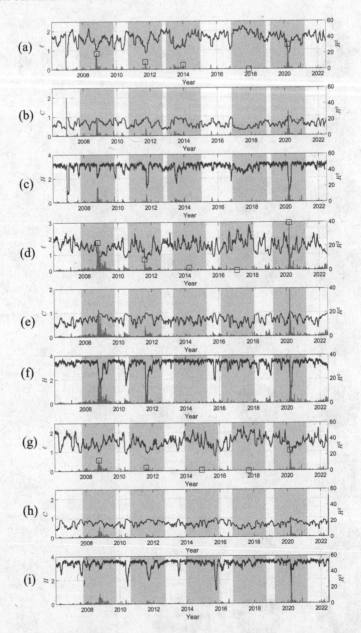

Fig. 4. Graphs exhibit dynamical signals of stock networks with $\theta = 0.225$ for SET (a–c), $\theta = 0.343$ for NASDAQ (d–f) and $\theta = 0.311$ for FTSE (g-i); they are identified with the squared average returns: (a) (d), and (g) the characteristic path length, (b), (e), and (h) the clustering coefficient, and lastly (c), (f), and (i) the network entropy. The gray strips show the periods of the highest and lowest squared returns with 150 data points before and after its maximum peaks denoted by black square boxes in (a), (d), and (f) in 2008, 2011, 2014, 2017, and 2020. Here we consider the squared average return as the instantaneous volatility. (Color figure online)

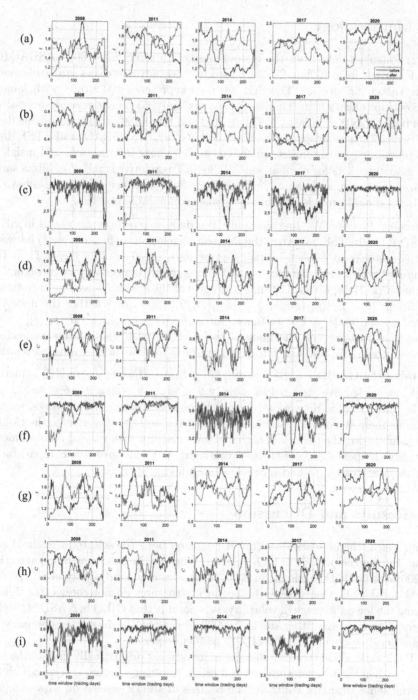

Fig. 5. Graphs represent zoom-in gray regions: (a–c) for SET, (d–f) for NASDAQ, and (g–i) for FTSE. Blue (red) denotes the signal before (after) a crisis. (Color figure online)

3 Methodology

Data: Financial data from three stock markets, namely SET100, NASDAQ100 and FTSE100 from 2006 to 2022 are used in this study. Data sets can be obtained from Yahoo Finance [3]. The SET100 data represents 100 stocks with largest market capitalization in Thailand; however, only 55 stocks provide accessible data back to 2008. Similarly, NASDAQ100 and FTSE100 provide accessible data of 74 and 77 stocks, respectively. The choices of SET, NASDAQ, and FTSE data sets represent a local market, where the authors have information, and insights, and global markets for comparison. While SET is relative small, it exhibits same stylized facts commonly found in stock markets [8]; hence, its collective behaviors of stocks should provide insights into other comparable markets.

Averaging Process: We first compute the daily log returns of stocks in SET, NASDAQ and FTSE, and then the Pearson correlation coefficients $\rho_{ij}(t)$ between any two stocks at time t by using prior data of 30 trading days. Then the computed correlation distance is used as weighted connectivity in constructing the network. Here, we use the time average θ of the ensemble-average correlation coefficient $(\theta = E_t\left[\langle\rho_{ij}(t)\rangle\right] = \sum_{t=1}^{T}\langle\rho_{ij}(t)\rangle/T)$ to decide whether the node i is linked to the node j. For instance, for a given threshold θ', if $|\rho_{ij}(t)| \geq \theta'$, there is a link between the nodes i and j. Otherwise, there is no connection between them. We also can filter the threshold from weak to strong correlation by using the standard deviation $\sigma_{\rho_{ij}} = \sqrt{E_t\left[\langle\rho_{ij}(t)\rangle^2\right] - \left[E_t\langle\rho_{ij}(t)\rangle\right]^2}$. In this work, we varied five thresholds, namely $\theta - 5\sigma_{\rho_{ij}}$, $\theta - 3\sigma_{\rho_{ij}}$, θ, $\theta + 3\sigma_{\rho_{ij}}$, and $\theta + 5\sigma_{\rho_{ij}}$ to visualize network structures. After constructing the networks, we can examine their dynamical features as mentioned. Lastly, we investigate the Holder exponent during the periods of high and low volatility with varying time windows to verify the persistence of the network signals. We focus on the comparison of signals in regression to and recovery from a financial crisis.

4 Results and Discussion

From Fig. 3, the network structure depends on both the selected threshold and time. During the periods of high volatility, such as the subprime mortgage crisis in 2008 and COVID pandemic in 2020, the stocks are bound tightly in SET, NASDAQ and FTSE. Conversely, the network structures do not connect as tightly during the lowest average of volatility. It is notable that GILD in NASDAQ yields larger distances from other stocks in 2020; it is a healthcare-related company, so this is sensible during the COVID-19 pandemic period.

   Panic selling and strategy changes in investment dramatically impact the hub sectors when the market approaches a crisis, resulting in other stocks experiencing the same effect. In contrast, the returns of stocks depend on slow-evolving information sharing, government policy, and news of each stock sector during the calm period. Hence, stocks interact weakly across different stock sectors. Similar dynamical behaviors of all features occur for every selected threshold; so we have depicted only the results for the threshold $\theta = 0.225$ for SET, $\theta = 0.343$ for NASDAQ, and $\theta = 0.311$ for FTSE.

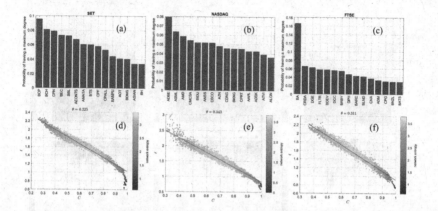

Fig. 6. (a)–(c) the top 15 highest probability of having a maximum degree of stock links with threshold $\theta = 0.225$ for SET, $\theta = 0.343$ for NASDAQ and $\theta = 0.311$ for FTSE; (d)–(f) the linear regression between the clustering coefficients and the characteristic path lengths; the red line shows the best linear fit $\ell = -1.624C + 2.645$ for SET, $\ell = -2.363C + 3.199$ for NASDAQ, and $\ell = -2.037C + 2.95$ for FTSE. (Color figure online)

To confirm the empirical results mentioned above, the dynamical features of network are investigated. As shown in Fig. 4, the characteristic path lengths and the clustering coefficients yield an opposite trend, see also Figs. 6(d)–(f), particularly during the high volatility (gray strips in Fig. 4) periods. This suggests that the network structure decreases in size, and every stock connects tightly due to the same behavior of prices dropping drastically as a result of a financial crisis. Furthermore, the network entropy shows less fluctuation over the period compared to the characteristic path lengths and clustering coefficients. Indeed, it shows a significant decay during a financial crisis, meaning that the network loses its complexity to sustain a diversity of degree distribution and becomes a more complete graph. This behavior is expected during a financial crisis because all stocks behave in the same trend.

To characterize the behaviors of a node, we rank the top 15 probability of having a maximum degree from the highest to the lowest, as shown in Figs. 6(a) for SET, 6(b) for NASDAQ, and 6(c) for FTSE. We find that the top 5 most probably connected stocks in SET are BCP, BCH, CPN, BEC, and BBL which are in the energy, the health care services, the property development, the media, and banking sectors, respectively. In NASDAQ, the most connected stocks are ADBE, ASML, AMD, CMCSA, and BIDU, which are semiconductor and software companies. For the FTSE, they are BA, CRDA, DGE, FLTR and BDEV for aerospace, chemical, alcoholic beverage, entertainment, and property development companies, respectively.

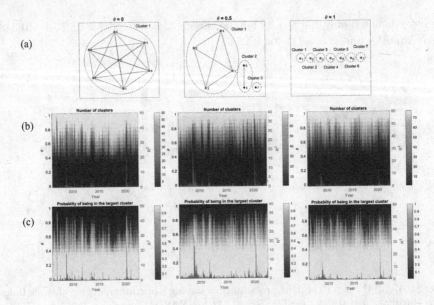

Fig. 7. (a) the number of clusters and the probability of a node being in the largest cluster (P_{clust}). For instance, $P_{clust} = 1, 4/7$, and $1/7$ for each panel, respectively; (b) the heat map of the selected threshold and the number of clusters; (c) the heat map of selected threshold and the probability of a node being in the largest cluster. The red signals represent the average squared returns. Each column of heat map represents data from left to right for SET, NASDAQ, and FTSE.

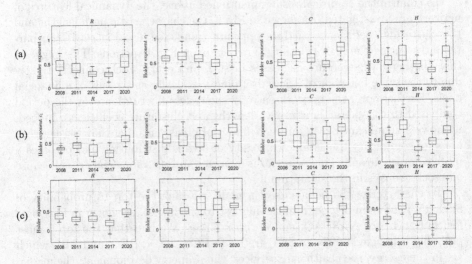

Fig. 8. Box plots of the Holder exponents with varying time windows between 300–500 trading days: (a) SET, (b) NASDAQ, and (C) FTSE.

Furthermore, we examine the relations of the selected threshold θ, the number of clusters, and the probability of a node being in the largest cluster, as shown in Fig. 7, in which panels (b) and (c) show that a small threshold contributes to a small number of clusters; nevertheless, it gives a high probability of a node being in the largest cluster. Also, the number of clusters tends to decrease during the period of high volatility (red peaks) as indicated by the blue area being greater than the yellow area. This implies that the stock network will shrink in size and tends to become a complete graph, as explained earlier. As a result, there is greater probability of a node being in the largest cluster. These results are consistent with those reported in Ref. [12] concerning the behavior of SET network structure when it evolves through a financial crisis. This work finds that similar behaviors are observed in NASDAQ and FTSE. Interestingly, as shown in Figs. 5(a)–(i), the signal of each feature before and after the peak of a crisis shows that it can recover the same value in about 50–100 days after the crisis, if there is no onset of another crisis. Moreover, network features in NASDAQ recovered more quickly in 2020 than those in SET and FTSE, except the network entropy. The concurrence of all three network features is notable.

Examining the Holder exponents of the network signals during the periods indicated by gray strips in Fig. 4 with varying time windows between 300–500 trading days, the variation of Holder exponents with box plots is presented in Figs. 8(a–c), where the periods of high volatility, particularly in 2020, show significantly higher Holder exponents, and are greater than 0.5 for all cases in SET. During such periods, the signals are more persistent than that in 2017 indicating the lowest volatility period. Thus, the network structure and entropy follow the increasing or decreasing trend from the previous signals. This can explain why there is a continuous drop or rise in the network's characteristic path lengths, clustering coefficients, and entropy during a high volatility period. However, the Holder exponents of NASDAQ and FTSE in Figs. 8(b–c) do not exhibit a distinct change at each period. We believe this is because the network features of NASDAQ and FTSE inherit more fluctuation than those of SET, in the sense that the features from SET have smoother trends in signals. Consequently, such fluctuation affects the behaviors of the Holder exponents.

5   Conclusion

We have constructed a dynamical network to examine how the stock network evolves during a financial crisis. The financial returns during a financial crisis show large changes when transiting through a crisis, known as high volatility clustering. This behavior persists for a while before it becomes more stable. From network perspectives, the results suggest that the stock network changes its structures and loses its stability to sustain a diversity of degrees. Each node exhibits increased probability of being in the largest cluster during a financial crisis. Furthermore, the Holder exponent during a period of high volatility is significantly higher than during that of low volatility, suggesting that the network structure and entropy become more persistent. However, we note that the size of selected time window and high fluctuation network signal can influence the behaviors of the Holder exponents. From the network features such as characteristic lengths, clustering coefficients and entropy, the networks of stocks recover the same values approximately 50–100 days after a crisis. Although the network features in SET and FTSE recovered faster than those in NASDAQ in 2008, they did slower in 2020, except for the network entropy.

References

1. List of economic crises. https://en.wikipedia.org/wiki/List_of_economic_crises
2. Wavelet leader multifractal analysis with Matlab. https://www.mathworks.com/help/wavelet/ug/multifractal-analysis.html
3. Yahoo Finance. https://finance.yahoo.com/. Accessed 15 Oct 2021
4. Boccaletti, S., Latora, V., Moreno, Y., Chavez, M., Hwang, D.U.: Complex networks: structure and dynamics. Phys. Rep. **424**(4), 175–308 (2006)
5. Freitas Cruz, I., Sampaio, J.: Multifractal analysis of movement behavior in association football. Symmetry **12**(8), 1287 (2020)
6. Jacob, R., Harikrishnan, K.P., Misra, R., Ambika, G.: Measure for degree heterogeneity in complex networks and its application to recurrence network analysis. Roy. Soc. Open Sci. **4**(1), 160757 (2017)
7. Jaffard, S., Lashermes, B., Abry, P.: Wavelet leaders in multifractal analysis. In: Qian, T., Vai, M.I., Xu, Y. (eds.) Wavelet Analysis and Applications. Applied and Numerical Harmonic Analysis, pp. 201–246. Birkhäuser, Basel (2007)
8. Jaroonchokanan, N., Termsaithong, T., Suwanna, S.: Dynamics of hierarchical clustering in stocks market during financial crises. Phys. A **607**, 128183 (2022)
9. Li, B., Pi, D.: Analysis of global stock index data during crisis period via complex network approach. PLOS One **13**(7), 1–16 (2018)
10. Nie, C.X., Song, F.T.: Constructing financial network based on PMFG and threshold method. Phys. A **495**, 104–113 (2018)
11. Nobi, A., Lee, S., Kim, D.H., Lee, J.W.: Correlation and network topologies in global and local stock indices. Phys. Lett. A **378**(34), 2482–2489 (2014)
12. Saichaemchan, S., Bhadola, P.: Evolution, structure and dynamics of the Thai stock market: a network perspective. J. Phys. Conf. Ser. **1719**(1), 012105 (2021)
13. Siegenfeld, A., Bar-Yam, Y.: An introduction to complex systems science and its applications. Complexity **2020**, 1–16 (2020)
14. Thitaweera, N., Sinthupinyo, S.: Correlation network analysis in the stock exchange of Thailand (SET). In: 6th International Conference on Machine Learning Technologies, p. 170–176. Association for Computing Machinery (2021)
15. Wang, B., Tang, H., Guo, C., Xiu, Z.: Entropy optimization of scale-free networks' robustness to random failures. Phys. A **363**(2), 591–596 (2006)
16. Wendt, H., Abry, P., Jaffard, S.: Bootstrap for empirical multifractal analysis. IEEE Signal Process. Mag. **24**(4), 38–48 (2007)
17. Yang, M.Y., Ren, F., Li, S.P.: Stock network stability after crashes based on entropy method. Front. Phys. **8**, 163 (2020)

A Community Detection Algorithm Using Random Walk

Rajesh Vashishtha[1], Anurag Singh[1(✉)], and Hocine Cherifi[2]

[1] National Institute of Technology Delhi, New Delhi, India
anuragsg@nitdelhi.ac.in
[2] University of Burgundy, Dijon, France
hocine.cherifi@u-bourgogne.fr

Abstract. Community structure plays an essential role in analyzing networks. Various algorithms exist to find the community structure that scores high on a graph clustering index called Modularity. In divisive community structure algorithms, initially, all the nodes belong to a single community. Each iteration divides the nodes into two groups, and finally, each node belongs to a single community. The main disadvantage of a divisive algorithm is that it is not able to find whether to divide the community further or not. A divisive community detection algorithm is proposed based on the graph spectra that give the termination method for community detection. We rely on Weighted Spectral Distribution (WSD) to divide the network into small sub-network or not. Experiments with various real-world networks show that the proposed method constantly compares favorably with the popular Girvan Newman's community detection algorithm.

Keywords: Community structure · Complex network · Graph spectra · Random walk

1 Introduction

In the last decades, there has been a great deal of research to understand the structure of real-world networks. Network modeling emphasizes the structural properties of the network, e.g., diameter, degree distribution, clustering coefficient, etc. Many real-world networks, such as social and scientific collaboration networks, follow a power-law degree distribution $p(d) \sim d^{-\gamma}$ where d denotes the degree of the node. The value of the exponent γ usually lies within the range, $2 \leq \gamma \leq 3$. The Barabasi Albert (BA) model initially introduces the preferential attachment and growth process. One can also achieve the desired goal using random walks, where the transition probability is proportional to the degree of node reached through a random edge Furthermore, Biased Random Walks allow defining the structural centrality of a node in a network [1–5]. Community structure is another property of paramount importance in real-world networks. Although there is no universal definition of communities in networks, one generally admits

---

The original version of this chapter was revised: a new acknowledgement was added. The correction to this chapter is available at
https://doi.org/10.1007/978-3-031-26303-3_27

T. N. Dinh and M. Li (Eds.): CSoNet 2022, LNCS 13831, pp. 227–235, 2023.
https://doi.org/10.1007/978-3-031-26303-3_20

that relationships within a community are dense while the relationships between different communities are rare [6]. This study focuses on finding communities in complex networks. In this setting, we examine the community detection issue by considering the network as a collection of nodes. Communities are formed by grouping the nodes into a community based on similarity measures.

This work proposes a method for discovering the communities in a network using Graph Spectra. We compare the proposed method with the popular Girvan Newman (GN) algorithm. GN uses Edge Betweenness to find the community in networks. As the complexity of edge betweenness computation is high, it is unsuitable for large networks and does not give so much high Modularity. Experiments show that the proposed method outperforms the GN algorithm. Indeed, the community detection algorithm can handle large networks and exhibits higher Modularity. Section 2 briefly presents some essential background elements of the community structure. Section 3 describes the proposed method for the community detection algorithm based on the graph spectra. Section 4 shows the results and analysis for the proposed community detection algorithm against Girvan-Newman. Section 5 discusses the conclusions and the future scope.

2   Community Detection Algorithms

One can apprehend Community detection as the graph partitioning issue in graph theory. The nodes are divided into groups, so the similarity within the groups is high and low in the inter-groups. The community structure may be categorized into:

- **Overlapping Community structure:** If one object presents more than one groups then, these groups form overlapping community.
- **Non overlapping community structure:** If no objects present more than one groups then, these groups forms a non-overlapping community. It is our case of study.

There are numerous algorithms to find communities in social networks such as Graph partitioning [7,8], Hierarchical clustering [9], Partitional clustering [10], Spectral clustering [11]. This work focus on Modularity-based clustering. These algorithms divide the network into small sub-networks based on a community structure fitness function called Modularity [12]. Modularity is defined as follows:

$$Q = \frac{1}{2|e|} \left[A_{ij} - \frac{r_i r_i}{2|e|} \right] \delta(i,j) \tag{1}$$

There are mainly two types of Modularity based community detection algorithms:

- Agglomerative Algorithm: They are also known as the bottom-up hierarchical algorithm. In this algorithm, all the nodes are initially present in the different communities. Based on the similarity of the vertex, merge the vertex into a single community. One uses a dendrogram to represent the community structure in this type of method.

– Divisive Algorithm: They are also known as the top-down hierarchical algorithm. In this algorithm, all the nodes are initially present in a single community. Based on the dissimilarity of the vertex, they divide the vertex into different communities.

The proposed work uses a divisive approach. Some of the Divisive existing algorithms are,

– **Spectral Clustering Algorithm:** This algorithm is based on the fielder vector of the Laplacian matrix. Fielder vector is the eigenvector corresponding to the Laplacian matrix's second spectrum. The fielder vector is very useful for graph clustering.

Algorithm 1. Spectral clustering Algorithm

1: **Input:** Adjacency matrix of the graph A and diagonal matrix D.
2: **Output:** In node divides into single community.
3:
4: First calculate the Laplacian matrix from the adjacency matrix $L = D - A$.
5: **while** Until all nodes divides into separate cluster. **do**
6:     Find the second eigenvalue and corresponding eigenvector. Some of the entry in the second eigenvector is positive and some entries are negative.
7:     The positive entries corresponding nodes belongs to one cluster and the negative entries corresponding nodes belongs to the another cluster.
8:     Recursively call the same procedure for each cluster.
9: **end while**

– **Girvan Newman Algorithm for community detection.** This algorithm is a divisive algorithm. This algorithm works on the edge of betweenness centrality.

Algorithm 2. Girvan Newman Algorithm

1: **Input:** Adjacency matrix of the network A.
2: **Output:** In node divides into single community.
3: **Procedure:** Calculate the betweenness of the each edge in the network.
4: Select the edge with maximum betweenness and remove it.
5: **if** network is partition into two group. **then**
6:     repeats the procedure for each sub network.
7: **else**
8:     Go to the Step 3.
9: **end if**

---

3   Proposed Methodology

This section explains the proposed method for uncovering the communities in the network. The algorithm is based on graph spectra. It uses Weighted Spectral Distribution (WSD) to check whether the network requires partitioning and whether one uses edge weights to remove the network's edges. It also relies on some sub-procedure for finding the communities.

3.1   Weight Spectral Distribution (WSD) Value for the Network

The WSD metric reflects the structure of the network based on the distribution of its random walk cycle. WSD understands the network structure in a better way [13]. Therefore, it helps in evaluating whether the sub-networks form a better community or not. One calculates the WSD value of a network with the help of the Laplacian matrix. We use the following method to calculate the WSD value for the network.

Step 1: Find the value of t, which is the maximum number of cycles of any length in the network. For example, if a network has three 3-length cycles and one 4-length cycle, the value of t is 3.

Step 2: Find the normalized Laplacian matrix of the network.

$$L(N) = I - D^{-1/2}AD^{-1/2}$$

where, $L(N)$ is the Laplacian matrix of the network. D is the diagonal matrix of the network.

$$L(N) = I - Y$$

where, $Y = D^{-1/2}AD^{-1/2}$. Let, the eigenvalue of the matrix $L(N)$ is λ_j. Hence, the eigenvalue of the matrix y is $1-\lambda_i$, and the value of the WSD of the network is $\sum_j (1-\lambda_j)^t$. WSD value is equivalent to the trace of the matrix Y^t. For example, we calculate the WSD value for the toy network shown in Fig. 1. There are four cycles, one is 4-length, and three are 3-length cycles. Hence, the value of t is 3. The value $t = 3$ is related to the clustering coefficient, and the value of $t = 4$

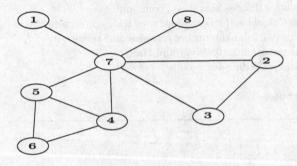

Fig. 1. WSD network

is related to the density of the graph. Density represents low diameter and high connectivity. In Fig. 1, there are eight nodes. The WSD value for the network 1 is $\sum_{j=1}^{8}(1 - \lambda_j)^3 = 0.693$.

3.2 Method for Finding the Centralities of Nodes in the Network

We explain the method used to find the centrality of each node in the network. We associate a sub-network with k-hop neighbors for each node, i.e., all nodes within k distance from the node. Then, we form the Laplacian matrix of this sub-network. Finally, we compute the second smallest eigenvalue of the Laplacian matrix to find the node's centrality using the following equation:

$$c_{node} = \frac{\lambda_2}{log_2(Deg_N)} \tag{2}$$

where λ_2 is the second smallest eigenvalues of the Laplacian matrix. Deg_N be the degree of node N. Let, D be the node in the toy network shown in the Fig. 2(a). The centrality of node D is calculated as follows: we take the small sub-network that is two-hop neighbour of node D shown in Fig. 2.

The Laplacian of sub network is as follows,

$$L = \begin{pmatrix} 2 & -1 & -1 & 0 & 0 & 0 \\ -1 & 2 & -1 & 0 & 0 & 0 \\ -1 & -1 & 3 & -1 & 0 & 0 \\ 0 & 0 & -1 & 2 & -1 & 0 \\ 0 & 0 & 0 & -1 & 2 & -1 \\ 0 & 0 & 0 & 0 & -1 & 1 \end{pmatrix}$$

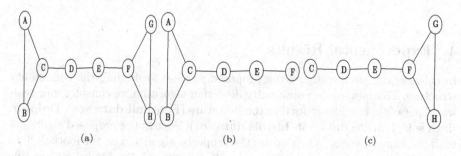

Fig. 2. (a) Toy Network 2, (b) Sub-network with respect to node D, (c) Sub-network with respect to node E

3.3 Edge Weight Computation Using Graph Spectra

Algorithm 3 explains how to calculate the edge weight for each edge in the network. This edge weight is used to find the best removable edge for partitioning the network.

Algorithm 3. Calculating the edge weights using Centrality Method

1: **Input:** Adjacency matrix of the given network.
2: **Output:** Edge weight matrix of the corresponding input network.
3: Calculate the centrality of each node.
4: For each edge in the network. Let, i and j be the end vertices and centrality of node i and j is c_i and c_j then the edge weight W_{ij} of that edge between i and j is calculated as follows:
$$W_{ij} = \sqrt{c_i * c_j}$$
5: Return the edge weight matrix for the input network.

3.4   Community Detection Algorithm Using WSD (CDAW)

Algorithm 4 is used to uncover the non-overlapping community structure of the network using Algorithm 3.

Algorithm 4. Algorithm CDAW

1: Find the WSD value for the given input network.
2: If the WSD value of the network is greater than 1, then go to step 3; else no need to partition the network.
3: Calculate the edge weight for each edge present in the remaining network using Algorithm 3.
4: Remove the edge with minimum edge weight and calculate the second eigenvalue of the Laplacian matrix for the remaining network. If the second spectral root of the Laplacian matrix is zero then go to **step 5** else go to step **step 3**.
5: If the second spectral root of the Laplacian matrix is zero, i.e., the network has at least two connected components.
6: Repeat **step 1-5** for each sub-networks.

4   Experimental Results

In this section, we investigate the proposed method for finding the community structure. To evaluate the community detection algorithm we consider four real-world networks commonly used in the literature (**Football data set , Dolphin data set , Karate data-set Lismis data set**). We test the proposed algorithm with the four data sets. When we use the proposed algorithm on the football data set, it divides the football data set into 12 groups, and the Modularity of this community structure is 0.598. When we apply the proposed community detection algorithm to the Dolphin data set, it divides the dolphin data set into three groups, and the Modularity of this community structure is 0.52. When we apply the proposed community detection algorithm to the Karate data set, it divides the Karate data set into four groups, and the Modularity of this community structure is 0.42. When we apply the proposed community detection algorithm to the Lesmis data set, it divides the Lesmis data set into 14 groups, and the Modularity of this community structure is 0.54.

4.1   Comparison with Girvan Newman Algorithm

The proposed algorithm is compared with the Girvan Newman algorithm on the same data sets. The outcomes are shown the Table 1. In this table, we show the Modularity and the number of communities.

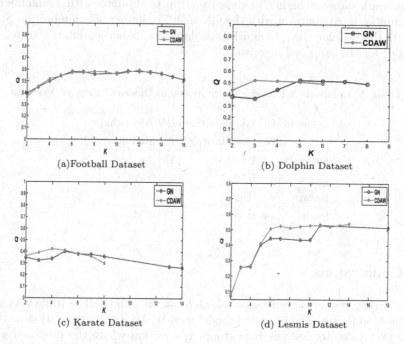

(a)Football Dataset

(b) Dolphin Dataset

(c) Karate Dataset

(d) Lesmis Dataset

Fig. 3. Comparison of the number of communities (K) against the modularity (Q) proposed algorithm with the GN community detection method.

The proposed algorithm provides a better results for getting the good modularity score against the number of communities regarding the number of communities and the Modularity of the community structure. In the football data set, the numbers of communities are 12. GN algorithm gives the maximum Modularity when it divides data into 11 communities. The proposed algorithm provides maximum Modularity when it divides the data into 12 communities, as shown in Fig. 3(a). In the Dolphin data set, the numbers of communities are 3. GN algorithm gives the maximum Modularity when it divides data into five communities. The proposed algorithm gives maximum Modularity when it divides the data into three communities, as shown in Fig. 3(b). In the Karate data set, the number of the community is 2. GN algorithm gives the maximum Modularity when it divides data into five communities. The proposed algorithm provides maximum Modularity when it divides the data into four communities, as shown in Fig. 3(c). In the Lesmis data set, the GN algorithm gives the maximum Modularity when it divides data into 11 communities. The proposed algorithm offers maximum Modularity when it divides the data into three communities, as shown in Fig. 3(d).

The Table 1 contains the conclusion of the Fig. 3. The higher modularity scores are found using the proposed CDAW algorithm for community detection algorithm in comparison to the benchmark Girvan-Newman algorithm. In Football network data-set higher modularity score is obtained with 12 number of communities in comparison with the GN with 11 number of communities. Similarity in the Dolphin, Karate and Lismis data-sets higher modularity scores are obtained for the proposed algorithm.

Table 1. Comparison between the proposed method and the GN Alorithm

Datasets	CDAW algorithm		GN Algorithm	
	K	Modularity Q	K	Modularity Q
Football	12	0.598	11	.58
Dolphin	3	0.52	5	0.51
Karate	4	0.42	5	0.401
lesmis	14	0.54	11	0.53

5   Conclusions

A new divisive algorithm is proposed for finding the communities while maximizing the Modularity. The proposed model uses WSD as the termination method WSD that helps to find the best stopping situation where the proposed algorithm terminates. Preliminary experiments show that it outperforms the GN Algorithm. In the future, another parameter of a real-world network, e.g., average diameter, may be varied. Indeed, one can control the average diameter of a network with a biased random walk. We plan to incorporate the proposed community detection algorithm into a recommendation system.

Acknowledgements. This work is supported by Science and Engineering Research Board (SERB), DST, Government of India under MATRICS project (Fill No. MTR/2019/000631).

References

1. Courtain, S., Leleux, P., Kivimäki, I., Guex, G., Saerens, M.: Randomized shortest paths with net flows and capacity constraints. Inf. Sci. **556**, 341–360 (2021)
2. Chakraborty, D., Singh, A., Cherifi, H.: Immunization strategies based on the overlapping nodes in networks with community structure. In: Nguyen, H.T.T., Snasel, V. (eds.) CSoNet 2016. LNCS, vol. 9795, pp. 62–73. Springer, Cham (2016). https://doi.org/10.1007/978-3-319-42345-6_6

3. Kumar, M., Singh, A., Cherifi, H.: An efficient immunization strategy using overlapping nodes and its neighborhoods. In: 2018 Companion Proceedings of the The Web Conference, pp. 1269–1275 (2018)
4. Ibnoulouafi, A., El Haziti, M., Cherifi, H.: M-centrality: identifying key nodes based on global position and local degree variation. J. Stat. Mech: Theory Exp. **2018**(7), 073407 (2018)
5. Rajeh, S., Savonnet, M., Leclercq, E., Cherifi, H.: Interplay between hierarchy and centrality in complex networks. IEEE Access **8**, 129717–129742 (2020)
6. Orman, G.K., Labatut, V., Cherifi, H.: Towards realistic artificial benchmark for community detection algorithms evaluation. arXiv preprint arXiv:1308.0577 (2013)
7. Barnes, E.R.: An algorithm for partitioning the nodes of a graph. SIAM J. Algebraic Discrete Methods **3**(4), 541–550 (1982)
8. Shi, J., Malik, J.: Normalized cuts and image segmentation. Departmental Papers (CIS), p. 107 (2000)
9. Friedman, J., Hastie, T., Tibshirani, R.: The Elements of Statistical Learning. Springer Series in Statistics, vol. 1. Springer, New York (2001)
10. Wagstaff, K., Cardie, C., Rogers, S., Schrödl, S., et al.: Constrained k-means clustering with background knowledge. In: ICML, vol. 1, pp. 577–584 (2001)
11. Ng, A.Y., Jordan, M.I., Weiss, Y.: On spectral clustering: analysis and an algorithm. In: Advances in Neural Information Processing Systems, pp. 849–856 (2002)
12. Newman, M.E.J.: Fast algorithm for detecting community structure in networks. Phys. Rev. E **69**(6), 066133 (2004)
13. Fay, D., et al.: Weighted spectral distribution for internet topology analysis: theory and applications. IEEE/ACM Trans. Netw. (ToN) **18**(1), 164–176 (2010)

Learning Heuristics for the Maximum Clique Enumeration Problem Using Low Dimensional Representations

Ali Baran Taşdemir⬛, Tuna Karacan⬛, Emir Kaan Kırmacı⬛,
and Lale Özkahya(✉)⬛

Department of Computer Engineering, Hacettepe University, Ankara, Turkey
alibaran@tasdemir.us, tunakx@gmail.com, ekaankirmaci@gmail.com,
lale.ozkahya@gmail.com

Abstract. Approximate solutions to various NP-hard combinatorial optimization problems have been found by learned heuristics using complex learning models. In particular, vertex (node) classification in graphs has been a helpful method towards finding the decision boundary to distinguish vertices in an optimal set from the rest. By following this approach, we use a learning framework for a pruning process of the input graph towards reducing the runtime of the maximum clique enumeration problem. We extensively study the role of using different vertex representations on the performance of this heuristic method, using graph embedding algorithms, such as Node2vec and DeepWalk, and representations using higher-order graph features comprising local subgraph counts. Our results show that Node2Vec and DeepWalk are promising embedding methods in representing nodes towards classification purposes. We observe that using local graph features in the classification process produce more accurate results when combined with a feature elimination process. Finally, we provide tests on random graphs to show the robustness and scalability of our method.

Keywords: Maximum clique enumeration · Node classification · Node embedding

1 Introduction

Graphs are natural objects to represent complex data in real-life relations, such as social networks, molecular networks, and finance networks. Most challenging problems arise in solving combinatorial optimization questions in a time-efficient way, since most of them are known to be NP-hard. The maximum clique enumeration problem (MCE) seeks to enumerate all subgraphs of maximum size, where all vertices are neighbors of each other. The MCE problem is NP-hard

T. Karacan and E. K. Kırmacı—Equal contribution.
Supported by the TÜBİTAK Project 118E283.

and it is a strengthening of the maximum clique problem, which aims to find the size of the maximum clique only. The maximum clique problem is a well-studied problem and known to be NP-complete among other strong hardness results [3, 25]. As many NP-complete problems defined on graphs arise as real-life problems on networks, the MCE problem is also applicable in the analysis of social network [24], behavioral networks [1], financial networks [2], and citation and dynamic networks [22].

Recently, there has been studies such as [6, 10] finding estimate solutions for these problems using a two-stage framework: 1) embedding the graph vertices into low-dimensional vectors; 2) using machine learning to reduce the size of the input graph via vertex classification for finding an approximate solution efficiently. This approach is shown to be applicable for very dense and large real-world networks and scalable to different domains.

Our Contribution: In this work, we build upon this approach by studying the role of using different vertex representations towards estimating the solution of the MCE problem. We conduct our experiments two-fold, representing vertices using: 1) graph embedding algorithms: Node2vec [11], DeepWalk [20], and GraphSage [12], 2) local frequencies of higher-order graph structures as learning features. We make use of these methods above to represent the vertices in the input graph as low-dimensional vectors. Later these vectors are used in a binary classification of the vertices to predict which vertices are in a maximum clique. This helps to remove the vertices that are least likely to be in a maximum clique from the graph so that the input size is reduced.

Our results show that Node2Vec and DeepWalk are promising embedding methods in representing nodes towards classification purposes. We also observe that using local counts of graphs in Fig. 1 as vertex features provides high classification accuracy when combined with a feature elimination process. Our method is tested on real world networks to show the high accuracy of our results. We show the robustness and scalability of our results on random graphs.

2   Framework

2.1   Low Dimensional Vertex Representations

As the initial step, we obtain the vector representation of each vertex in the input graph. Graph embeddings are used to transform the vertices of a graph to vectors and a good embedding is expected to capture graph topology, such as degree distribution, and clustering coefficient. Thus, we use 3 different embedding methods, Node2vec [11], DeepWalk [20], and GraphSage [12], to represent the vertices as vectors. The first two use a strategy based on random walks preserving the connectivity information between a vertex and its neighbors. GraphSage is an algorithm focusing on inductive learning tasks and represents vertices by aggregating features from their neighbors. In addition to embedding techniques, we use the local frequencies of all 5-vertex induced subgraphs (graphlets) in Fig. 1 as vertex features, using the state-of-the-art algorithm Evoke [18]. As seen in Fig. 1, mostly there is more than one role/position of a vertex in a graphlet as distinguished by different colors in Fig. 1. Each of these positions is called an *orbit* and each vertex occurs in one of the orbits in a graphlet. The *orbit frequency* at a vertex is the number of times a vertex occurs in one of these orbits and each of these frequencies is considered as a distinct vertex feature.

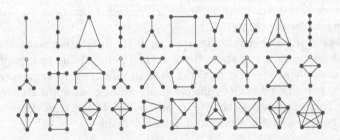

Fig. 1. The 5-vertex subgraphs and orbits whose local counts are used as graph features using the Evoke algorithm in [18].

2.2 Classification Procedure

The vertices are classified according to their presence in a maximum clique. If the vertex is in a maximum clique, we label that vertex with label-1. If the vertex is not in a maximum clique, we label that vertex with label-0. The purpose of this classification is to find the vertices that will be pruned from the graph. The goal is to train a model by obtaining a mapping $\gamma : V \rightarrow \{0, 1\}$ using the training set as $T = \{<f(v_i), y_i>\}_{i=1}^{L}$ with L samples, where $f(v_i)$ indicates the feature vector associated with $v_i \in V$ and $y_i \in \{0, 1\}$ is the class label for for that vertex.

In this learning process, a probabilistic classifier P is used to obtain a probability distribution over $\{0, 1\}$ for every given $f(u)$, $u \in V$. With the help of this and defining a *confidence threshold* $q \in [0, 1]$, the vertex set is pruned to reduce the size of the graph as follows. By choosing a well-performing confidence threshold q, the vertices to prune are defined by the set $V' = \{u \in V : P(u = 1) \leq q\}$. In that sense, picking a higher value of q implies a higher pruning ratio. We experiment with various classifiers via scikit-learn that is an open-sourced library providing tools for supervised and unsupervised machine learning [19]. To obtain the confidence values, we use predict-proba method in the scikit-learn implementation which outputs the probability that a vertex is in a maximum clique.

Vertex Classification: For each network G_i in the training set, we list all maximum cliques $\mathcal{C}_i = \{C_1, C_2, ..., C_n\}$ in G_i, and label the vertices contained in these cliques with 1. To create a balanced dataset, we use the in-sample approach and randomly pick 1.5 times of this many vertices from $G_i \backslash \mathcal{C}_i$ and label them with 0. We use mainly two classifier algorithms based on their performance in the experiments: logistic regression and random forest. For logistic regression, we train with a 5-fold cross-validation for a maximum of 100 epochs, using one of L1 or L2 regularizers and 0.5 or 1 as the regularization term multiplier determined by a grid search. For random forest, we train with a 5-fold cross-validation with 5, 25, or 100 estimators and using "gini" or "entropy" split quality function determined by a grid search.

Evaluation Metrics: The *pruning ratio* P describes the overall ratio of the number of vertices predicted to not be in a solution with probability at least q. The *clique number* of a graph G is the number of vertices in a maximum clique of G, denoted by $\omega(G)$. The *clique accuracy* is the overall ratio of the number of graphs G with $\omega(G) = \omega(G')$ in a sampling set, where G' is the graph obtained by pruning.

3   Experimental Results

In the experiments, we use the state-of-the-art algorithm cliquer [16] to find cliques, which uses the branch-and-bound algorithm of [17]. To calculate clique counts on graphs, we use a computer with Amd Epyc 7B12 (2.25 GHz) processor and 102 GB memory, running Debian 9. Cliquer requires large memory for big networks. Hence, alternatively, we also use the igraph-python library [5] to enumerate maximum cliques, which is a software implementation of a modified Bron-Kerbosch algorithm [7]. For that process, we use a computer with an Intel Core i7 processor and 32 GB of memory.

3.1   Results on Real-World Graphs

For testing the performance of our framework on real-world networks, we chose three network domains from [21]: biological networks, social networks, and web networks, that contain 26, 93, and 12 networks, respectively. We train a distinct classifier for each particular representation method and particular network domain. The classifiers for the biological networks, social networks, and web networks use a sample of 25, 90, and 10 networks, respectively. The F1 scores for all classifiers are above 80% (around 90% in average over this group) except classifiers trained using Graphsage embedding vectors and Evoke vertex feature vectors on the Social networks domain, for which the F1 score is around 55%.

Node Embedding Specifications: We use python implementations of Node2vec and DeepWalk in [4] and [23]. For both DeepWalk and Node2vec, the representation sizes of all vertices, the walk number, the walk length, the skip-gram window size, and the number of parallel processes are set to 128, 5, 10, 10 and 8, respectively. The value of the return parameter p is chosen as 0.25 and the in-out parameter q as 0.75. For GraphSage, we perform unsupervised learning by using the mean aggregator. We set the representation size for embedding to 50 and use two layers with the size of 50 and L2 normalization after each layer. We use randomly initiated features with a length of 32. The parameters listed above are not optimized and fine tuning the values would help to improve the accuracy and runtime.

Table 1. The clique number (ω), pruning ratio (P), runtime in seconds by igraph (T_i) results on real-world networks (with $|V|$-$|E|$) using $q = 0.4$. Representation methods as Node2Vec (N2V), Deepwalk (DW), GraphSage (GS), and graph features obtained via Evoke (Ev.), respectively.

	bio-WormNet-v3 (16K - 763K)			web-wikipedia2009 (2M - 5M)			web-google-dir (876K - 5M)		
	ω	P	T_i	ω	P	T_i	ω	P	T_i
orig.	121	-	7044	31	-	312	44	-	63
N2V	**111**	**0.77**	5324	25	0.92	267	**34**	**0.85**	54
DW	78	0.6	0.2	**30**	**0.88**	275	30	0.88	52
GS	104	0.40	448	19	0.83	262	29	0.81	52
Ev.	61	0.94	3.2	9	0.97	160	20	0.98	35
[15]	90	0.90	-	31	0.99	-	44	0.97	-

	socfb-A-anon (3M - 24M)			socfb-B-anon (3M - 21M)			socfb-Texas84 (36K - 2M)		
	ω	P	T_i	ω	P	T_i	ω	P	T_i
orig.	25	-	1317	24	-	1166	51	-	98
N2V	11	0.99	1287	18	0.99	921	**49**	**0.93**	23
DW	**18**	**0.94**	992	**23**	**0.75**	935	42	0.38	3.2
GS	18	0.73	1081	16	0.74	865	27	0.72	0.9
Ev.	16	0.69	698	17	0.71	683	29	0.77	2.3
[15]	23	0.94	-	23	0.94	-	44	0.97	-

We present the results on real-world networks in Table 1, where the computation cost of enumerating all maximum cliques in the pruned graph is listed as T_i. The optimal confidence threshold is decided to be $q = 0.4$ considering the trade-off between the accuracy and pruning ratio. We present the performance of four different representation methods and compare it with the state-of-the-art method used in [15] together with the exact values listed as 'original'. For each test network, the best performing model using our method is highlighted. We observe that classifiers using graph embeddings by Node2Vec and DeepWalk as feature vector perform better compared to the other models used for our method. Our method differs from the state-of-the-art method in [15] also by not having a preprocessing stage, which uses a degree-method to eliminate the vertices with relatively low degrees. In that sense, Node2Vec and DeepWalk are observed as promising embedding methods in representing nodes towards classification purposes.

We also observe that the use of higher-order node features needs fine tuning of its parameters such as the subset of features used. In the following section, we introduce a feature elimination method that greatly improves the performance of the classifiers when using the local node features introduced above.

3.2 Feature Elimination Procedure

In order to improve the performance of the results obtained by orbit counts as features, we conduct a feature elimination process. We apply a combination of different feature selection steps, which are RFE (Random Feature Elimination), Univariate Selection, and Pearson Correlation, using the scikit-learn library. Higher-order graph features are selected by using a combination of these techniques. Features are ranked by using each method from 1 to n in descending order, n being the total number of features. The scores given by these different techniques are averaged to obtain a combined score for each feature. Initially, we experiment with feature vectors without ranking them by their performance in each network domain, called "f1", "f2", and "f3". We determine three threshold scores s_1, s_2 and s_3 such that $s_1 > s_2 > s_3$. Then, fi ($i \in \{1, 2, 3\}$) contains all vectors with a score above s_i. Hence, we have $f1 \subset f2 \subset f3 \subset f0$, where being a subset means containment relation in terms of the set of features in the corresponding vector and $f0$ is the vector of all features. Next, we redo these evaluations obtaining scores when the performance of the features are restricted to a particular domain. We introduce the domain-specific vectors as $fx1$, $fx2$, and $fx3$, where x takes the value b, w, and s for the biology, web, and social networks domain, respectively.

Table 2. The information loss vs. pruning ratio for the corresponding feature vectors. The information loss is given by the error percentage in the predicted clique number.

Network	f0		f1		f2		f3		fx1		fx2		fx3	
bio-WormNet-v3	0.50	0.94	**0.00**	**0.80**	**0.00**	**0.81**	**0.00**	**0.81**	0.00	0.75	**0.00**	**0.81**	**0.00**	**0.81**
web-wikipedia-2009	0.71	0.97	0.68	0.96	0.45	0.87	**0.35**	**0.97**	**0.39**	**0.97**	0.35	0.95	0.68	0.98
web-google-dir	0.55	0.98	**0.27**	**0.94**	0.52	0.96	0.52	0.97	**0.27**	**0.97**	0.55	0.97	0.52	0.97
socfb-A-anon	0.36	0.69	0.20	0.72	**0.12**	**0.73**	0.12	0.71	0.16	0.73	0.16	0.58	**0.08**	**0.71**
socfb-B-anon	0.29	0.71	0.17	0.71	**0.17**	**0.72**	0.13	0.69	**0.17**	**0.71**	0.17	0.62	0.21	0.60
socfb-Texas84	0.43	0.77	**0.27**	**0.75**	0.29	0.75	0.27	0.73	0.51	0.81	0.20	0.63	0.31	0.62

In Table 2, we observe that feature elimination procedure increases the accuracy in the predicted clique number when the values are compared with those under $f0$. In this table, we highlight the cases that produce slightly better results than the reference case $f0$ by reducing the information loss remarkably while having comparable pruning ratios. Moreover, the domain-independent feature vectors $f1$, $f2$ and $f3$ seem to provide results that are at least as accurate as the domain-specific feature vectors.

3.3   Robustness and Scalability

The experiments presented below aim to show that our method is able to accurately enumerate maximum cliques even in the cases when the training is done on small instances of graphs or when there is only one maximum clique in the input graph. For that, we generate random graphs for training and testing our model by planting a maximum clique of a particular size in each of them.

The random graph $G(n, p)$, also known as Erdős-Renyi graph [8] is a graph on n vertices, where every edge exists with probability p. We sample all random graphs using the Erdős-Renyi model with $G(n, 1/2)$. To plant a clique of size k, we sample k vertices uniformly at random and insert edges between all vertex pairs in this set. Thus, we guarantee that each network in the dataset contains at least one clique with size at least k.

For every n, the value of the expected clique number k is $2 \log(n)$. Hence, in the training stage, we sample $G(n, 1/2)$ with a planted clique of the order close to k and generate 300 different graphs for the training set corresponding to each different pair (n, k). We use the same process for each of the three embedding methods.

Vertex Classification: The vertices are represented by the feature vectors comprising the local frequencies of all orbits listed in Fig. 1. In the training process, we use all vertices with label-1 (in max-clique) and sample 1.5 times of this amount from vertices with label-0 (not in max-clique). Although efficient algorithms exist [9] for finding these cliques, it is still an open problem to efficiently find the cliques whose order are in the range between $\sqrt{n}$ and $2 \log(n)$. Hence, to show that our method is also able to produce accurate and efficient results in that range, we pick values in our experiments as $(n, k) \in \{(128, 12), (256, 13), (512, 15)\}$, using sample graphs with the same value of n and k.

Table 3. Robustness with fixed n and increasing k, using random forest as the classifier algorithm and $q = 0.4$. For each planted clique of order $k + 1$, $k + 2$ and $k + 3$, the pruning ratio, P, and the clique accuracy, A_C, show average values.

Training sets		k + 1		k + 2		k + 3	
n	k	P	A_C	P	A_C	P	A_C
128	12	0.94	0.73	0.93	0.96	0.93	1.0
256	13	0.89	0.34	0.89	0.48	0.89	0.54
512	15	0.87	0.05	0.87	0.15	0.87	0.20

In the testing stage, we use random graphs with a growing planted clique size as $k' = k + 1, k + 2, k + 3$, where for each value of k' 100 samples are used, shown in Table 3. The clique accuracy indicates the ratio of the sample graphs whose maximum clique size does not change after the pruning stage.

As expected, the performance of the classifiers improves for larger values of k'. This is also observed on the performance of the algorithms solving the planted clique problem for large k [13,14]. For larger values of n the accuracy drops, since existence of a planted clique causes a more significant deviation from the expected value of ω as n increases. We observe speedups of the order 100x when $q = .40$ is used.

4   Conclusions

We propose a heuristic method to find an approximate solution of the Maximum Clique Enumeration problem by narrowing the search space. Our experiments show that graph embedding algorithms Node2Vec and DeepWalk are suitable choices for representing the vertices in this framework as well as using higher-order graph features. Our method is also shown to be robust and scalable by extensive experiments on real world and random networks. The accuracy and runtime results can be further improved by optimizing the hyperparameters in the embedding process and incorporating preprocessing techniques.

Acknowledgements. This research was supported in part by the TÜBİTAK Project 118E283.

References

1. Bernard, H.R., Killworth, P.D., Sailer, L.: Informant accuracy in social network data IV: a comparison of clique-level structure in behavioral and cognitive network data. Soc. Netw. **2**(3), 191–218 (1979)
2. Boginski, V., Butenko, S., Pardalos, P.M.: Statistical analysis of financial networks. Comput. Stat. Data Anal. **48**(2), 431–443 (2005)
3. Chen, J., Huang, X., Kanj, I.A., Xia, G.: Strong computational lower bounds via parameterized complexity. J. Comput. Syst. Sci. **72**(8), 1346–1367 (2006)
4. Cohen, E.: Node2vec (2020). https://github.com/eliorc/node2vec
5. Csardi, G., Nepusz, T., et al.: The Igraph software package for complex network research. InterJ. Complex Syst. **1695**(5), 1–9 (2006)
6. Dutta, S., Lauri, J.: Finding a maximum clique in dense graphs via $\chi 2$ statistics. In: Proceedings of the 28th ACM International Conference on Information and Knowledge Management, pp. 2421–2424 (2019)
7. Eppstein, D., Löffler, M., Strash, D.: Listing all maximal cliques in sparse graphs in near-optimal time. In: Cheong, O., Chwa, K.-Y., Park, K. (eds.) ISAAC 2010. LNCS, vol. 6506, pp. 403–414. Springer, Heidelberg (2010). https://doi.org/10.1007/978-3-642-17517-6_36
8. Erdős, P., Rényi, A.: On random graphs i. Publ. Math. (Debrecen) **6**, 290–297 (1959)
9. Feldman, V., Grigorescu, E., Reyzin, L., Vempala, S.S., Xiao, Y.: Statistical algorithms and a lower bound for detecting planted cliques. J. ACM (JACM) **64**(2), 1–37 (2017)
10. Grassia, M., Lauri, J., Dutta, S., Ajwani, D.: Learning multi-stage sparsification for maximum clique enumeration. arXiv preprint arXiv:1910.00517 (2019)

11. Grover, A., Leskovec, J.: node2vec: scalable feature learning for networks. In: Proceedings of the 22nd ACM SIGKDD International Conference on Knowledge Discovery and Data Mining, pp. 855–864 (2016)
12. Hamilton, W.L., Ying, R., Leskovec, J.: Inductive representation learning on large graphs. In: Proceedings of the 31st International Conference on Neural Information Processing Systems, pp. 1025–1035 (2017)
13. Jerrum, M.: Large cliques elude the metropolis process. Random Struct. Algorithms **3**(4), 347–359 (1992)
14. Kučera, L.: Expected complexity of graph partitioning problems. Discret. Appl. Math. **57**(2–3), 193–212 (1995)
15. Lauri, J., Dutta, S.: Fine-grained search space classification for hard enumeration variants of subset problems. In: Proceedings of the AAAI Conference on Artificial Intelligence, vol. 33, pp. 2314–2321 (2019)
16. Niskanen, S., Östergård, P.R.: Cliquer user's guide: version 1.0. Helsinki University of Technology Helsinki, Finland (2003)
17. Östergård, P.R.: A fast algorithm for the maximum clique problem. Discret. Appl. Math. **120**(1–3), 197–207 (2002)
18. Pashanasangi, N., Seshadhri, C.: Efficiently counting vertex orbits of all 5-vertex subgraphs, by evoke. In: Proceedings of the 13th International Conference on Web Search and Data Mining, pp. 447–455 (2020)
19. Pedregosa, F., et al.: Scikit-learn: machine learning in python. J. Mach. Learn. Res. **12**, 2825–2830 (2011)
20. Perozzi, B., Al-Rfou, R., Skiena, S.: DeepWalk: online learning of social representations. In: Proceedings of the 20th ACM SIGKDD International Conference on Knowledge Discovery and Data Mining, pp. 701–710 (2014)
21. Rossi, R.A., Ahmed, N.K.: The network data repository with interactive graph analytics and visualization. In: AAAI (2015). http://networkrepository.com
22. Stix, V.: Finding all maximal cliques in dynamic graphs. Comput. Optim. Appl. **27**(2), 173–186 (2004)
23. THUNLP: OpenNE (2018). https://github.com/thunlp/OpenNE
24. Wasserman, S., Faust, K., Iacobucci, D.: Social Network Analysis: Theory and Methods (1995)
25. Zuckerman, D.: Linear degree extractors and the inapproximability of max clique and chromatic number. In: Proceedings of the Thirty-Eighth Annual ACM Symposium on Theory of Computing, pp. 681–690 (2006)

Optimization

Competitive Influence Maximisation with Nonlinear Cost of Allocations

Sukankana Chakraborty$^{(\boxtimes)}$ (iD) and Sebastian Stein (iD)

University of Southampton, Highfield Campus, Southampton SO17 1BJ, UK
{sc8n15,ss2}@ecs.soton.ac.uk

Abstract. We explore the competitive influence maximisation problem in the voter model. We extend past work by modelling real-world settings where the strength of influence changes nonlinearly with external allocations to the network. We use this approach to identify two distinct regimes—one where optimal intervention strategies offer significant gain in outcomes, and the other where they yield no gains. The two regimes also vary in their sensitivity to budget availability, and we find that in some cases, even a tenfold increase in the budget only marginally improves the outcome of an intervention in a population.

Keywords: Influence maximisation · Nonlinear model · Voter dynamics

1 Introduction

Our interactions with our peers often impact our personal choices, behaviours and opinions. Social networks have therefore attracted considerable attention as a medium to control collective behaviours in populations through external interventions [17]. A key challenge in this aspect is to determine the optimal allocation of external influence on the network that can maximise the outcome of an intervention in a population. The influence maximisation approach addresses this problem by exploiting interpersonal ties in a social network, to maximise the adoption of an innovation or a behaviour in a population [11]. It is typically framed as an optimisation problem that identifies the most influential individuals in a network who can maximise the spread of a desired behaviour in the

This research was sponsored by the U.S. Army Research Laboratory and the U.K. Ministry of Defence under Agreement Number W911NF-16-3-0001. The views and conclusions contained in this document are those of the authors and should not be interpreted as representing the official policies, either expressed or implied, of the U.S. Army Research Laboratory, the U.S. Government, the U.K. Ministry of Defence or the U.K. Government. The U.S. and U.K. Governments are authorized to reproduce and distribute reprints for Government purposes notwithstanding any copyright notation hereon. The authors would like to thank Dr. Markus Brede for the insightful discussions. The authors acknowledge the use of the IRIDIS High Performance Computing Facility at the University of Southampton for the completion of this work.

T. N. Dinh and M. Li (Eds.): CSoNet 2022, LNCS 13831, pp. 247–258, 2023.
https://doi.org/10.1007/978-3-031-26303-3_22

rest of the population. In the past, influence maximisation has been categorically studied using diffusion models [11]. While these models aptly describe how decisions and behaviours virally spread in social systems, their representation of individual states reflect long-term commitments such as buying a car. The one-off, immutable nature of individual states in these models make them unsuitable for studying settings where individual choices (or opinions) are transient and free of abiding commitments [13]. In contrast, dynamical models are used to study instances where individuals frequently change their states as they interact with their social neighbourhood, and thus is a more appropriate model when studying behaviour and opinion dynamics in social networks[1].

In this paper, we consider a dynamical model known as the voter model to capture influence flow in a population [5,9]. This paradigmatic model is characterised by its simple but effective approach to study reality-based social dynamics [16]. The voter model has attracted considerable attention within influence maximisation research [12,23]. Most of this work however, mimics the traditional setting where a limited budget is used to convert ("seed" or activate) a small number of individuals who subsequently influence the rest of the population. Such an approach focuses on identifying the most influential individuals in the network and abstracts all other information about how the intervention budget (e.g. marketing budget) should be used. In the real world, influence maximisation efforts are typically led with resources such as time and money, and a strategy detailing the optimal distribution of these resources would be a useful result. With this in mind, we assume continuous allocations of resources (e.g. time and money) on the network, where individuals are targeted with varying intensities based on their importance in the influence spread process. We focus on studying the influence maximisation problem in competitive settings, as dynamics in the voter model either converge to an ordered consensus, or reach a fragmented state at equilibrium [2], and consensus is rarely ever achieved in the real world.

It is important to note that when considering traditional methods, the intervention budget constrains the number of nodes "seeded" at the start of the dynamics. In the continuous approach however, resources are spread heterogeneously over the network and thus an explicit relationship between the amount of resources allocated to a node and the strength of influence experienced by them needs to be defined. Past work in this area has strictly assumed this relationship to be linear [4,18], i.e. the amount of allocated resources is directly proportional to the strength of influence experienced by the node. However, this assumption may not consistently apply to all real-world settings. For instance, some interventions take longer to be understood or adopted (e.g. adoption of green technologies), and hence require more resources [6]. Similarly, studies from marketing research show that more resources is not always better, and that the duration and complexity of advertisements often have a *diminishing returns* effect on customer engagement and interest [7]. With this in mind, here we present a novel model that considers nonlinear relationships between the strength of influence and external allocations. So far nonlinearity has only been considered in the

---

[1] For a comprehensive review see [2].

spread dynamics by introducing noise such as in the q-voter model [3] or by adding contrarians to the network [20], and to the best of our knowledge, it has never been studied in terms of allocations to the network.

2   Model

We consider a population of N individuals, connected via a social network. The structure of the network is given by a graph $G(V, E)$, where vertices V represent individuals and edges E the relationships between them. Any vertex $i \in V = \{1, 2, \ldots, N\}$ is connected to other individuals in the network $\{j \in V; j \neq i\}$ through a subset of E. Edges between individuals are indicated using weights w_{ij}. Here we consider unweighted and undirected graphs, where binary weights are used to capture the structure of the network, such that $w_{ij} = 1$, if i and j have an edge between them, else $w_{ij} = 0$. Additionally, W here is symmetric as we consider undirected graphs.

We explore a setting where two controllers (A and B) compete to maximise their influence (or opinions) in the population. At any given point in time, individuals in the network strictly adhere to one of two opinions (A or B), corresponding to each controller. Opinions are characterised using binary state variables $\sigma_{A,i}(t) \in \{0, 1\}$, where $\sigma_{A,i}(t) = 1$ implies node i is in state A, or in state B ($\sigma_{A,i}(t) = 0$) at time t. From here, it follows that $\sigma_{B,i}(t) = 1 - \sigma_{A,i}(t)$. Controllers maximise their opinion shares in the population by influencing the network externally. Here we assume a nonlinear relation between allocations and the strength of influence experienced by the node. For any node i, external influence from controllers A and B are $p_{A,i}$ and $p_{B,i}$, when $p_{A,i}^\gamma$ and $p_{B,i}^\gamma$ amount of resources are allocated to it. Influence over the entire network is described using non-negative vectors, $p_A \in \mathbb{R}_+^N$ and $p_B \in \mathbb{R}_+^N$ which are nonlinearly constrained by controller budgets B_A and B_B, as $\sum_i p_{A,i}^\gamma = B_A$ and $\sum_i p_{B,i}^\gamma = B_B$.

Nodes update their opinions using voter dynamics [9], at every time step, a node is selected uniformly at random to update their opinion state where they copy the state of a neighbouring node j with the probability $w_{ji}/(\sum_{j \in \mathcal{N}_i} w_{ji} + p_{A,i} + p_{B,i})$ where $\mathcal{N}_i$ is the immediate social neighbourhood of i, or copy the state of an external controller (say A) with the probability $p_{A,i}/(\sum_{j \in \mathcal{N}} w_{ji} + p_{A,i} + p_{B,i})$. As opinions are stochastic, we approximate the global behaviour in the system by assuming that $x_{A,i}$ is the probability a node i is in state $\sigma_{A,i} = 1$, which gives us the rate at which it chooses to remain in opinion state A as,

$$\frac{dx_{A,i}}{dt} = (1 - x_{A,i}) \frac{\sum_j w_{ji} x_{A,j} + p_{A,i}}{\sum_j w_{ji} + p_{A,i} + p_{B,i}} - x_{A,i} \frac{\sum_j w_{ji}(1 - x_{A,j}) + p_{B,i}}{\sum_j w_{ji} + p_{A,i} + p_{B,i}}. \quad (1)$$

Here the terms $\frac{\sum_j w_{ji} x_{A,j} + p_{A,i}}{\sum_j w_{ji} + p_{A,i} + p_{B,i}}$ and $\frac{\sum_j w_{ji}(1 - x_{A,j}) + p_{B,i}}{\sum_j w_{ji} + p_{A,i} + p_{B,i}}$ quantify the total influence a node i experiences from their immediate neighbourhood and from external controllers in favour of opinions A and B respectively. We estimate the global behaviour of the population by estimating the total share of opinions obtained by each controller at equilibrium. We determine steady-state conditions by setting $\frac{du_{A,i}}{dt} = 0$ in Eq. (1), which for an arbitrary network of size N yields $[L + diag(p_A + p_B)]x_A = p_A$, where L is the Laplacian of the network given by a $N \times N$ matrix with diagonal elements representing the total strength of all edges on a node ($L_{ii} = \sum_j w_{ji}$) and off-diagonal elements are $L_{ij} = -w_{ij}$. The total vote-share obtained by controller A at equilibrium is then given by $X_A = \frac{1}{N}\mathbf{1}^T x_A = \frac{1}{N}\mathbf{1}^T[L + diag(p_A + p_B)]^{-1}p_A$, which can be used to present the formal optimisation problem as,

$$p_A^* = \arg\max_{p_A \in \mathcal{P}} X_A^*(L, p_B), \tag{2}$$

where $\mathcal{P}$ is a set of all possible allocations p_A such that $\sum_{i=1}^{N} p_{A,i}^\gamma = B_A.^2$

3   Methods

We now propose approaches to solve the optimisation problem given in Eq. (2). Depending on the value of γ, the constraint space changes yielding the following *two* settings: (i) when $\gamma > 1$, we observe *diminishing returns* in terms of the strength of influence experienced by nodes as allocations are increased, and (ii) when $0 < \gamma < 1$, we find that individuals take longer (or more resources) to respond to external influence and hence illustrates a *delayed influence* effect. Below we present numerical approaches for each case.

3.1   Numerical Methods

Case $\gamma > 1$: We first consider the *diminishing returns* instance where $\gamma > 1$. The strength of influence experienced by a node is given by the γ-th root of allocated resources $p_{A,i}^\gamma$ and thus the norm of the allocation vector is constrained by $B_A^{1/\gamma}$, implying that the constraint set here is convex in nature. Given that the vote-share X_A is a concave function of allocations p_A [18], the problem we have at hand is therefore a convex optimisation problem. A common approach taken to solve this type of constrained optimisation problem uses the Lagrange method where the objective function is maximised by iteratively stepping in the optimal direction, within the constraint space. To employ this method in our work we follow the approach discussed in [21] which yields the optimal direction as,

$$p_{A,i}(t+1) \leftarrow p_{A,i}(t) + \eta\frac{\nabla_{p_A(t)}X_{A,i}(t)}{p_{A,i}^{\gamma-2}(t)}, \tag{3}$$

---

² The total vote-share is a function of the network structure L, competitor allocations p_B and controller allocations p_A. The maximum vote-share X_A^* is achieved by the optimal allocation vector p_A^*.

where η is the step-length and $\nabla_{p_A(t)} X_{A,i}(t)$ is the gradient of the vote-share function wrt to allocations[3]. To solve the optimisation problem we first initialise a random, feasible allocation vector which then uses Eq. (3) to iteratively update allocations until the total vote-share can no longer be improved, or alternatively we obtain a μ-approximation of the optimal allocation configuration. Note that the allocation vector is also normalised by scaling the entries in p_A at every time step to satisfy the budget constraint $\sum_i p_{A,i}^\gamma = B_A$. The step-length is adjusted using back-tracking[4] to ensure convergence, and given that the problem at hand is a convex optimisation problem, the algorithm is guaranteed to converge to the global maximum.

Case $0 < \gamma < 1$: Next, we consider the *delayed influence* setting where $0 < \gamma < 1$. The constraint set in this case is clearly nonconvex, and hence the Lagrange method cannot be employed here. In general, nonconvex optimisation problems are difficult to solve. In some cases however, the structure of the objective function and the constraint function can be exploited to design polynomial-time algorithms that yield near-optimal solutions [10]. Given that our objective function here is similar to the one shown in [18] and the constraint space is in the shape of an ℓ_p-norm ball where $0 < p < 1$, we employ the projected gradient ascent algorithm proposed in [18], and modify the projection step to meet the nonlinear budget constraint considered in our work. For the projection method, we use the IRBP algorithm[5] which is an instance of a majorisation-minimisation algorithm, where the algorithm iteratively alternates between a majorisation step and a minimisation step[6] until it converges [22]. As this is a nonconvex problem, we do not have any theoretical guarantees of reaching the optimal solution, and therefore we run the algorithm for multiple initialisations of the allocation vector and consider the mean result obtained over all simulations.

4   Results

We now use the above approaches to study the problem in synthetic and real world networks. We also present analytical approaches in synthetic networks to provide benchmarks for our numerical results.

4.1   Analytical Benchmark

First we propose an analytical method to determine optimal allocations. Analytical methods typically apply to simplified network structures, and here we consider the core-periphery network which has a core of highly connected nodes

---

[3] Given by $\nabla_{p_A} X_A = \frac{1}{N} \mathbf{1}^T [L + diag(p_A + p_B)]^{-1} (I - diag(x_A))$.

[4] When $X_A(t+1) < X_A(t)$, the solution $p_A(t+1)$ at time step $(t+1)$ is rejected and the allocation vector is optimised again with an updated step-length $\eta(t+1) = \frac{\eta(t)}{2}$.

[5] See https://github.com/Optimizater/Lp-ball-Projection for more details.

[6] The majorisation step relaxes and linearises the ℓ_p ball to obtain a weighted ℓ_1 ball, and the minimisation step obtains the projection of the point on the ℓ_1 ball.

and other sparsely connected peripheral nodes. The bimodal degree-distribution resembles many real-world leader-follower type networks, and the simplified network structure limits the degrees of freedom which allows us to apply analytical methods to the problem at hand. We employ a degree-based mean-field approximation to determine optimal allocations analytically. The approximation method assumes that nodes with the same degrees have similar behaviours. Nodes are grouped based on their degrees, and the behaviour of the population is approximated by averaging over the behaviours of each group of nodes. Although such an approximation is not always effective, it has been observed to work well in networks where there are no degree correlations [15]. We obtain a degree-based mean-field approximation by following the approach taken in [18], and modifying it to reflect our budget constraint as follows,

$$X_{MF} = \frac{\left(\sum_k \frac{P_k k}{k+a_k^{\frac{1}{\gamma}}+b_k^{\frac{1}{\gamma}}}\right)\left(\sum_k \frac{P_k k a_k^{\frac{1}{\gamma}}}{k+a_k^{\frac{1}{\gamma}}+b_k^{\frac{1}{\gamma}}}\right)}{\sum_k \frac{P_k k (a_k^{\frac{1}{\gamma}}+b_k^{\frac{1}{\gamma}})}{k+a_k^{\frac{1}{\gamma}}+b_k^{\frac{1}{\gamma}}}} + \sum_k \frac{P_k a_k^{\frac{1}{\gamma}}}{k+a_k^{\frac{1}{\gamma}}+b_k^{\frac{1}{\gamma}}}. \qquad (4)$$

Here P_k is the degree-distribution of the network and k is the degree of nodes in the network. Additionally, b_k is the competitor's allocation to the group of nodes with degree k and a_k is the controller's allocation to the same group of nodes. The influence experienced by nodes with degree k from both controllers are $a_k^{1/\gamma}$ and $b_k^{1/\gamma}$, and are uniform across all nodes in a given group. We can now use Eq. (4) to determine the optimal allocation patterns in any large, arbitrary core-periphery network structure.

4.2 Core-Periphery Network

Core-periphery networks of size $N = 1000$ are considered, with a core formed by $P_1 = p_r = 0.25$ (or 25% of the total nodes in the network). Nodes in the highly clustered core have a degree of $k_1 = 30$ and the sparsely connected peripheral nodes have degree $k_2 = 3$. We examine two settings, one where the competitor targets the core and another where they target the peripheral nodes.

Competitor Allocations to the Core: We first consider the instance where B targets the hub nodes. The competitor budget is distributed uniformly across all nodes in the core of the network. Assuming ϵ_A is the fraction of the total budget allocated to hub nodes by controller A, we obtain an expression for total vote-share using Eq. (4), and use semi-analytical methods to determine the optimal allocation ϵ_A^* that maximise the total vote-shares. We also use the computational methods described in Sect. 3.1 to determine optimal allocations numerically. For numerical results, we consider 10 instances of core-periphery networks, each of size $N = 1000$, $p_r = 0.25$, $k_1 = 30$ and $k_2 = 3$. Networks are generated using the configuration model [14]. We consider three budget scenarios: (i) insufficient budget $B_A/B_B = 0.1$, (ii) equal budget $B_A/B_B = 1$ and finally (iii) excess budget $B_A/B_B = 10$. In each case, $B_B = N$.

Figure 1a illustrates the fraction of total resources allocated to the hub nodes in the network, as γ is varied. Figure 1b shows the corresponding vote-shares obtained by the controller. We find a stark contrast in optimal allocations and vote-shares between the two regimes: (i) $0 < \gamma < 1$ and (ii) $\gamma > 1$. We find that allocation strategies are highly sensitive to controller budgets when allocations have a *delayed effect* on influence i.e. $0 < \gamma < 1$. Whereas when allocations yield *diminishing returns* on influence ($\gamma > 1$), optimal strategies do not change significantly even when budgets differ considerably. A similar phenomenon is also reflected in Fig. 1b, where controller budgets significantly affect vote-shares when $0 < \gamma < 1$, and have significantly less effect on vote-shares when $\gamma \gg 1$. Observe that the linear case, $\gamma = 1$, clearly acts the transition point between the two regimes. Taking a closer look at the region where $0 < \gamma < 1$, we find that optimal allocations oscillate between discrete and continuous configurations (i.e. hub nodes are exclusively targeted or entirely avoided) particularly for limited budget. These fluctuations are more abrupt in analytical results, which maybe due to the *all-or-none* strategy adopted in the mean-field approximation, i.e. all nodes with the same degree are uniformly targeted in the analytical app-roach. This assumption also results in discrepancies between the analytical and the numerical results. For instance, when $B_A/B_B = 0.1$ we find that while the analytical solution opts for a discrete strategy (targets only the hub nodes) when $0.3 \leq \gamma \leq 0.5$, the numerical solution does not allocate any resources to the hub nodes in these settings and only targets a fraction of the peripheral nodes. Whereas from Fig. 1b we find that the numerical method yields higher vote-shares than the analytical results in this region, thus highlighting the limita-tions of the mean-field approximation in such instances. Finally, we also observe inconsistencies in Fig. 1a for very high values of γ, i.e. $\gamma > 4$ likely caused by numerical instabilities [1].

Competitor Allocations to the Periphery: We now consider the instance where competitor B targets the periphery, and we illustrate our results in Fig. 1. Once again, we find that controller budgets heavily impact optimal allocations and vote-shares in the region where $0 < \gamma < 1$. The variation in optimal alloca-tions decreases as γ crosses over into the region where $\gamma \geq 1$. We also find that, for the most part numerical results closely replicate analytical results, with some exceptions. As before, numerical inconsistencies are observed when $\gamma > 4$. We also observe discrepancies between the analytical and the numerical results in the region $0.4 \leq \gamma \leq 0.6$, and argue that such disparities exist as allocations are artificially constrained in the analytical method, whereas numerical approaches are more flexible, thus yielding higher vote-shares.

4.3   Real-World Collaboration Network

Next, to explore how these results apply to the real world, we explore a col-laboration network among network scientists consisting of $N = 379$ scientists

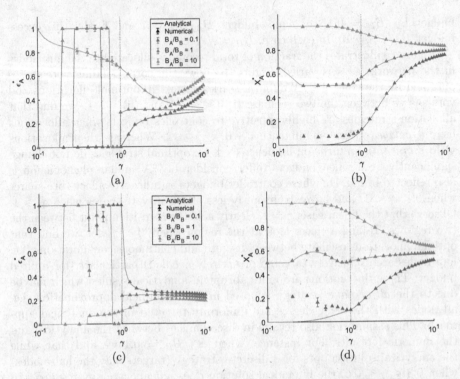

Fig. 1. Figure showing analytical and numerical results for optimal configurations of allocations in a core-periphery network of size $N = 1000$ and average degree $\langle k \rangle = 9.75$. Hub nodes here have degree $k_1 = 30$ while peripheral nodes have degree $k_2 = 3$. The top panel ((a) and (b)) shows results against competitor allocations to the hub nodes. Figure (a) shows the optimal fraction ϵ_A^* of the total budget allocated to hub nodes as γ is varied. Figure (b) shows the corresponding optimal vote-shares X_A^* obtained by the controller. The bottom panel ((c) and (d)) shows results against competitor allocations to the periphery. Figure (c) shows optimal allocations ϵ_A^* to the hub nodes while (d) shows the corresponding vote-shares. Numerical results obtained are a μ-approximation of the optimal solutions where $\mu = 10^{-10}$. Results are averaged over 100 simulations, where algorithms for both the convex and nonconvex optimisation problem are run 10 times with random initialisations of p_A (and $\eta(0) = 1$) on 10 realisations of core-periphery networks. Errorbars show the 95% confidence intervals. The result for $\gamma = 0.1$ and $B_A = 10N$ is missing due to runtime errors caused by numerical overflows.

connected through coauthorship of papers [8,19]. Given that both competitor allocation settings yield similar results in synthetic networks, here for the sake of brevity we focus on the instance where B targets hub nodes. Identifying hub nodes in a heterogeneous network is not a straight-forward process. Here for simplicity, we use the degree centrality measure to distinguish between hubs and peripheral nodes. Given that the average degree of the network is $\langle k \rangle = 4.8$, we assume nodes with degrees above $k > 5$ are hubs, and those with $k \leq 5$

form the periphery of the network. We find that this method classifies nearly 30% of the network as hubs and the rest as the periphery. Since it is infeasible to apply analytical methods to highly heterogeneous network structures, here we rely on numerical approaches to determine optimal allocations. Optimal allocations are determined as the nonlinearity constraint of the allocation vector changes between $0.1 \leq \gamma \leq 10$ and the controller budget is varied between $0.1 \leq B_A/B_B \leq 10$, where $B_B = N$. Results are averaged over 10 simulations, and shown in Fig. 2.

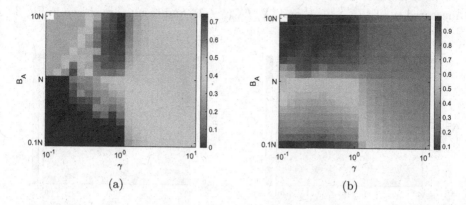

(a)   (b)

Fig. 2. Figure showing optimal allocations and vote-shares in a real-world network depicting collaborations among network scientists ($N = 379$ and average degree $\langle k \rangle = 4.8$). Here the competitor B targets hubs nodes. Figure (a) illustrates the optimal fraction of total budget that is allocated to the hub nodes, and Figure (b) shows the corresponding optimal vote-shares for varying γ and B_A. Numerical results obtained are a μ–approximation of the optimal solutions where $\mu = 10^{-10}$. Results are averaged over 10 simulations, where algorithms for both the convex and nonconvex optimisation problems are run 10 times with random initialisations of p_A and $\eta(0) = 1$. The missing values for $\gamma = 0.1$ and $B_A = 10N$ across the heatmaps are due to numerical overflows.

Figure 2a illustrates the optimal fraction of the total budget allocated to the hub nodes when the competitor targets the hubs. Here too we observe that optimal allocation patterns are highly sensitive to change in budget conditions in the region $0 < \gamma < 1$, whereas optimal allocation patterns are more uniform when $\gamma > 1$. Furthermore, we observe that under low budget conditions, optimal allocation configurations allocate less resources to nodes targeted by the competitor and focus more resources on nodes avoided by the competitor. As the budget increases $B_A > N$, we find that more resources are allocated to the nodes targeted by the competitor. Figure 2b illustrates the optimal vote-shares obtained in this instance. Similar to results in Sect. 4.2, we find that vote-shares vary significantly with budgets when $0 < \gamma < 1$, but not as much when $\gamma \geq 1$.

So far we have observed patterns of optimal allocations in settings with nonlinear budget constraints. We now examine how much vote-share a controller gains from optimally targeting a network as opposed to employing a naïve strategy. For comparison, we consider two simple heuristics: (i) the degree-based approach and (ii) uniform allocation. We consider the collaboration network among network scientists for our simulations. We vary γ and B_A, and in each instance determine the optimal allocation and vote-shares numerically. For comparison, we define the allocation vector for the degree-based approach as $p_{A,i} \propto k_i$, where k_i is the degree of a node i, normalised to meet the budget constraint. Additionally, the allocation vector for uniform allocation is given by $p_{A,i} = (B_A/N)^{1/\gamma}$, $1 \le i \le N$. The corresponding vote-shares are determined for each instance as X_A^{deg} and X_A^{uni} respectively. Gain in vote-shares is then measured as $[X_A^*/X_A^{deg} - 1]$ and $[X_A^*/X_A^{uni} - 1]$, and results are shown in Fig. 3.

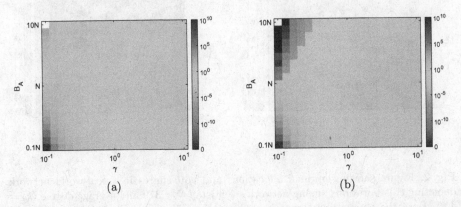

(a)   (b)

Fig. 3. Figures showing the gain in vote-shares obtained when employing optimal allocation strategies in a real-world collaboration network ($N = 379$ and $\langle k \rangle = 4.8$), as opposed to simple heuristics such as (i) degree-based targeting or (ii) uniform allocations. The competitor targets the hub nodes. Figure (a) shows the gain in vote-shares $[X_A^*/X_A^{deg} - 1]$ when the optimal strategy is compared to degree-based targeting, whereas Figure (b) shows the gain in vote-shares $[X_A^*/X_A^{uni} - 1]$ when the optimal strategy is compared to uniform allocations. Numerical results obtained are a μ-approximation of the optimal solutions where $\mu = 10^{-10}$. Results shown are mean values obtained over 10 simulations, where algorithms for both the convex and non-convex optimisation problems are run 10 times with random initialisations of p_A and $\eta(0) = 1$. The missing values for $\gamma = 0.1$ and $B_A = 10N$ across the heatmaps are due to numerical overflows.

Figure 3a and 3b illustrate gain in vote-shares against competitor allocations to the hub. We find that the sensitivity of gain in vote-shares to controller budget is more in the *delayed influence* setting, as compared to the *diminishing returns* setting. In particular, we observe that the controller can gain significant vote-shares ($\approx 10^{10}$ times) by targeting the network optimally for low values of γ and a low budget B_A. This implies that the effectiveness of the optimal strategy in comparison to other heuristics is higher when individuals take longer (or more resources) to experience or respond to external influence under low budget conditions. We further observe that for larger budgets and low γ, the strategy of targeting the network uniformly yields near-optimal results.

5   Conclusions

Here we explore the competitive influence maximisation problem with continuous allocations in the voter model. Contrary to traditional methods, where nodes are typically targeted in a binary fashion, here we consider continuous allocation of influence to the network where an array of nodes are targeted with varying intensities. We assume that two controllers compete to maximise their vote-shares in the network. Traditionally, the influence maximisation problem has been studied in a linear setting where the cost of influence (or allocations) is analogous to the strength of influence experienced by individuals, and thus the relationship between the cost and effect of influence was expressed using a linear function. However, assuming a linear cost function may be an over-simplification that may not apply to all real-world settings. Thus, here we study the competitive influence maximisation problem for nonlinear cost functions, and we consider settings where the effect of influence varies nonlinearly with the cost of influence. Specifically, we consider two scenarios: (i) where increasing allocations diminishes the marginal strength of influence experienced by nodes, and (ii) where nodes take longer or more allocations to start experiencing the effect of influence (observed as the delayed effect of influence). We find that optimal allocations and vote-shares are highly sensitive to budget conditions where allocations have a delayed effect on influence experienced by nodes. On the contrary, when allocations have a diminishing effect on influence, optimal allocations and vote-shares show limited sensitivity to budget conditions. We further show that targeting the network optimally under low budget conditions and under nonlinear budget constraints can result in significant gain in vote-shares, when compared to more naïve approaches. Our results consider optimal allocations only for known competitor allocations, and an interesting future direction for this work would be to study the problem in a game-theoretic framework that assumes incomplete knowledge of competitor allocations.

References

1. Barbero, A., Sra, S.: Modular proximal optimization for multidimensional total-variation regularization. arXiv preprint arXiv:1411.0589 (2014)
2. Castellano, C., Fortunato, S., Loreto, V.: Statistical physics of social dynamics. Rev. Mod. Phys. **81**(2), 591 (2009)
3. Castellano, C., Muñoz, M.A., Pastor-Satorras, R.: Nonlinear q-voter model. Phys. Rev. E **80**(4), 041129 (2009)
4. Chakraborty, S., Stein, S., Brede, M., Swami, A., de Mel, G., Restocchi, V.: Competitive influence maximisation using voting dynamics. In: Proceedings of the 2019 IEEE/ACM International Conference on Advances in Social Networks Analysis and Mining, pp. 978–985 (2019)

5. Clifford, P., Sudbury, A.: A model for spatial conflict. Biometrika **60**(3), 581–588 (1973)
6. Darko, A., Chan, A.P.: Review of barriers to green building adoption. Sustain. Dev. **25**(3), 167–179 (2017)
7. Goldstein, D.G., McAfee, R.P., Suri, S.: The effects of exposure time on memory of display advertisements. In: Proceedings of the 12th ACM Conference on Electronic Commerce, pp. 49–58. ACM (2011)
8. Guimera, R., Danon, L., Diaz-Guilera, A., Giralt, F., Arenas, A.: Self-similar community structure in a network of human interactions. Phys. Rev. E **68**(6), 065103 (2003)
9. Holley, R.A., Liggett, T.M.: Ergodic theorems for weakly interacting infinite systems and the voter model. Ann. Probab. **3**(4), 643–663 (1975)
10. Jain, P., Kar, P., et al.: Non-convex optimization for machine learning. Found. Trends Mach. Learn. **10**(3–4), 142–363 (2017)
11. Kempe, D., Kleinberg, J., Tardos, É.: Maximizing the spread of influence through a social network. In: Proceedings of the Ninth ACM SIGKDD International Conference on Knowledge Discovery and Data Mining, pp. 137–146 (2003)
12. Masuda, N.: Opinion control in complex networks. New J. Phys. **17**(3), 033031 (2015)
13. Moon, S.: Analysis of Twitter unfollow: how often do people unfollow in Twitter and why? In: Datta, A., Shulman, S., Zheng, B., Lin, S.-D., Sun, A., Lim, E.-P. (eds.) SocInfo 2011. LNCS, vol. 6984, pp. 7–7. Springer, Heidelberg (2011). https://doi.org/10.1007/978-3-642-24704-0_6
14. Newman, M.: Networks. Oxford University Press, Oxford (2018)
15. Pastor-Satorras, R., Vespignani, A.: Epidemic spreading in scale-free networks. Phys. Rev. Lett. **86**(14), 3200 (2001)
16. Redner, S.: Reality-inspired voter models: a mini-review. C R Phys. **20**(4), 275–292 (2019)
17. Rogers, E.M.: Diffusion of Innovations. Simon and Schuster, New York (2010)
18. Romero Moreno, G., Chakraborty, S., Brede, M.: Shadowing and shielding: effective heuristics for continuous influence maximisation in the voting dynamics. PLoS ONE **16**(6), e0252515 (2021)
19. Rossi, R.A., Ahmed, N.K.: The network data repository with interactive graph analytics and visualization. In: AAAI (2015). http://networkrepository.com
20. Tanabe, S., Masuda, N.: Complex dynamics of a nonlinear voter model with contrarian agents. Chaos Interdisc. J. Nonlinear Sci. **23**(4), 043136 (2013)
21. Tanaka, T., Aoyagi, T.: Optimal weighted networks of phase oscillators for synchronization. Phys. Rev. E **78**(4), 046210 (2008)
22. Yang, X., Wang, J., Wang, H.: Towards an efficient approach for the nonconvex ℓ_p ball projection: algorithm and analysis. arXiv preprint arXiv:2101.01350 (2021)
23. Yildiz, E., Ozdaglar, A., Acemoglu, D., Saberi, A., Scaglione, A.: Binary opinion dynamics with stubborn agents. ACM Trans. Econ. Comput. (TEAC) **1**(4), 1–30 (2013)

Frank Wolfe Algorithm for Nonmonotone One-Sided Smooth Function Maximization Problem

Hongxiang Zhang[1], Chunlin Hao[1], Wenying Guo[2], and Yapu Zhang[1(✉)]

[1] Institute of Operations Research and Information Engineering,
Beijing University of Technology, Beijing 100124, People's Republic of China
zhanghx010@emails.bjut.edu.cn, {haochl,zhangyapu}@bjut.edu.cn
[2] School of Statistics, Capital University of Economics and Business,
Beijing 100070, People's Republic of China
guowy@cueb.edu.cn

Abstract. In this paper, we study the problem of maximizing a nonmonotone one-sided-η smooth (OSS for short) function $\psi(x)$ under a downwards-closed convex polytope constraint. The concept of OSS was first proposed by Mehrdad et al. [1,2] to express the properties of multilinear extension of some set functions. It is a generalization of the continuous DR submodular function. The OSS property guarantees an alternative bound based on Taylor expansion. If the objective function is nonmonotone diminishing return (DR) submodular, Bian et al. [3] gave a $1/e$ approximation algorithm with a regret bound $O(\frac{LD^2}{2K})$. On general convex sets, Dürr et al. [4] gave a $\frac{1}{3\sqrt{3}}$ approximation solution with $O(\frac{LD^2}{(\ln K)^2})$ regrets. In this paper, we consider maximizing the more general OSS function, and by adjusting the iterative step of the Jump-Start Frank Wolfe algorithm, an approximation of $1/e$ can still be obtained in the case of a larger regret bound $O(\frac{L(\mu D)^2}{2K})$. (where L, μ, D are some parameters, see Table 1). The larger the parameter η we choose, the more regrets we will receive, because of $\mu = \left(\frac{\beta}{\beta+1}\right)^{-2\eta}$ ($\beta \in (0,1]$).

Keywords: Approximation algorithm · One-sided smooth · Nonmonotone · Frank Wolfe

1 Introduction

The concept of one-sided-smooth OSS was first proposed by Mehrdad et al. [1,2] to express the properties of multilinear continuous extension of set functions or nonconvex functions. For example, the diminishing return (DR) submodular function is a case where $\eta = 0$, the multilinear extension [5] of proportionally submodular functions is the case of $\eta = 1$, the multilinear extension of the diversity functions is a more general case ($\eta = 1, 2, 3...$) [1]. One-sided η-smooth functions have an important position in many fields, such as web search [6], machine learning [7,8], document aggregation [9], recommender systems [10,11].

© The Author(s), under exclusive license to Springer Nature Switzerland AG 2023
T. N. Dinh and M. Li (Eds.): CSoNet 2022, LNCS 13831, pp. 259–267, 2023.
https://doi.org/10.1007/978-3-031-26303-3_23

A function ψ is one-sided-smooth if

$$\frac{1}{2}v^T\nabla^2\psi(x)v \le \eta \cdot \frac{\|v\|_1}{\|x\|_1}v^T\nabla\psi(x)$$

for all $x, v \ge 0, x \ne 0$. In this paper, we discuss a nonmonotone OSS function maximization problem:

$$\max\psi(x) \quad s.t.\ x \in [0,1]^n, x \in \mathbb{P} \tag{1}$$

where $\mathbb{P}$ is a downwards-closed convex polytope, and it has a upper bound vector ϖ (i.e. any $x \in \mathbb{P}, x \le \varpi$). $\psi : [0,1]^n \to R_+$ is a nonnegative nonmonotone normalized OSS function. Next, we define the regret function for the OSS maximization problem

$$r \cdot \max_{x\in\mathbb{P}}\psi(x) - \psi(x_K) \le \pi(K), \tag{2}$$

where r is the approximate ratio, $\pi(K)$ is the regret function, K is the number of iterations.

Similar to Lipschitz smoothness [12], one-sided-smoothness can control the approximation ratio and the complexity of related algorithms. The main method to solve the above problem is continuous greedy. The core of continuous greedy is to maximize the multilinear extension function. Submodularity has some beautiful properties for the multilinear continuous extension. For example, monotone concavity along a fixed direction. The property was often used to bound a Taylor expansion in algorithm analysis. Since nonsubmodular multilinear extensions don't have this property, Mehrdad et al proposed a "OSS property" condition to propose an alternative bound based on Taylor series. When the OSS function is monotone, Mehrdad et al. [2] provided a tight $(1 - 1/e^{(1-\beta)(\beta/(\beta+1))^{2\eta}})$ approximation for the maximization problem under downwards-closed convex polytope constraint. For the nonmonotone OSS maximization problem, the research of the algorithm mainly focus on the nonmonotone DR-submodular maximization problem, that is $\eta = 0$. It is difficult to maximize a non-monotone continuous DR-submodular functions. Bian et al and Niazadeh et al. [13,14] have given methods to maximize non-monotone diminishing return (DR) submodular functions and they got the same 1/2-approximation guarantee. Both algorithms come from the double greedy frameworks in [15,16]. In 2019, Bian et al. [3] provided a 1/e approximation algorithm on downwards-closed convex sets. For general convex sets containing origin, Dürr et al. [4] gave a $\frac{1}{3\sqrt{3}}$ approximation solution.

In order to optimize the more general problems (i.e. $\eta > 0$), we propose a nonmonotone Frank Wolfe method. This method comes from the technique of solving convex optimization problems [17,18], and has been widely introduced into the research of submodular optimization and machine learning problems [12,19]. The core of the Frank Wolfe is to approximate the original function by a linear gradient function at each iteration point x_k. If the objective function is monotone, then the gradient function is nonnegative and it was often used to analyze the approximate solution [20]. However, it is not useful for the nonmonotone case. To overcome this problem, we use several optimization tools that include the OSS function is (θ, ϖ)-continuous, gradient is μ-bounded. The above two optimization tools help us establish the connection between two adjacent iteration points and the optimal solution (i.e. $\psi(x^{k+1}) \ge (1-\rho)\psi(x^k)+\rho(1-\rho\mu^2)^{t^k/\rho\mu^2}\psi(z^*) - \frac{L(D\rho\mu)^2}{2}$, The notations used see Table 1). This connection is very helpful for our design of nonmonotone Frank Wolfe algorithm for OSS maximization problems.

Main Contributions: This paper mainly studies whether the approximate ratio $1/e$ can be guaranteed theoretically under the general OSS maximization problem. We design the nonmonotone Jump-Start Frank Wolfe (JSFK) algorithm by applying several techniques, including Jump-Start, nonmonotone Frank-Wolfe, (θ, ϖ)-continuous, μ-bounded gradient. JSFK algorithm iterates by constantly accessing the optimal linear gradient function $d^T \nabla \psi(x)$ ($d \in \mathbb{P}, d \leq \varpi - x$), and finally outputs the solution. We theoretically prove that JSFK algorithm has a $1/e$ approximation ratio with a regret $O(\frac{L(\mu D)^2}{2K})$ for any OSS function maximization problem, and it needs at least $O(K)$ iterations. Because $\mu = \left(\frac{\beta}{\beta+1}\right)^{-2\eta}$, we choose the larger parameter η, the algorithm will receive the more regret (i.e., the result of the algorithm will get worse as the parameter increases).

Main Differences and Limitations: In this paper, our main work is to extend the study of non-monotone continuous DR-submodular functions under the same approximation ratio guarantee. The analysis idea of the algorithm comes from the Algorithm 2 (in [3]) and Algorithm 1 (in [4]), but there are essential differences and limitations.

- Main differences:
    - The analysis of Algorithm 2 (in [3]) and Algorithm 1 (in [4]) strictly depends on the concaveness in the nonnegative direction of the continuous DR-submodular function.
    - Our algorithm iterates from a non-zero point. (It is possible to set the polyhedron constraint to not contain the origin, unless for some k, $x^k \geq z^*$ hold, where z^* denotes the optimal solution).
- Limitations:
    - The analysis of our algorithm requires stronger gradient conditions: μ-bounded continuous.
    - Compared with Algorithm 2 (in [3]), the Jump-Start Frank Wolfe algorithm need to pay more regrets to obtain the same approximate ratio.
    - If the parameter η is too large, the optimization model and its algorithm that we study are meaningless

1.1 Organization

The remainder of this paper is organized as follows. Section 2 introduced some definitions and necessary lemmas for the later algorithm designs. In Sect. 3, a nonmonotone algorithm are proposed for the deterministic nonmonotone OSS problem. The corresponding theoretical analysis for their efficiency are also provided. The last section concludes this work.

2 Preliminaries

The notations used in this paper are listed in Table 1.

In this section, we would offer some notations and definitions, which are used throughout the whole paper. Let ψ be a normalized, nonnegative OSS function. For $\forall u, u' \in [0,1]^n$, $u \leq u'$ if and only if $u_i \leq u'_i$ holds. $u \vee u'$ denotes the coordinate-wise maximum of u and u', and $u \wedge u'$ denotes the coordinate-wise minimum.

Definition 1. Down-Closed: For $\forall x \in \mathbb{P}$, if $0 \leq x' \leq x$, then $x' \in \mathbb{P}$.

To obtain a constant approximation ratio in polynomial time, the constraint set needs to have the property of downwards-closed. Therefor the downwards-closed condition is very important for approximation algorithms analysis process.

Lemma 1 ([2]). $\psi : [0,1]^n \to \mathbb{R}_+$ is OSS on $[x, x + \epsilon v]$, then we have

$$v^T \nabla \psi(x + \epsilon v) \leq \left(\frac{\|x + \epsilon v\|_1}{\|x\|_1}\right)^{2\eta} v^T \nabla \psi(x). \tag{3}$$

where $x, v \in [0,1]^n, x \neq 0$ and $\epsilon > 0$ such that $(x + \epsilon v) \in [0,1]^n$.

Table 1. Symbol description

β, η, ϵ:	Some parameters greater than zero are given in advance;
μ:	$\mu = \left(\frac{\beta}{\beta+1}\right)^{-2\eta}$
$\mathbb{P}$:	Polyhedron: such as linear function polyhedron $Ax \leq b$, matrix polyhedron, etc.;
ϖ:	The upper bound vector of $x \in P$;
u:	A vector belongs to $[0,1]^n$;
D:	The diameter of P, where $D := \max_{x,x' \in \mathbb{P}} \|x - x'\|$, and $D \leq \|\varpi\|$;
L:	The Lipschitz parameter

Definition 2 *([21])*. **Lipschitz smooth:** *A gradient function* $\psi : [0,1]^n \to \mathbb{R}$ *is Lipschitz smooth if for all* $x, y \in [0,1]^n$, *it holds that*

$$\|\nabla\psi(x) - \nabla\psi(x')\| \leq L\|x - x'\|. \tag{4}$$

Lemma 2. *A gradient function* $\psi : [0,1]^n \to \mathbb{R}$ *has L-Lipschitz gradients if for all* $x, x' \in [0,1]^n$, *then we have*

$$|\psi(x') - \psi(x) - \langle \nabla\psi(x), x' - x \rangle| \leq \frac{L}{2}\|x' - x\|^2. \tag{5}$$

We defer to the proofs to the full version.

In order to obtain the convergence approximation solution, we need the gradient function to be Lipschitz continuous. In addition, in order to overcome the difficulties caused by the nonmonotone property of the objective function, the objective function also needs to satisfy the following two conditions.

Definition 3. (θ, ϖ)**-continuous:** *Given* $\theta \in [0, \varpi]$ *(where* ϖ *is the upper bound vector of* $x \in \mathbb{P}$). *Then a function* $\psi : [0,1]^n \to \mathbb{R}_+$ *is* (θ, ϖ)-*continuous if for any* $x \in [0, \theta]$, $y \in \mathbb{P}$, *the following inequality*

$$\psi(x \vee y) \geq \left(1 - \left[\min_{i \in [n]} \frac{\varpi_i}{\theta_i}\right]^{-1}\right)\psi(y). \tag{6}$$

holds.

Definition 4. μ**-bounded continuous:** *A gradient function* $\nabla\psi(x) : [0,1]^n \to \mathbb{R}$ *is* μ-*bounded continuous if for any* $x, y, y' \in \mathbb{P}$, *the following inequality holds*

$$|\langle \nabla\psi(x), y \rangle| \leq \mu \cdot |\langle \nabla\psi(x), y' \rangle| \tag{7}$$

where $\mu = \left(\frac{\beta}{\beta+1}\right)^{-2\eta}$.

The gradient function with a μ-bound provides a bound for the variation of the gradient function to avoid the function being too singular.

Definition 5 *([22])*. *Continuous DR-submodular functions: A continuously twice differentiable function* $\psi : \mathbb{R}_{\geq 0}^n \to \mathbb{R}_+$ *is DR-submodular if it satisfies*

$$\psi(ke_i + x) - \psi(x) \geq \psi((k+l)e_i + x) - \psi(le_i + x),$$

where $k, l \in \mathbb{R}_+$ *and* $x, (ke_i + x), ((k+l)e_i + x) \in \mathbb{R}_{\geq 0}^n$.

Submodularity has some beautiful properties for the multilinear continuous extension. For example, monotone concavity along a fixed direction. The property was often used to bound a Taylor expansion in algorithm analysis. Next we need consider some questions: no monotonicity? no submodularity? no unidirectional concave? Mehrdad et al. proposed a "OSS-property" condition which guarantees an alternative bound based on Taylor series. But they did not consider the case where the function is nonmonotone.

3   Jump-Start Frank Wolfe Algorithm for Nonmonotone Setting

For the nonmonotone OSS problems, we propose Jump-Start Frank Wolfe algorithm, our algorithm mainly uses the Frank Wolfe skill in convex optimization. That is, the following linear optimization problem is solved in each iteration of the algorithm

$$\max_{d \in \mathbb{P}} d^T \nabla \psi(x_k).$$

The algorithm can be briefly described in three parts: (i) Choose an initial feasible solution $x^0 = \beta \arg\max_{x \in P, x \leq \varpi} \|x\|_1, \beta \to 0, \beta \in [0,1)$; (ii) Identify the Iteration direction by $Frank\ Wolfe$ skill; (iii) Travel an acceptable distance in the selected direction.

In Algorithm 1, t^k is the cumulative step size. When $t^k = 1$, x^K is the convex combination of feasible solutions. x^K must be a feasible solution. ϖ is used to bound the growth of x^k (For any $k < K$, we can get $x^k \leq \varpi[1 - (1 - \rho\mu^2)^{t^k/\rho\mu^2}]$), where $\mu = \left(\frac{\beta}{\beta+1}\right)^{-2\eta}$. The following is a specific analysis.

Algorithm 1. JSFW: Jump-Start Frank Wolfe Algorithm($\psi, \mathbb{P}, \beta, \eta, K, \varpi$)

Input: η-OSS function $\psi : [0,1]^n \cap \mathbb{P} \to \mathbb{R}_+$; $\mathbb{P}$: down-closed convex polytope; ϖ: the given bounded vector (i.e. for any $x \in \mathbb{P}$, $x \leq \varpi$ holds).
Parameter: $\beta \in (0,1], \eta \geq 0$, step size $\rho = 1/K, K \approx O(n^2)$. Let $\mu = (\beta/\beta+1)^{-2\eta}$.
Output: x^K.

1: $t^0 \to \beta$.
2: $k = 0$.
3: $x^0 = \beta \arg\max_{x \in \mathbb{P}, x \leq \varpi} \|x\|_1$.
4: **while** $t^k < 1$, **do**
5:    $d^k = \arg\max_{d \in \mathbb{P}, d \leq \varpi - x^k} d^T \nabla \psi(x^k)$, $\rho_k = \min\{\rho, 1 - t^k\}$.
6:    $x^{k+1} = x^k + \mu^2 \rho_k d^k$, $t^{k+1} = t^k + \mu^2 \rho_k$.
7:    $k \leftarrow k + 1$.
8: **end while**

Lemma 3. *Assume $x^0 = \beta \arg\max_{x \in P, x \leq \varpi} \|x\|_1$. For any $k < K$, it holds,*

$$x^k \leq \varpi \left[1 - \left(1 - \rho\mu^2\right)^{t^k/\rho\mu^2} \right], \tag{8}$$

where $\mu = \left(\frac{\beta}{\beta+1}\right)^{-2\eta}$.

We defer to the proofs to the full version.

Lemma 4. *Let z^* denote the optimal solution, then for all x^k in Algorithm 1, we have*

$$\psi(x^k \vee z^*) - \psi(x^k) \leq \mu |(x^k \vee z^* - x^k)^T \nabla \psi(x^k)|. \tag{9}$$

We defer to the proofs to the full version.

Lemma 5. *For any $k < K$, the following inequality holds*

$$\psi(x^{k+1}) \geq (1-\rho)\psi(x^k) + \rho(1-\rho\mu^2)^{t^k/\rho\mu^2} \psi(z^*) - \frac{L(D\rho\mu)^2}{2}, \tag{10}$$

where $\mu = \left(\frac{\beta}{\beta+1}\right)^{-2\eta}$, z^* denotes the optimal solution.

Proof. Let $l^k := x^k \vee z^* - x^k$, then the following two conditions hold: 1) $l^k \leq \varpi - x^k$; 2) $l^k \in \mathbb{P}$, since property of downwards-closed. So l^k is a feasible solution in Algorithm 1). Because $\nabla\psi(x)$ is Lipschitz continuous, from Lemma 2, we have

$$\psi(x^{k+1}) - \psi(x^k) \geq \rho\mu^2 \langle \nabla\psi(x^k), d^k \rangle - \frac{L(\rho\mu^2)^2}{2}\|d^k\|^2$$

(D : The diameter of $\mathbb{P}$)

$$\geq \rho\mu^2 \langle \nabla\psi(x^k), d^k \rangle - \frac{L(\rho\mu^2)^2}{2}D^2$$

(d^k is an ascending direction)

$$\geq \rho\mu^2 |\langle \nabla\psi(x^k), d^k \rangle| - \frac{L(\rho\mu^2)^2}{2}D^2$$

(By the μ − bounded)

$$\geq \rho\mu |\langle \nabla\psi(x^k), l^k \rangle| - \frac{L(\rho\mu)^2}{2}D^2$$

$$= \rho\mu |\langle \nabla\psi(x^k), x^k \vee z^* - x^k \rangle| - \frac{L(\rho\mu)^2}{2}D^2$$

(By the Lemma 4)

$$\geq \rho\mu [\psi(x^k \vee z^*) - \psi(x^k)] \cdot \mu^{-1} - \frac{L(\rho\mu)^2}{2}D^2$$

(By the (θ, ϖ) − continuity and

$\lambda = \min_{i \in [n]} \frac{\varpi_i}{\theta_i}$)

$$\geq \rho[\left(1 - \frac{1}{\lambda}\right)\psi(z^*) - \psi(x^k)] - \frac{L(\rho\mu)^2}{2}D^2$$

$\left(\theta := \varpi\left[1 - (1-\rho\mu^2)^{t^k/\rho\mu^2}\right]\right)$

$$= \rho[(1-\rho\mu)^{t^k/\rho\mu^2}\psi(z^*) - \psi(x^k)] - \frac{L(\rho\mu^2)^2}{2}D^2.$$

Hence

$$\psi(x^{k+1}) \geq (1-\rho)\psi(x^k) + \rho(1-\rho\mu^2)^{t^k/\rho\mu^2}\psi(z^*) - \frac{L(D\rho\mu)^2}{2}.$$

$\square$

Theorem 1. *Consider Algorithm 1 with uniform step size ρ. For $k = 1, ..., K$ it holds that*

$$\psi(x^{k+1}) \geq t^{k+1}e^{-t^{k+1}}\psi(z^*) - \frac{(k+1)L}{2}(\rho\mu D)^2 - O(\rho^2)\psi(z^*). \tag{11}$$

The larger the parameter η we choose, the more regrets we will receive, because of $\mu = \left(\frac{\beta}{\beta+1}\right)^{-2\eta}$ ($\beta \in (0,1]$).

Proof. Firstly, it holds when $k = 0$ (notice that $t^0 = \beta \to 0$). Assume that it holds for k. Then for $k+1$, from $0 \leq \rho\mu \leq t \leq 1$ and Lemma 5, we get

$$\psi(x^{k+1}) \geq (1-\rho)\psi(x^k) + \rho(1 - \rho\mu^2)^{t^k/\rho\mu^2}\psi(z^*) - \frac{LD^2}{2}\rho^2\mu^2$$

$$(Because\ e^{-t} - O(\rho\mu^2) \leq (1 - \rho\mu^2)^{t/\rho\mu^2})$$

$$\geq (1-\rho)\psi(x^k) + \rho\left[e^{-t^k} - O(\rho\mu^2)\right]\psi(z^*) - \frac{LD^2}{2}\rho^2\mu^2$$

$$\geq (1-\rho)\left[t^k e^{-t^k}\psi(z^*) - \frac{kL}{2}(\rho\mu D)^2 - O(\rho^2)\psi(z^*)\right]$$

$$+\rho\left[e^{-t^k} - O(\rho\mu^2)\right]\psi(z^*) - \frac{LD^2}{2}\rho^2\mu^2$$

$$= \left[(1-\rho)t^k e^{-t^k} + \rho e^{-t^k}\right]\psi(z^*) - \frac{(\rho\mu D)^2 L}{2}[(1-\rho)k + 1]$$

$$-[(1-\rho)O(\rho^2) + \rho O(\rho\mu^2)]\psi(z^*)$$

$$\geq \left[(1-\rho)t^k e^{-t^k} + \rho e^{-t^k}\right]\psi(z^*) - \frac{(k+1)(\rho\mu D)^2 L}{2}$$

$$-O(\rho^2)\psi(z^*).$$

Next, let $g(t) = te^{-t}$, the function is monotonically increasing in $[0, 1]$ and $g(t+\rho) - g(t) \leq \rho g'(t)$. Then we get

$$\left[(1-\rho)t^k e^{-t^k} + \rho e^{-t^k}\right]\psi(z^*)$$

$$\geq (t^k + \rho)e^{-(t^k+\rho)}\psi(z^*)$$

$$\geq (t^k + \rho\mu^2)e^{-(t^k+\rho\mu^2)}\psi(z^*)$$

$$= (t^{k+1})e^{-(t^{k+1})}\psi(z^*).$$

So the Theorem 1 holds, that is

$$\psi(x^{k+1}) \geq t^{k+1}e^{-t^{k+1}}\psi(z^*) - \frac{(k+1)L}{2}(\rho\mu D)^2 - O(\rho^2)\psi(z^*)$$

The algorithm termination condition is $t = 1$, and we need about $O(Ln^2)$ number of iterations, when the algorithm terminates, we can get the following solution

$$\psi(x^{out}) \geq e^{-1}OPT - \frac{L(\mu D)^2}{2K} - O(\rho^2)\psi(z^*).$$

Because $\mu = \left(\frac{\beta}{\beta+1}\right)^{-2\eta}$, we choose the larger parameter η, the algorithm will receive the more regret (i.e., the result of the algorithm will get worse as the parameter increases). Most of OSS problems are continuous, and our algorithms can be applied directly. However, if the optimization problems are discrete, then we need to round the corresponding non-discrete results. Generally, the traditional rounding techniques mentioned in the paper [2] are all available, the rounding solution will have a loss.

4   Conclusion

In this paper, we design a nonmonotone Jump Start Frank Wolfe algorithm for nonmonotone OSS maximization problem with a down-closed convex polytope constraint. Our algorithm obtains a e^{-1}-approximation solution with a $\frac{L(\mu D)^2}{2K}$ (where $\mu = \left(\frac{\beta}{\beta+1}\right)^{-2\eta}$) regret. If $\eta = 0$, the result is same as the nonmonotone DR-submodular maximization problem in [5] and the result may not be tight. In order to get the convergent approximation solution, the algorithm requires at least $O(K)$ iterations. We choose the larger parameter η, the algorithm will receive more regret. Our paper also provides a good tool to maximize the multilinear extension of some nonmonotone set functions.

Acknowledgements. The first author is supported by National Natural Science Foundation of China (No. 12131003) and General Research Projects of Beijing Educations Committee in China under Grant (No. KM201910005013).

References

1. Mehrdad, G., Richard, S., Bruce, S.: Beyond submodular maximization, pp. 1–60. arXiv preprint arXiv:1904.09216 (2019)
2. Mehrdad, G., Richard, S., Bruce, S.: Beyond submodular maximization via one-sided smoothness. In: Proceedings of SODA, pp. 1006–1025 (2021)
3. Bian, A., Levy, K., Krause, A., Buhmann, J.: Non-monotone continuous DR-submodular maximization: structure and algorithms. In: Proceedings of NeurIPS, pp. 487–497 (2018)
4. Dürr, C., Thăng, N., Srivastav, A., Tible, L.: Non-monotone DR-submodular maximization over general convex sets. In: Proceedings of IJCAI, pp. 2148–2154 (2021)
5. Chandra, C., Quanrud, K.: Submodular function maximization in parallel via the multilinear relaxation. In: Proceedings of the SODA, pp. 303–322 (2019)
6. Radlinski, F., Dumais, S.: Improving personalized web search using result diversification. In: Proceedings of ACM SIGIR, pp. 691–692 (2006)
7. Ghadiri, M., Schmidt, M.: Distributed maximization of submodular plus diversity functions for multilabel feature selection on huge datasets. In: Proceedings of AISTATS, pp. 2077–2086 (2019)
8. Zadeh, S., Ghadiri, M., Mirrokni, V., Zadimoghaddam, M.: Scalable feature selection via distributed diversity maximization. In: Proceedings of AAAI, pp. 2876–2883 (2017)
9. Abbassi, Z., Mirrokni, V., Thakur, M.: Diversity maximization under matroid constraints. In: Proceedings of ACM SIGKDD, pp. 32–40 (2013)
10. Xin, D., Cheng, H., Yan, X., Han, J.: Extracting redundancy-aware top-k patterns. In: Proceedings of ACM SIGKDD, pp. 444–453 (2006)
11. Carbonell, J., Goldstein, J.: The use of MMR, diversity-based reranking for reordering documents and producing summaries. In: Proceedings of ACM SIGIR, pp. 335–336 (1998)

12. Bian, A., Levy, K., Krause, A., Buhmann, J.: Continuous DR-submodular maximization: structure and algorithms. In: Proceedings of NeurIPS, pp. 486–496 (2017)
13. Bian, Y., Buhmann, J., Krause, A.: Optimal continuous DR-submodular maximization and applications to provable mean field inference. In: Proceedings of ICML, pp. 644–653 (2019)
14. Niazadeh, R., Roughgarden, T., Wang, J.: Optimal algorithms for continuous non-monotone submodular and DR-submodular maximization. J. Mach. Learn. Res. **21**, 1–31 (2020)
15. Buchbinder, N., Feldman, M.: Deterministic algorithms for submodular maximization problems. ACM Trans. Algorithms **14**, 1–20 (2018)
16. Buchbinder, N., Feldman, M., Seffi, J., Schwartz, R.: A tight linear time (1/2)-approximation for unconstrained submodular maximization. SIAM J. Comput. **44**, 1384–1402 (2015)
17. Martin, J.: Revisiting Frank-Wolfe: projection-free sparse convex optimization. In: Proceedings of ICML, pp. 427–435 (2013)
18. Freund, R., Grigas, P.: New analysis and results for the Frank-Wolfe method. Math. Program. **155**, 199–230 (2016). https://doi.org/10.1007/s10107-014-0841-6
19. Zhang, M., Shen, Z., Mokhtari, A., Hassani, H., Karbasi, A.: One sample stochastic Frank-Wolfe. In: Proceedings of ICAIS, pp. 4012–4023 (2020)
20. Feldman, M., Naor, J., Schwartz, R.: A unified continuous greedy algorithm for submodular maximization. In: Proceedings of FOCS, pp. 570–579 (2011)
21. Nesterov, Y.: Introductory Lectures on Convex Optimization: A Basic Course, vol. 87. Springer, New York (2003). https://doi.org/10.1007/978-1-4419-8853-9
22. Bian, Y., Buhmann, J., Krause, A.: Continuous submodular function maximization, pp. 1–64. arXiv preprint arXiv:2006.13474 (2020)

Non-monotone k-Submodular Function Maximization with Individual Size Constraints

Hao Xiao, Qian Liu, Yang Zhou, and Min Li[✉]

School of Mathematics and Statistics, Shandong Normal University,
Jinan 250014, People's Republic of China
{zhouyang,liminemily}@sdnu.edu.cn

Abstract. In the problem of maximizing non-monotone k-submodular function f under individual size constraints, the goal is to maximize the value of k disjoint subsets with size upper bounds $B_1, B_2, \ldots, B_k$, respectively. This problem generalized both submodular maximization and k-submodular maximization problem with total size constraint. In this paper, we propose two results about this kind of problem. One is a $\frac{1}{B_m+4}$-approximation algorithm, where $B_m = \max\{B_1, B_2, \ldots, B_k\}$. The other is a bi-criteria algorithm with approximation ratio $\frac{1}{4}$, where each subset is allowed to exceed the size constraint by up to B_m, and in the worst case, only one subset will exceed B_m.

Keywords: Approximation algorithm · Non-monotone k-submodular function · Individual size constraint

1 Introduction

Let V denote a finite set. A function $f : 2^V \to \mathbb{R}_+$ is submodular if for any $S, T \subseteq V$, we have $f(S) + f(T) \geq f(S \cup T) + f(S \cap T)$. It is generalized to k-submodular function by firstly defining the new domain as the set of k-tuples consisting of disjoint subsets in V, i.e., $(k+1)^V := \{(X_1, \ldots, X_k) \mid X_i \subseteq V(\forall i \in \{1, \ldots, k\}), X_i \cap X_j = \varnothing(\forall i \neq j)\}$. Then the k-submodular function $f : (k+1)^V \to \mathbb{R}_+$ is defined to satisfy

$$f(\boldsymbol{x}) + f(\boldsymbol{y}) \geq f(\boldsymbol{x} \sqcup \boldsymbol{y}) + f(\boldsymbol{x} \sqcap \boldsymbol{y}),$$

for any elements $\boldsymbol{x} = (X_1, \ldots, X_k), \boldsymbol{y} = (Y_1, \ldots, Y_k)$ in $(k+1)^V$, where

$$\boldsymbol{x} \sqcup \boldsymbol{y} = \left(X_1 \cup Y_1 \backslash (\bigcup_{i \neq 1} (X_i \cup Y_i)), \ldots, X_k \cup Y_k \backslash (\bigcup_{i \neq k} (X_i \cup Y_i)) \right),$$

and

$$\boldsymbol{x} \sqcap \boldsymbol{y} = \left(X_1 \cap Y_1, \ldots, X_k \cap Y_k \right).$$

Supported by Natural Science Foundation of Shandong Province of China (Nos. ZR2020MA029, ZR2021MA100) and National Science Foundation of China (No. 12001335).

T. N. Dinh and M. Li (Eds.): CSoNet 2022, LNCS 13831, pp. 268–279, 2023.
https://doi.org/10.1007/978-3-031-26303-3_24

Obviously, when $k = 1$, the k-submodular function is reduced to submodular. For general k, the k-submodular function is more complex, since there are more conditions for the order or position of the items. Moreover, k-submodular function also has very important research value and broad application prospects, such as maximizing influence, sensor placement, and so on [4,6].

For monotone unconstraint k-submodular maximization problem, Ward and Živný [18] proposed a deterministic $\frac{1}{2}$-approximation algorithm based on greedy technique. It was improved to $\frac{k}{2k-1}$ based on randomized algorithm by Iwata et al. [3]. When the function is non-monotone, Ward and Živný presented a $\max\{\frac{1}{3}, \frac{1}{1+a}\}$-approximation algorithm, where $a = \max\{1, \sqrt{\frac{k-1}{4}}\}$. Iwata et al. [3] and Oshima [7] improved this approximate guarantee to $\frac{1}{2}$ and $\frac{k^2+1}{2k^2+1}(k \geq 3)$, respectively. Ohsaka and Yoshida [6] used greedy method to give the approximation ratio of $\frac{1}{2}$ for monotone k-submodular maximization subject to a total size constraint. They also designed a $\frac{1}{3}$-approximation algorithm for monotone k-submodular function subject to individual size constraints, and showed that the time complexity of all these algorithms is $O(nkB)$, where n denotes the cardinality of V and B denotes a parameter about the size constraint. Ene and Nguyen [1] designed a new streaming algorithm for the latter case based on primal and dual technique. Wang and Zhou [17] preserved the same approximation ratios for above results with constraints by introducing the multilinear extension and using continuous greedy method. Recently, Nguyen and Thai [5] proposed some streaming algorithms for the total size constraint under noise. They also showed that there is a $\frac{1}{3}$-approximation algorithm for the non-monotone case. For the problem of maximizing non-monotone k-submodular function with two kinds of constraints, Shi et al. [13] proposed the approximation ratios of $1 - e^{-\frac{1}{k}} - \epsilon$ (total size and individual size difference constraints) and $1 - 1/e - \epsilon$ (individual size constraint), while both the time complexities $O(\frac{n(k-1)!}{\epsilon} \log \frac{n}{\epsilon} + k(k - 1))$ and $O(\frac{nk!}{\epsilon} \log \frac{n}{\epsilon})$ depend on the factorial of k. There are also many results about the k-submodular maximization or minimization with other constraints, such as matroid, knapsack, and so on [2,8–12,14–16,19].

In this paper, we are interested in the problem of maximizing non-monotone k-submodular function under individual size constraints. Given k integers $B_1, B_2, \ldots, B_k$ denoting the individual size bound, we select one $\boldsymbol{x} = (X_1, X_2, \ldots, X_k)$ in $(k+1)^V$ to maximize $f(\boldsymbol{x})$, where $|X_i| \leq B_i$ for each $i \in \{1, 2, \ldots, k\}$. That is, the number of elements in each subset does not exceed the corresponding individual size constraint. In the process of research, we find a property of the solution according to the pairwise monotonicity of k-submodular function. That is, when the solution cannot include any further elements, at least $k - 1$ subsets of the solution satisfy the individual size constraints. By using this property, we present our main contribution about two approximation algorithms. The first one is a $\frac{1}{B_m+4}$-approximation algorithm with complexity $O(nkB)$, where $B_m = \max\{B_1, \ldots, B_k\}$ and $B = \sum_{i=1}^{k} B_i$. This approximation ratio depends on the size bounds, so it will not be so good as the size increases. The other result is given in the sense of bi-criteria, that is, we present a solution $\hat{\boldsymbol{x}} = (X_1, X_2, \ldots, X_k)$ with performance guarantee $\frac{1}{4}$ in polynomial time, but this solution may be against the individual size constraints with $|X_i| \leq B_i + B_m, i = 1, 2, \ldots, k$.

The structure of this paper is as follows. In Sect. 2, we introduce some notations and some related properties of k-submodular function. In Sect. 3, we propose two algorithms for the problem, and give the corresponding results and some proof processes. The omitted proof process is supplemented in the Appendix. In Sect. 4, we list the summary and future research direction.

2 Preliminaries

In order to describe k-submodular function in detail, we introduce several notations. Assume that $V = \{e_1, e_2, \ldots, e_n\}$ is the ground set, $k \geq 1$, $[k] := \{1, \ldots, k\}$ and $(k+1)^V := \{(X_1, \ldots, X_k) \mid X_i \subseteq V(\forall i \in [k]), X_i \cap X_j = \varnothing(\forall i \neq j)\}$ means the set of k-tuples consisting of disjoint subsets in V. $f : (k+1)^V \to \mathbb{R}_+$ is k-submodular function if

$$f(x) + f(y) \geq f(x \sqcup y) + f(x \sqcap y),$$

for any elements $x = (X_1, \ldots, X_k)$, $y = (Y_1, \ldots, Y_k)$ in $(k+1)^V$, where

$$x \sqcup y = \left(X_1 \cup Y_1 \backslash (\bigcup_{i \neq 1}(X_i \cup Y_i)), \ldots, X_k \cup Y_k \backslash (\bigcup_{i \neq k}(X_i \cup Y_i))\right),$$

and

$$x \sqcap y = \left(X_1 \cap Y_1, \ldots, X_k \cap Y_k\right).$$

For any $x = (X_1, X_2, \ldots, X_k)$ in $(k+1)^V$, it can always correspond to an n-dimensional vector $\bar{x}$ containing $\{0, 1, \ldots, k\}$ as follows,

$$\bar{x}_i = \begin{cases} j & \text{if } e_i \in X_j, \\ 0 & \text{if } e_i \notin supp(x), \end{cases}$$

where $supp(x) = \cup_{j \in [k]} X_j$ is the support set composed of all elements in x. So we donnot distinguish the difference between these two symbols and denote them by x. In addition, for a given element $e \in V$ and a position $i(i \in \{0, \ldots, k\})$, we define a new notation $1^{(e,i)}$ to represent a special vector, in which only the component corresponding to the element e is i and the rest components are 0. For $e \notin supp(x)$, we use the notation $x + 1^{(e,i)}$ to represent that the component corresponding to the element e in x changes from 0 to i, and the other components remain unchanged. Similarly, for $e \in X_i$, the notation $x - 1^{(e,i)}$ only changes the component corresponding to the element e in x from i to 0. Therefore, for any $x \in (k+1)^V$, we can define

$$\Delta_{e,i} f(x) = f(x + 1^{(e,i)}) - f(x), \ e \notin supp(x),$$

as the difference between the function value of vector x and $x + 1^{(e,i)}$. f is monotone means

$$\Delta_{e,i} f(x) \geq 0, \ e \notin supp(x), \forall i \in [k].$$

In this paper, f is not needed to be monotone, so we cannot obtain that any element added to any position can increase the marginal value. But there is an important property of k-submodular function named pairwise monotonicity [18],

$$\Delta_{e,i} f(\boldsymbol{x}) + \Delta_{e,j} f(\boldsymbol{x}) \geq 0, \ e \notin supp(\boldsymbol{x}), \forall i, j \in [k], i \neq j.$$

The pairwise monotonicity implies that there is at most one decreasing position when $e \notin supp(\boldsymbol{x})$ is added to $\boldsymbol{x}$. In fact, this property is very important for us to design the algorithms. Moreover, $\forall \boldsymbol{x}, \boldsymbol{y} \in (k+1)^V$, the following orthant submodularity holds [18],

$$\Delta_{e,i} f(\boldsymbol{y}) \leq \Delta_{e,i} f(\boldsymbol{x}), \ \boldsymbol{x} \preceq \boldsymbol{y}, \ e \notin supp(\boldsymbol{y}), \forall i \in [k],$$

where $\boldsymbol{x} \preceq \boldsymbol{y}$ means $X_i \subseteq Y_i$ for all $i \in [k]$. The orthant submodularity means that the marginal return of a smaller vector is not less than the one of bigger vector.

Given a non-monotone k-submodular function f and k integer size bounds $B_1, B_2, \ldots, B_k$, the problem of maximizing non-monotone k-submodular function subject to individual size constraints (denoted by **MkISC** for short) is to find a vector $\boldsymbol{s} \in (k+1)^V$ to maximize $f(\boldsymbol{s})$, where $\boldsymbol{s}$ satisfies the individual size constraints, i.e.,

$$\max_{\boldsymbol{s} \in (k+1)^V} f(\boldsymbol{s}) \quad \text{s. t. } |S_i| \leq B_i \ \forall i \in [k].$$

For any feasible solution $\boldsymbol{s} = (S_1, S_2, \ldots, S_k)$ satisfying $|S_i| = B_i$ for each $i \in [k]$, we call it a *full solution*. Different from the monotone case, we can not decide whether an optimal solution is full or not in non-monotone case. In fact, even for the feasible solution, we do not know whether it is full or not. But, if there are two or more subsets with elements less than the given bounds, we can try to add the element not in its support to one of these subsets. Based on the pairwise monotonicity, there should exist one case that does not decrease the function value. Thus, we can show the following main property.

Property 1. For any feasible solution $\boldsymbol{s} = (S_1, S_2, \ldots, S_k)$ of **MkISC** with maximal support set, it is a full solution or there exists at most one position not full. That is,

$$|\{i \in [k] : |S_i| < B_i\}| \leq 1.$$

3   Main Results

In this part, we will propose two algorithms to solve **MkISC** based on Property 1. The first algorithm introduced in Subsect. 3.1 is to directly and greedily select the elements and their positions, which can ensure the maximum gain of f after adding the element in each iteration. This algorithm mainly follows the algorithm k-Greedy-IS in [6]. But there are many differences from its analysis, especially about the meaning of definition o^j. The second algorithm shown in Subsect. 3.2 is a bi-criteria algorithm by using a combination of two greedy methods to select elements and their positions. Although the latter algorithm leads to violation of the individual size constraints, it can further improve the approximation factor.

Algorithm 1. Greedy algorithm for MkISC

Input: A non-monotone k-submodular function $f : (k+1)^V \to \mathbb{R}_+$ and integers $B_1, \ldots, B_k \in \mathbb{Z}_+$.

output: A vector $s = (S_1, \ldots, S_k)$ with $|S_i| \leq B_i$ for $i \in [k]$.

1: $s \leftarrow \mathbf{0}$ and $B \leftarrow \sum_{i=1}^{k} B_i$
2: **for** j from 1 to B **do**
3:     $I^j \leftarrow \{i \mid |S_i| < B_i, \forall i \in [k]\}$
4:     $(e, i) \leftarrow \arg\max_{e \in V \setminus supp(s), i \in I^j} \Delta_{e,i} f(s)$
5:     **if** $\Delta_{e,i} f(s) \geq 0$ **then**
6:         $s \leftarrow s + \mathbf{1}^{(e,i)}$
7:     **else**
8:         break the for-loop
9:     **end if**
10: **end for**
11: **return** s

3.1 Greedy Algorithm for MkISC

Algorithm 1 is designed based on greedy method. Firstly, in each iteration j, we can see that a greedily selected element is added to the set of indicators in I^j. From the definition of I^j, it can be known that the number of added elements will not exceed the size constraint of any subset. Secondly, considering the process of selected element in each iteration, we divide them into two cases to analyze. One is $|I^j| \geq 2$ and the other is $|I^j| = 1$, where $|I^j|$ refers to the number of elements in I^j. If $|I^j| \geq 2$, from Property 1, there is $i \in I^j$, making $\Delta_{e,i} f(s) \geq 0$, so the elements can be added. Thus, the elements can be added until $k - 1$ subsets reach the size constraints. If $|I^j| = 1$, there is only one subset with elements less than the given bound. If there is an element making the gain of f positive, we add the element to the subset of indicator in I^j. Otherwise, we terminate the algorithm. Therefore, there are at least $k - 1$ subsets of the final output solution of Algorithm 1 meet the size constraints. Assume that s is the final output solution of Algorithm 1, and let $r = |supp(s)|$, we have $B - B_m \leq r \leq B$, where $B_m = \max\{B_1, \ldots, B_k\}$.

For $j \in [r]$, in the j-th iteration of Algorithm 1, let e^j, i^j be the chosen element and position, and s^j be the solution after this iteration, i.e., $s^j = s^{j-1} + \mathbf{1}^{(e^j, i^j)}$, where $s^0 = \mathbf{0}$. Use $o^0 = o$ to denote an optimal solution. For $i \in [k]$, let $S^{(i,j)} := O_i^{j-1} \setminus supp(s^{j-1})$ represent the set composed of the remaining elements in the i-th subset of $o^{j-1} = (O_1^{j-1}, O_2^{j-1}, \ldots, O_k^{j-1})$ after removing the elements from $supp(s^{j-1})$. We denote

$$o^j = \begin{cases} e^j & \text{if } S^{(i^j, j)} = \varnothing \text{ or } e^j \in S^{(i^j, j)}, \\ \text{arbitrary element } e \in S^{(i^j, j)} & \text{otherwise,} \end{cases}$$

and

$$o^{j-1/2} = \begin{cases} o^{j-1} - \mathbf{1}^{(o^j, o^{j-1}(o^j))} & \text{if } S^{(i^j, j)} = \varnothing \text{ or } e^j \in S^{(i^j, j)}, \\ o^{j-1} - \mathbf{1}^{(o^j, i^j)} - \mathbf{1}^{(e^j, o^{j-1}(e^j))} & \text{otherwise,} \end{cases}$$

where $o^{j-1}(o^j)$ and $o^{j-1}(e^j)$ represent the indexes of subsets of elements o^j and e^j in o^{j-1}. At the same time, we define o^j as the constructing solution after the j-th iteration of Algorithm 1, i.e., $o^j = o^{j-1/2} + 1^{(e^j, i^j)}$. From these definitions, one can easily obtain that

$$s^{j-1} \preceq o^{j-1/2}, \forall j \in [r]. \tag{1}$$

We can show the following lemmas.

Lemma 1. *For any iteration $j \in [r]$ of the Algorithm 1, we have the following results:*

    (a) If $|I^j| \geq 2, f(o^{j-1}) - f(o^j) \leq 3\Delta_{e^j, i^j} f(s^{j-1})$,

    (b) If $|I^j| = 1, f(o^{j-1}) - f(o^j) \leq 2f(s^j) - f(s^{j-1})$.

Proof. (a) We will give the proof in three cases according to the construction forms of the vector o^j.

    Case a1. $S^{(i^j, j)} = \varnothing$.

In this case, we have $o^j = e^j, o^{j-1/2} = o^{j-1} - 1^{(o^j, o^{j-1}(o^j))}$, and $o^j = o^{j-1/2} + 1^{(e^j, i^j)}$. Then we give the analysis depending on o^j belonging to the support set of o^{j-1} or not.

If $o^j \in supp(o^{j-1})$, we have $o^{j-1}(o^j) \neq 0$. Moreover, we also have $o^{j-1}(o^j) \in I^j \backslash \{i^j\}$. If not, the position $o^{j-1}(o^j)$ has been full before the iteration j. So, $e^l = o^j$ should belong to $supp(s^l)$ for some $l < j$, which contradicts that it can be chosen during the j-th step. Since $|I^j| \geq 2$, taking any $i' \in I^j \backslash \{i^j\}$, by pairwise monotonicity, we have

$$f(o^{j-1/2} + 1^{(e^j, i')}) + f(o^j) - 2f(o^{j-1/2}) \geq 0. \tag{2}$$

Then, we can get

$$
\begin{aligned}
f(o^{j-1}) - f(o^j) &= f(o^{j-1}) + f(o^{j-1/2} + 1^{(e^j, i')}) - 2f(o^{j-1/2}) \\
&\quad - [f(o^j) + f(o^{j-1/2} + 1^{(e^j, i')}) - 2f(o^{j-1/2})] \\
&\leq \Delta_{o^j, o^{j-1}(o^j)} f(o^{j-1/2}) + \Delta_{e^j, i'} f(o^{j-1/2}) \\
&\leq \Delta_{o^j, o^{j-1}(o^j)} f(s^{j-1}) + \Delta_{e^j, i'} f(s^{j-1}) \\
&= f(s^{j-1} + 1^{(o^j, o^{j-1}(o^j))}) + f(s^{j-1} + 1^{(e^j, i')}) - 2f(s^{j-1}) \\
&\leq 2[f(s^j) - f(s^{j-1})],
\end{aligned}
$$

where the first inequality is due to (2), the second inequality is due to orthant submodularity together with (1), and the third inequality is due to Algorithm 1.

Otherwise, o^j does not belong to the support set of o^{j-1}, i.e., $o^{j-1}(o^j) = 0$. Furthermore, $o^{j-1/2} = o^{j-1}$. Since $|I^j| \geq 2$, taking any $i' \in I^j \backslash \{i^j\}$, similarly, we have

$$
\begin{aligned}
f(o^{j-1}) - f(o^j) &= \Delta_{e^j, i'} f(o^{j-1/2}) - [\Delta_{e^j, i^j} f(o^{j-1/2}) + \Delta_{e^j, i'} f(o^{j-1/2})] \\
&\leq f(o^{j-1/2} + 1^{(e^j, i')}) - f(o^{j-1/2}) \\
&\leq f(s^{j-1} + 1^{(e^j, i')}) - f(s^{j-1}) \\
&\leq f(s^j) - f(s^{j-1}).
\end{aligned}
$$

Case a2. $S^{(i^j,j)} \neq \varnothing$ and $e^j \in S^{(i^j,j)}$.

In this case, we have $o^j = e^j$ and $o^j = o^{j-1}$, then

$$f(o^{j-1}) - f(o^j) = 0 \leq \Delta_{e^j,i^j} f(s^{j-1}).$$

Case a3. $S^{(i^j,j)} \neq \varnothing$ and $e^j \notin S^{(i^j,j)}$.

In this case, we have $o^j \neq e^j$, $o^j \in S^{(i^j,j)}$, $o^{j-1/2} = o^{j-1} - 1^{(o^j,i^j)} - 1^{(e^j,o^{j-1}(e^j))}$, and $o^j = o^{j-1/2} + 1^{(e^j,i^j)}$. Depending on e^j in the support set of o^{j-1} or not, we similarly give the following proof. Firstly, assume that $o^{j-1}(e^j) \neq 0$, and by the similar analysis in **Case a1**, we have $o^{j-1}(e^j) \in I^j \backslash \{i^j\}$. Then specifying $i' \in I^j \backslash \{i^j\}$ because of $|I^j| \geq 2$, we have,

$$f(o^{j-1}) - f(o^j)$$
$$\leq f(o^{j-1}) + f(o^{j-1/2} + 1^{(e^j,i')}) - 2f(o^{j-1/2}) \quad \text{(pairwise monotonicity)}$$
$$\leq \Delta_{e^j,o^{j-1}(e^j)} f(o^{j-1/2}) + \Delta_{o^j,i^j} f(o^{j-1/2}) + \Delta_{e^j,i'} f(o^{j-1/2})$$
$$\text{(orthant submodularity and } o^{j-1/2} \preceq o^{j-1/2} + 1^{(o^j,i^j)})$$
$$\leq f(s^{j-1} + 1^{(e^j,o^{j-1}(e^j))}) + f(s^{j-1} + 1^{(o^j,i^j)}) + f(s^{j-1} + 1^{(e^j,i')}) - 3f(s^{j-1})$$
$$\leq 3\Delta_{e^j,i^j} f(s^{j-1}).$$

Secondly, assume that $o^{j-1}(e^j) = 0$. Since $|I^j| \geq 2$, taking $i' \in I^j \backslash \{i^j\}$, we obtain

$$f(o^{j-1}) - f(o^j) = \Delta_{e^j,i'} f(o^{j-1/2}) + \Delta_{o^j,i^j} f(o^{j-1/2})$$
$$- [f(o^j) + f(o^{j-1/2} + 1^{(e^j,i')}) - 2f(o^{j-1/2})]$$
$$\leq f(o^{j-1/2} + 1^{(e^j,i')}) + f(o^{j-1}) - 2f(o^{j-1/2})$$
$$\leq \Delta_{o^j,i^j} f(s^{j-1}) + \Delta_{e^j,i'} f(s^{j-1})$$
$$\leq 2[f(s^j) - f(s^{j-1})].$$

(b) We have $I^j = \{i^j\}$. Similar to the analysis of (a), we show the proof in three cases.

Case b1. $S^{(i^j,j)} = \varnothing$.

In this case, we have $o^j = e^j$, $o^{j-1/2} = o^{j-1} = s^{j-1}$, and $o^j = o^{j-1/2} + 1^{(e^j,i^j)} = s^{j-1} + 1^{(e^j,i^j)} = s^j$, then

$$f(o^{j-1}) - f(o^j) = f(s^{j-1}) - f(s^j) \leq 0 \leq \Delta_{e^j,i^j} f(s^{j-1}) = f(s^j) - f(s^{j-1}).$$

Case b2. $S^{(i^j,j)} \neq \varnothing$ and $e^j \in S^{(i^j,j)}$.

In this case, the proof process is similar to **Case a2**.

Case b3. $S^{(i^j,j)} \neq \varnothing$ and $e^j \notin S^{(i^j,j)}$.

In this case, we have $o^j = e \in S^{(i^j,j)}$, $o^{j-1/2} = o^{j-1} - 1^{(o^j,i^j)}$, and $o^j = o^{j-1/2} + 1^{(e^j,i^j)}$.

If there exists $i^a \in [k]$ not equal to i^j, satisfying $\Delta_{e^j,i^a} f(s^{j-1}) \leq \Delta_{e^j,i^j} f(s^{j-1})$, we have

$$
\begin{aligned}
f(o^{j-1}) - f(o^j) &\leq \Delta_{o^j,i^j} f(o^{j-1/2}) + \Delta_{e^j,i^a} f(o^{j-1/2}) \\
&\leq f(s^{j-1} + 1^{(o^j,i^j)}) + f(s^{j-1} + 1^{(e^j,i^a)}) - 2f(s^{j-1}) \\
&\leq 2[f(s^j) - f(s^{j-1})],
\end{aligned}
$$

where the first inequality is due to pairwise monotonicity, the second inequality is due to orthant submodularity, and the third inequality is due to Algorithm 1.

Otherwise, for any $i^b \in [k] \backslash I^j$, $\Delta_{e^j,i^b} f(s^{j-1}) > \Delta_{e^j,i^j} f(s^{j-1}) \geq 0$, we have

$$
\begin{aligned}
f(o^{j-1}) - f(o^j) &\leq f(o^{j-1/2} + 1^{(e^j,i^b)}) + f(o^{j-1}) - 2f(o^{j-1/2}) \\
&\leq f(s^{j-1} + 1^{(o^j,i^j)}) + f(s^{j-1} + 1^{(e^j,i^b)}) - 2f(s^{j-1}) \\
&\leq f(s^{j-1} + 1^{(o^j,i^j)}) - f(s^{j-1}) + f(1^{(e^j,i^b)}) \\
&\leq f(s^j) - f(s^{j-1}) + f(s^1) \\
&\leq 2f(s^j) - f(s^{j-1}),
\end{aligned}
$$

where the first inequality is due to pairwise monotonicity, the second inequality and the third inequality are due to orthant submodularity, and the forth inequality and the fifth inequality are due to Algorithm 1. $\qquad\square$

Lemma 2. *For o^r, we have $f(o^r) \leq f(s)$.*

Proof. It is trivial if $o^r = s$. In particular, if s is a full solution, and by the process of constructing of o^r, we know that $o^r = s$. Therefore, $f(o^r) = f(s)$.

In the following discussion, we assume that s is not full in the position i and in this case $s \preceq o^r$. Denote $t = |supp(o^r)| - r$. Also by the ways to construct the sequence o^j for $j \in [r]$, we know that the positions of these t items in $supp(o^r) \backslash supp(s)$ must be the same and coincide the position i. Suppose that

$$
o^r = s + 1^{(v_1,i)} + \cdots + 1^{(v_t,i)}.
$$

And for each $j \in [t]$, denote

$$
o^{r+j} = o^{r+j-1} - 1^{(v_j,i)}.
$$

By Algorithm 1, for any $j \in [t]$, the element v_j out of support set of s in position i gives negative marginal return, i.e., $f(s + 1^{(v_j,i)}) < f(s)$. Therefore, by using orthant submodularity, we can obtain the relationship between o^{r+j-1} and o^{r+j} directly

$$
f(o^{r+j-1}) - f(o^{r+j}) \leq f(s + 1^{(v_j,i)}) - f(s) < 0.
$$

Thus, we have

$$
f(o^r) - f(s) = \sum_{j=1}^{t} [f(o^{r+j-1}) - f(o^{r+j})] < 0.
$$

That is

$$f(o^r) < f(s).$$

□

Then by the above two lemmas, we can give our main results in the following theorem.

Theorem 1. *The $1/(B_m + 4)$-approximate solution can be output by the Algorithm 1 with complexity $O(nkB)$, where $B_m = \max\{B_1, \ldots, B_k\}$.*

Proof. It is trivial to obtain the complexity. Without loss of generality, we assume that the position of the last insertion in Algorithm 1 is i. Let r_1 be the number of iterations when $k - 1$ subsets reach the size constraints, then we have

$$
\begin{aligned}
f(o) - f(o^r) &= \sum_{j=1}^{r} [f(o^{j-1}) - f(o^j)] \\
&= \sum_{j=1}^{r_1} [f(o^{j-1}) - f(o^j)] + \sum_{j=r_1+1}^{r} [f(o^{j-1}) - f(o^j)] \\
&\leq 3 \sum_{j=1}^{r_1} [f(s^j) - f(s^{j-1})] + \sum_{j=r_1+1}^{r} [2f(s^j) - f(s^{j-1})] \\
&= 3 \sum_{j=1}^{r_1} [f(s^j) - f(s^{j-1})] + 2 \sum_{j=r_1+1}^{r} [f(s^j) - f(s^{j-1})] + \sum_{j=r_1+1}^{r} f(s^{j-1}) \\
&\leq 3 \sum_{j=1}^{r} [f(s^j) - f(s^{j-1})] + \sum_{j=r_1+1}^{r} f(s^{j-1}) \\
&\leq 3 \sum_{j=1}^{r} [f(s^j) - f(s^{j-1})] + \sum_{j=r_1+1}^{r} f(s^r) \\
&= 3[f(s) - f(0)] + (r - r_1) \cdot f(s) \\
&\leq 3[f(s) - f(0)] + B_i \cdot f(s) \\
&\leq 3f(s) + B_m \cdot f(s) \quad (B_m = \max\{B_1, \ldots, B_k\}),
\end{aligned}
$$

where the first inequality is due to Lemma 1 (a), the second inequality and the third inequality are due to Algorithm 1. Then together with Lemma 2, we have

$$f(o) \leq (B_m + 4) \cdot f(s).$$

□

Algorithm 2. Bi-criteria algorithm for MkISC

Input: A non-monotone k-submodular function $f : (k+1)^V \rightarrow \mathbb{R}_+$ and integers $B_1, \ldots, B_k \in \mathbb{Z}_+$.

Output: A vector $s = (S_1, \ldots, S_k)$ with $|S_i| \leq B_i + B_m$ for $\forall i \in [k]$, where $B_m = \max\{B_1, \ldots, B_k\}$.

1:   $s \leftarrow \mathbf{0}, I^1 \leftarrow \{1, 2, \ldots, k\}, j = 1$
2:   **while** $|I^j| \geq 2$ **do**
3:     $(e, i) \leftarrow \arg\max_{e \in V \setminus supp(s), i \in I^j} \Delta_{e,i} f(s)$
4:     $s \leftarrow s + \mathbf{1}^{(e,i)}$
5:     $j \leftarrow j + 1$
6:     $I^j \leftarrow \{i \mid |S_i| < B_i, \forall i \in [k]\}$
7:   **end while**
8:   $\{l\} \leftarrow I^j$
9:   **for** p from 1 to $B_l - |S_l|$ **do**
10:    $(e, i) \leftarrow \arg\max_{e \in V \setminus supp(s), i \in [k]} \Delta_{e,i} f(s)$
11:    $s \leftarrow s + \mathbf{1}^{(e,i)}$
12:    $I^{j+1} \leftarrow I^j$
13: **end for**
14: **return** s

3.2 Bi-criteria Algorithm for MkISC

In each iteration of Algorithm 2, we can see that the scope of greed can be limited to I^j or $[k]$. When the scope of greed is limited in I^j, by pairwise monotonicity, there is $i \in I^j$, such that $\Delta_{e,i} f(s) \geq 0$. So the elements can be added to the subset represented in I^j. This situation can continue until the $k - 1$ subsets of the algorithm solution reach the size constraints, i.e., only one subset does not reach the size constraint. When this case happens, the greedy range of the element will be limited to $[k]$. By pairwise monotonicity, there is $i \in [k]$, such that $\Delta_{e,i} f(s) \geq 0$. Therefore, the algorithm 2 can select the element and subset that maximize the gain of $\Delta_{e,i} f(s)$, then put the selected element in this subset, and continue the iterative process until the number of elements of the algorithm solution reaches $B = \sum_{i=1}^{k} B_i$. Thus, the final output solution s of Algorithm 2 satisfies $|supp(s)| = B$ and $|S_i| \leq B_i + B_m$ for all $i \in [k]$, where $B_m = \max\{B_1, \ldots, B_k\}$. In fact, from Steps 8–12 in Algorithm 2, one can see that at most B_m items are put in the "wrong" position but there may be $k - 1$ individual size constraints are violated. So we relax our violated constraint bounds to $B_i + B_m, \forall i \in [k]$.

The main result is given as follows, and its proof will be shown in Appendix.

Theorem 2. *The Algorithm 2 provides an approximation guarantee of $1/4$ with complexity $O(knB)$, where $B = \sum_{i=1}^{k} B_i$. But this solution may be against the individual size constraints, and it just satisfies $|S_i| \leq B_i + B_m, \forall i \in [k]$, where $B_m = \max_{i \in [k]}\{B_i\}$.*

4 Discussion

In this paper, we study the maximization non-monotone k-submodular function problem with individual size constraints. Two deterministic algorithms are proposed. Next, we will study the stochastic and streaming algorithm to improve the deterministic algorithm and reduce its complexity.

References

1. Ene, A., Nguyen, H.: Streaming algorithm for monotone k-submodular maximization with cardinality constraints. In: Proceedings of ICML, pp. 5944–5967 (2022)
2. Huber, A., Kolmogorov, V.: Towards minimizing k-submodular functions. In: Mahjoub, A.R., Markakis, V., Milis, I., Paschos, V.T. (eds.) ISCO 2012. LNCS, vol. 7422, pp. 451–462. Springer, Heidelberg (2012). https://doi.org/10.1007/978-3-642-32147-4_40
3. Iwata, S., Tanigawa, S., Yoshida, Y.: Improved approximation algorithms for k-submodular function maximization. In: Proceedings of SODA, pp. 404–413 (2016)
4. Krause, A., Singh, A., Guestrin, C.: Near-optimal sensor placements in Gaussian processes: theory, efficient algorithms and empirical studies. J. Mach. Learn. Res. **9**(8), 235–284 (2008)
5. Nguyen, L., Thai, M.: Streaming k-submodular maximization under noise subject to size constraint. In: Proceedings of ICML, pp. 7338–7347 (2020)
6. Ohsaka, N., Yoshida, Y.: Monotone k-submodular function maximization with size constraints. In: Proceedings of NeurIPS, pp. 694–702 (2015)
7. Oshima, H.: Improved randomized algorithm for k-submodular function maximization. SIAM J. Discrete Math. **35**(1), 1–22 (2021)
8. Pham, C.V., Vu, Q.C., Ha, D.K.T., Nguyen, T.T.: Streaming algorithms for budgeted k-submodular maximization problem. In: Mohaisen, D., Jin, R. (eds.) CSoNet 2021. LNCS, vol. 13116, pp. 27–38. Springer, Cham (2021). https://doi.org/10.1007/978-3-030-91434-9_3
9. Pham, C., Vu, Q., Ha, D., Nguyen, T., Le, N.: Maximizing k-submodular functions under budget constraint: applications and streaming algorithms. J. Comb. Optim. **44**, 723–751 (2022). https://doi.org/10.1007/s10878-022-00858-x
10. Qian, C., Shi, J., Tang, K., Zhou, Z.: Constrained monotone k-submodular function maximization using multiobjective evolutionary algorithms with theoretical guarantee. IEEE Trans. Evol. Comput. **22**, 595–608 (2018)
11. Rafiey, A., Yoshida, Y.: Fast and private submodular and k-submodular functions maximization with matroid constraints. In: Proceeding of ICML, pp. 7887–7897 (2020)
12. Sakaue, S.: On maximizing a monotone k-submodular function subject to a matroid constraint. Discrete Optim. **23**, 105–113 (2017)
13. Shi, G., Gu, S., Wu, W.: k-submodular maximization with two kinds of constraints. Discrete Math. Algorithms Appl. **13**(4), 2150036 (2021)
14. Sun, Y., Liu, Y., Li, M.: Maximization of k-submodular function with a matroid constraint. In: Du, D.Z., Du, D., Wu, C., Xu, D. (eds.) TAMC 2022. LNCS, vol. 13571, pp. 1–10. Springer, Cham (2022). https://doi.org/10.1007/978-3-031-20350-3_1
15. Tang, Z., Wang, C., Chan, H.: Monotone k-submodular secretary problems: cardinality and knapsack constraints. Theor. Comput. Sci. **921**, 86–99 (2022)
16. Tang, Z., Wang, C., Chan, H.: On maximizing a monotone k-submodular function under a knapsack constraint. Oper. Res. Lett. **50**(1), 28–31 (2022)

17. Wang, B., Zhou, H.: Multilinear extension of k-submodular functions. arXiv:2107.07103 (2021)
18. Ward, J., Živný, S.: Maximizing k-submodular functions and beyond. ACM Trans. Algorithms **12**(4), 1–26 (2016). Article 47
19. Zheng, L., Chan, H., Loukides, G., Li, M.: Maximizing approximately k-submodular functions. In: Proceeding of SDM, pp. 414–422 (2021)

A Heuristic Algorithm for Student-Project Allocation Problem

Nguyen Thi Uyen[1,2], Giang L. Nguyen[2], Canh V. Pham[3], Tran Xuan Sang[4], and Hoang Huu Viet[1(✉)]

[1] School of Engineering and Technology, Vinh University, Vinh City, Vietnam
{uyennt,viethh}@vinhuni.edu.vn
[2] Institute of Information Technology, VAST, Hanoi, Vietnam
nlgiang@ioit.ac.vn
[3] ORLab, Faculty of Computer Science,, Phenikaa University, Hanoi, Vietnam
canh.phamvan@phenikaa-uni.edu.vn
[4] Cyber School, Vinh University, Vinh, Nghe An, Vietnam
sangtx@vinhuni.edu.vn

Abstract. The Student-Project Allocation problem with lecturer preferences over Students with Ties (SPA-ST) is to find a stable matching of students and projects to satisfy the constraints on student preferences over projects, lecturer preferences over students, and the maximum number of students given by each project and lecturer. This problem has attracted many researchers because of its wide applications in allocating students to projects at many universities worldwide. However, the main weakness of existing algorithms is their high computational cost. In this paper, we propose a heuristic algorithm to improve solution quality and execution time for solving the SPA-ST problem of large sizes. Experimental results on randomly generated datasets show that our algorithm outperforms the state-of-the-art algorithm regarding solution quality and execution time.

Keywords: Student-Project Allocation Problem · Heuristic algorithm · Blocking pairs · Stable matching · MAX-SPA-ST · Large sizes

1 Introduction

The problem of allocating students to projects based on their preferences, called SPA [2,7,18,19,21], is to find a stable matching of the students and the projects to satisfy the constraints on their preference lists. This problem has played an important role at many universities in the world [6,10,14,17]. However, it has a strict constraint on preference lists: students and lecturers must rank given projects in a certain order that cannot cover a reality case: two projects are ranked in the same order of preference in their lists. Abraham et al. [1] proposed a new variant of the SPA problem, called preferences over Students containing Ties (SPA-ST), with an adjustment that helps students have more options

© The Author(s), under exclusive license to Springer Nature Switzerland AG 2023
T. N. Dinh and M. Li (Eds.): CSoNet 2022, LNCS 13831, pp. 280–291, 2023.
https://doi.org/10.1007/978-3-031-26303-3_25

when choosing projects from the lecturers: the list of preferences of lecturers and students can contain equality relations.

In the SPA-ST problem, there are three stability criteria of matching: *weakly stable*, *strongly stable*, or *super-stable* matching [19,20]. Irving et al. [22] proved that a super-stable matching is strongly stable, and a strongly stable matching is a weakly stable matching. If a super-stable matching is found, all weakly stable matchings have the same sizes. Besides, they also showed that weakly stable matching always exist and have different sizes [11]. Therefore, the problem of finding a maximum size weakly stable matching is known as the NP-hard problem [22]. In addition, strongly stable or super-stable matchings, whose goal is to find a stable matching with a maximum number of matched students, may not exist because the constraints are too tight [15,20,22].

Practically, the problem of finding a maximum weakly stable matching is the most suitable for real-life applications because it focuses on assigning as many students as possible to projects. This problem is known as the MAX-SPA-ST problem and it has attracted much attention from the research community because of its application in education optimization problems. Some universities applied the MAX-SPA-ST problem to optimize the assignment of projects for lecturers to satisfy the constraints of large-scale problems, such as the School of Computer Science, the University of Glasgow [14], Faculty of Science, University of Southern Denmark [6], Department of Computer Science, York University, and elsewhere [3–5,8]. Unfortunately, finding an efficient algorithm to solve the MAX-SPA-ST of large sizes is still a main challenge for the research community. The approximation algorithm is one of the popular methods for solving the SPA-ST problem [1,12,16,21]. Cooper et al. [7] first proposed a 3/2-approximation algorithm named APX for solving the MAX-SPA-ST based on Király's idea [13]. Latter, Manlove et al. [15] presented an Integer Programming (IP) model to find a strongly stable matching for SPA-ST. Recently, Olaosebikan et al. [19] provided a polynomial-time algorithm to find a super-stable matching. However, they proved that it might not exist for the SPA-ST problem.

In this paper, we call a *weakly stable* matching a *stable* matching. Accordingly, we propose a new heuristic algorithm for solving the MAX-SPA-ST problem. The main difference between our algorithm from the others is that we design *two heuristic functions* to estimate reasonable solutions: the first is used to choose the best project to match with students, while the second is to discard the worst student from the matching when the project or lecturer is *over-subscribed*. By combining two functions, our algorithm can find a suitable solution within a reasonable time. Experimental results show that our proposed algorithm performs better in terms of the average execution time and percentage of perfect matchings compared to the most recent APX algorithm [7] for solving the MAX-SPA-ST problem of large sizes.

The rest of this paper is organized as follows. Section 2 presents preliminaries of the SPA-ST problem, Sect. 3 describes our proposed algorithm, Sect. 4 discusses our experimental results, and Sect. 5 concludes our work.

2   Preliminaries

An SPA-ST instance consists of a set $S = \{s_1, s_2, \cdots, s_n\}$ of students, a set $\mathcal{P} = \{p_1, p_2, \cdots, p_q\}$ projects, and a set $\mathcal{L} = \{l_1, l_2, \cdots, l_m\}$ of lecturers. Each student s_i ranks a set of acceptable projects in order of preference containing ties. Each lecturer $l_k \in \mathcal{L}$ offers a set of projects and ranks a set of students in a preference list containing ties. Each lecturer has a capacity $d_k \in \mathbb{Z}^+$ indicating the maximum number of students assigned to l_k. Each project $p_j \in \mathcal{P}$ is offered by one lecturer and has a capacity $c_j \in \mathbb{Z}^+$ indicating the maximum number of students that can be assigned to it. For example, an SPA-ST instance given in Table 1 consists of a set of $S = \{s_1, s_2, s_3, s_4, s_5, s_6, s_7\}$ of students, a set $\mathcal{P} = \{p_1, p_2, p_3, p_4, p_5, p_6, p_7, p_8\}$ of projects, and a set $\mathcal{L} = \{l_1, l_2, l_3\}$ of lecturers.

Table 1. An example of SPA-ST instance

Students' preferences	Lecturers' preferences	
s_1: $(p_1 \ p_7)$	l_1: $(s_7 \ s_4) \ s_1 \ s_3 \ (s_2 \ s_5) \ s_6$	l_1 offers p_1, p_2, p_3
s_2: $p_1 \ p_3 \ p_5$	l_2: $s_3 \ s_2 \ s_7 \ s_5$	l_2 offers p_4, p_5, p_6
s_3: $(p_2 \ p_1) \ p_4$	l_3: $(s_1 \ s_7) \ s_6$	l_3 offers p_7, p_8
s_4: p_2		
s_5: $p_1 \ p_4$		
s_6: $p_2 \ p_8$	Project capacities	$c_1 = 2, c_j = 1, (2 \leq j \leq 8)$
s_7: $(p_5 \ p_3) \ p_8$	Lecturer capacities	$d_1 = 3, d_2 = 2, d_3 = 2$

For any pair $(s_i, p_j) \in S \times \mathcal{P}$, where p_j is offered by l_k, we refer (s_i, p_j) as an *acceptable pair* if s_i and p_j both find each other acceptable, i.e., p_j is ranked by a student s_i and a lecturer l_k ranks s_i. Each student $s_i \in S$ has a set $A_i \subseteq \mathcal{P}$ of acceptable projects that they rank in the order of preference. We let $R_{s_i}(p_j)$ and $R_{l_k}(s_i)$ be the rank of p_i in s_i's ranks list and the rank of s_i in l_k's ranks list, respectively.

A matching $\mathcal{T}$ of an SPA-ST instance is a set of acceptable pairs (s_i, p_j) or $(s_i, \varnothing)$ such that $|\mathcal{T}(s_i)| \leq 1$ for all $s_i \in S$, $|\mathcal{T}(p_j)| \leq c_j$ for all $p_j \in \mathcal{P}$, and $|\mathcal{T}(l_k)| \leq d_k$ for all $l_k \in \mathcal{L}$, meaning that each s_i belongs to at most one pair. A project p_j is *under-subscribed*, *full*, or *over-subscribed* if $|\mathcal{T}(p_j)| < c_j$, $|\mathcal{T}(p_j)| = c_j$, or $|\mathcal{T}(p_j)| > c_j$, respectively. Similarly, a lecturer l_k is *under-subscribed*, *full*, or *over-subscribed* if $|\mathcal{T}(l_k)| < d_k$, $|\mathcal{T}(l_k)| = d_k$, or $|\mathcal{T}(l_k)| > d_k$, respectively. If an acceptable project and lecturer are *under-subscribed*, this project is a *potential* project. If $(s_i, p_j) \in \mathcal{T}$, then s_i is matched to p_j, denoted by $\mathcal{T}(s_i) = p_j$. If $\mathcal{T}(s_i) = \varnothing$, then s_i is *unassigned* in $\mathcal{T}$, and we let the set of *unsigned* students be $\mathcal{U}$.

Given a matching $\mathcal{T}$, a pair $(s_i, p_j) \in S \times \mathcal{P}$ is a blocking pair in $\mathcal{T}$ if it meets the conditions of (1), (2), and (3) as follows:

1. s_i and p_j find accept each other;
2. s_i prefers p_j to $\mathcal{T}(s_i)$ or $\mathcal{T}(s_i) = \varnothing$;

3. *either* (a), (b) *or* (c) *holds as follows*:
   (a) $|T(p_j)| < c_j$ and $|T(l_k)| < d_k$;
   (b) $|T(p_j)| < c_j$, $|T(l_k)| = d_k$, and *(i) either* $s_i \in T(l_k)$ *or (ii)* l_k *prefers* s_i to the worst student in $T(l_k)$;
   (c) $|T(p_j)| = c_j$ and l_k prefers s_i to the worst student in $T(p_j)$.

A matching T is *stable* if it admits no blocking pair. Otherwise, it is *unstable*. The size of a stable matching T, denoted by $|T|$, is the number of students assigned in T. If $|T| = n$, then T is a *perfect* matching. Otherwise, T is a *non-perfect* matching. $|U|$ is the number of unassigned students in a *non-perfect* matching T. If the size of a weakly stable matching is equal to n, denoted by $|T| = n$, then T is a *perfect* matching. Otherwise, T is *non-perfect*.

3   Proposed Algorithm

In this section, we introduce our proposed algorithm. The core of our algorithm is two *heuristic functions* that can improve execution time and solution quality for the MAX-SPA-ST problem. The first one helps students to choose a suitable project for matching, while the second one determines which students will be removed from the current matching when the lecturer or project is *over-subscribed*.

3.1   Heuristic Functions

This section presents two heuristic functions to guide students and lecturers to select appropriate projects and students, respectively, so that our algorithm can reach a perfect matching with a better solution quality and shorter execution time. To do so, we design two heuristic functions as follows:

(1) The heuristic function $h(p_j)$: For each student $s_i \in S$, we define the first heuristic function $h(p_j)$ for every project p_j in s_i's rank list, where p_j is offered by l_k, to choose the best project in terms of the minimum value of $h(p_j)$ in Eq. 1 as follows:

$$h(p_j) = R_{s_i}(p_j) - \min(d_k - |T(l_k)|, 1)/2 - (c_j - |T(p_j)|)/(2 \times c_j + 1). \quad (1)$$

If (s_i, p_j) is an *acceptable* pair, then $1 \le R_{s_i}(p_j) \le q$. If l_k is *full*, i.e., $d_k - |T(l_k)| = 0$, then $min(d_k - |T(l_k)|, 1)/2 = 0$. If l_k is *under-subscribed*, i.e., $|T(l_k)| < d_k$, then $min(d_k - |T(l_k)|, 1)/2 = 0.5$ for all $l_k \in L$. Besides, we have $0 \le c_j - |T(p_j)| \le c_j$, then $0 \le (c_j - |T(p_j)|)/(2 \times c_j + 1) < 0.5$ for all $p_j \in P$, meaning that $0 < h(p_j) \le q$.

(2) The heuristic function $g(s_t)$: For each student $s_i \in S$, when s_i offers to p_j, if p_j is *full* or l_k is *full*, then we choose the worst student in terms of the maximum heuristic value to remove her/his from the current matching. To do this, we define a heuristic function $g(s_t)$ in Eq. 2 as follows:

$$g(s_t) = R_{l_k}(s_t) + t(s_t) + r(s_t)/(q + 1). \quad (2)$$

For each student $s_t \in \mathcal{T}(l_k)$, we have $1 \leq R_{l_k}(s_t) \leq n$. Moreover, $t(s_t) = sum(min(d_z - |\mathcal{T}(l_z)|, 1) \times min(c_u - |\mathcal{T}(p_u)|, 1) \times n)$, where p_u is the same ties with $\mathcal{T}(s_t)$ in s_t's rank list and p_u is offered by l_z. If p_u is *under-subscribed*, i.e., $|\mathcal{T}(p_u)| < c_u$, then $1 \leq c_u - |\mathcal{T}(p_u)| \leq c_u$, thus we have $min(c_u - |\mathcal{T}(p_u)|, 1) = 1$. Otherwise, if p_u is *full*, i.e., $|\mathcal{T}(p_u)| = c_u$, then $c_u - |\mathcal{T}(p_u)| = 0$, thus we have $min(c_u - |\mathcal{T}(p_u)|, 1) = 0$. Similarly, we have $min(d_z - |\mathcal{T}(l_z)|, 1) = 1$ or 0, if l_z is *under-subscribed* or *full*, respectively. If p_u is a *potential* project, then we have $min(d_z - |\mathcal{T}(l_z)|, 1) \times min(c_u - |\mathcal{T}(p_u)|, 1) \times n = n$, otherwise, $min(d_z - |\mathcal{T}(l_z)|, 1) \times min(c_u - |\mathcal{T}(p_u)|, 1) \times n = 0$. Thus, we have $0 \leq t(s_t) \leq (q-1) \times n$. Let $r(s_t)$ be the number projects ranked by s_t, we have $1 \leq r(s_t) \leq q$, thus $0 < r(s_t)/(q+1) < 1$. This means that we have $1 < g(s_t) < (q \times n + 1)$ for all $s_t \in \mathcal{T}(l_k)$.

3.2 Our Algorithm

This section presents our proposed algorithm, called HAG (Algorithm 1). At the beginning, HAG assigns a matching $\mathcal{T} = \varnothing$ and a count variable $v(s_i) = 0$ for all students $s_i \in \mathcal{S}$. At each iteration, if there exists an *unassigned* $s_i \in \mathcal{S}$ such that s_i's rank list is non-empty, then HAG runs as follows. First, it calculates the heuristic function $h(p_j)$ for each project $p_j \in \mathcal{P}$ in s_i's rank list. Then, s_i proposes a project p_j corresponding to the minimum value of $h(p_j)$. Let l_k be a lecturer who offers p_j, we consider the following cases:

1. If both p_j and l_k are *under-subscribed*, HAG adds (s_i, p_j) into $\mathcal{T}$.
2. If p_j is *full*, HAG calculates $g(s_t)$ for all $s_t \in \mathcal{T}(p_j)$ and chooses a student s_t with maximal value of $g(s_t)$ by the function $Choose\_Student(\mathcal{T}(p_j), l_k)$. If $g(s_t) > n + 1$ or $R_{l_k}(s_t) > R_{l_k}(s_i)$, then HAG removes (s_t, p_j), adds (s_i, p_j) into $\mathcal{T}$ and deletes p_j in s_t's rank list if $g(s_t) < n+1$. Otherwise, HAG deletes p_j in s_i's rank list. Note that if $g(s_t) > n + 1$, then s_t contains a *potential* project with the same ties as p_j in s_t's rank list.
3. If l_k is *full*, HAG calculates $g(s_w)$ for all $s_w \in \mathcal{T}(l_k)$ and chooses a student s_w with maximal value of $g(s_w)$ by the function $Choose\_Student(\mathcal{T}(l_k), l_k)$. If $g(s_w) > n + 1$ or $R_{l_k}(s_w) > R_{l_k}(s_i)$, then HAG removes (s_w, p_u), where $p_u = \mathcal{T}(s_w)$, and adds (s_i, p_j) into $\mathcal{T}$. If $g(s_w) < n+1$, then HAG deletes p_u in s_w's rank list. Otherwise, it deletes p_j in s_i's rank list. When a project p_u is removed, it can form blocking pairs with students in $\mathcal{T}(l_k)$, then HAG calls the function $Repair(p_u, l_k)$ to break blocking pairs satisfying the condition of *(3bi)*.

The above process repeats until a stable matching $\mathcal{T}$ is found. If $|\mathcal{T}| = n$, then HAG returns a perfect matching. Otherwise, HAG calls the function $Escape(p_u, l_k)$ to assign *unassigned* students for the current stable matching. Our HAG stops when a *perfect* matching is found or all *unassigned* students cannot find any projects to assign them. In the latter case, the algorithm returns a stable matching of a maximum size found so far.

Algorithm 1: HAG algorithm for MAX-SPA-ST

Input: An SPA-ST instance I
Output: A stable matching $\mathcal{T}$

1. **function** HAG(I)
2. $\mathcal{T} := \varnothing$;
3. $v(s_i) := 0, \forall s_i \in \mathcal{S}$;
4.   **while** *true* **do**
5. $s_i :=$ an *unassigned* student that s_i's rank list is non-empty;
6.   **if** *there exists no student s_i* **then**
7.   **if** $|\mathcal{T}| = n$ **then** break;
8.   **else**
9. $\mathcal{T}' :=$ Escape($\mathcal{T}$);
10.   **if** $\mathcal{T}' = \mathcal{T}$ **then** break;
11. $\mathcal{T} := \mathcal{T}'$;
12.   continue;
13.   **end**
14.   **end**
15.   **for** *each $p_j \in A_i$* **do**
16. $l_k :=$ a lecturer who offers p_j;
17. $h(p_j) := R_{s_i}(p_j)\text{-}min(d_k\text{-}|\mathcal{T}(l_k)|, 1)/2\text{-}(c_j\text{-}|\mathcal{T}(p_j)|) /(2 \times c_j + 1)$;
18.   **end**
19. $p_j := \operatorname{argmin}(h(p_j) > 0), \forall p_j \in \mathcal{P}$;
20. $l_k :=$ a lecturer who offers p_j;
21.   **if** *p_j and l_k are under-subscribed* **then**
22. $\mathcal{T} := \mathcal{T} \cup \{(s_i, p_j)\}$;
23.   **else if** *p_j is full* **then**
24. $[s_t, g(s_t)] :=$ Choose_Student($\mathcal{T}(p_j), l_k$);
25.   **if** $g(s_t) > n + 1$ or $R_{l_k}(s_i) < R_{l_k}(s_t)$ **then**
26. $\mathcal{T} := \mathcal{T} \setminus \{(s_t, p_j)\} \cup \{(s_i, p_j)\}$;
27.   **if** $g(s_t) < n + 1$ **then**
28. $R_{s_t}(p_j) := 0$;
29.   **end**
30.   **else**
31. $R_{s_i}(p_j) := 0$;
32.   **end**
33.   **else**
34. $[s_w, g(s_w)] :=$ Choose_Student($\mathcal{T}(l_k), l_k$);
35.   **if** $g(s_w) > n + 1$ or $R_{l_k}(s_i) < R_{l_k}(s_w)$ **then**
36. $\mathcal{T} := \mathcal{T} \setminus \{(s_w, p_u)\} \cup \{(s_i, p_j)\}$, where $p_u = \mathcal{T}(s_w)$;
37.   Repair(p_u, l_k);
38.   **if** $g(s_w) < n + 1$ **then**
39. $R_{s_w}(p_u) := 0$;
40.   **end**
41.   **else**
42. $R_{s_i}(p_j) := 0$;
43.   **end**
44.   **end**
45.   **end**
46.   **return** $\mathcal{T}$;
47. **end function**

The function $Choose\_Student(\mathcal{T}(p_j), l_k)$ is used to choose a student with the maximum value of $g(s_t)$. Let X be the set of students such that $X = \mathcal{T}(p_j)$, where p_j is *full*, or $X = \mathcal{T}(l_k)$ where l_k is *full*. For each $s_t \in X$, the algorithm calculates the heuristic value of $g(s_t)$ and returns a student s_t with the maximum value of $g(s_t)$.

The function $Repair(p_u, l_k)$ is used to break blocking pairs satisfying the condition of *(3bi)*. When a project p_u is removed, for each $s_k \in \mathcal{T}(l_k)$, if $R_{s_k}(p_u) < R_{s_k}(p_z)$, where $p_z = \mathcal{T}(s_k)$, then the algorithm removes (s_k, p_z) and adds (s_k, p_u) into $\mathcal{T}$. This process repeats for each project which is removed until it cannot form blocking pairs.

If HAG reaches a *stable* matching but it is *non-perfect* matching, it gets stuck at a local minimum. At each iteration, for each student $s_i \in \mathcal{S}$, HAG proposes a project $p_j \in \mathcal{P}$ which is the minimum rank in s_i's rank list and adds (s_i, p_j) into matching $\mathcal{T}$. Since ties were given in l_k's rank list (p_j is offered by l_k), there exists a different student s_u who prefers the most project p_j or p_z (p_z is offered by l_k) where $R_{l_k}(s_u) = R_{l_k}(s_i)$. When s_u applies to p_j (p_j is *full*) or p_z (l_k is *full*), we keep (s_i, p_j) and reject (s_u, p_z) or (s_u, p_j), then HAG can result in a *non-perfect* matching. If we add (s_u, p_z) or (s_u, p_j) and remove (s_i, p_j), HAG can result in a *perfect* matching. This means that we can add (s_u, p_z) or (s_u, p_j) and remove (s_i, p_j), then s_i applies to other projects. Moreover, we can improve stable matchings' size by using the function $Escape(\mathcal{T})$. For each *unassigned* student $s_u \in \mathcal{S}$, the algorithm finds a project p_z such that there exists a student s_i where p_j is assigned to s_i and $v(s_u) \geq v(s_i)$. Then, the algorithm replaces (s_i, p_j) by (s_u, p_z) in $\mathcal{T}$ and increases the value of $v(s_u)$. When p_j is removed, it can form blocking pairs with other students in $\mathcal{T}(l_k)$, thus we call the function $Repair(p_j, l_k)$ to break blocking pairs. It should be noted that the condition $v(s_u) \geq v(s_i)$ means that the number of replacements of s_u is higher than that of s_i, i.e., s_u is prioritized to assign to $p_z \in l_k$. Accordingly, matching $\mathcal{T}$ is *unstable* and s_i is an *unassigned* student, then s_i proposes other projects in s_i's rank list.

4   Performance Evaluation

In this section, we present several experiments to evaluate the efficiency of our HAG algorithm. We compared the average execution time and solution quality found by HAG with those found by the approximation algorithm, named APX [7] for solving MAX-SPA-ST problem of large sizes. We implemented these algorithms by Matlab R2019a software on a system with Xeon-R Gold 6130 CPU 2.1 GHz computer with 16 GB RAM.

Datasets: To perform experiments, we adapted a random SMTI problem generator [9] to generate SPA-ST instances with seven parameters $(n, m, q, p_1, p_2, C, D)$, where n is the number of students, m is the number of lecturers, q is the number of projects, p_1 is the probability of incompleteness, p_2 is the probability of ties, C is the total capacity of projects, denoted by $C = \sum_{j=1}^{q} c_j$ where c_j is the capacity of project p_j offered by lecturer l_k, D is the total capacity of lecturers, denoted by $D = \sum_{k=1}^{m} d_k$ where d_k is the capacity of each lecturer l_k.

4.1 Evaluate the Variation of Ties and Incompleteness

In this experiment, we evaluate the influence of the probability of p_1 and p_2 on the solution quality and execution time of HAG in comparison with APX. We generated SPA-ST instances by letting parameters $(n, m, q, p_1, p_2, C, D)$, in which $n \in \{100, 200\}$, $m = 0.05n$, $q = 0.1n$, $p_1 \in [0.1, 0.8]$ and $p_2 \in [0.0, 1.0]$ with step of 0.1. The total capacities of projects and lecturers are $C = 1.2n$, and $D = 1.1n$, respectively. In addition, c_j is distributed for each project $p_j \in \mathcal{P}$ such that $0.6C/q \leq c_j \leq 1.4C/q$ and d_k is set of each lecturer $l_k \in \mathcal{L}$ such that $d_k = D/m$.

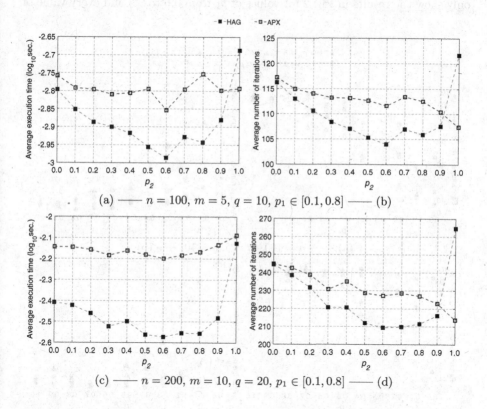

Fig. 1. Average execution time and average number of iterations of HAG vs. APX

We first compare the average execution time and the average number of iterations of HAG with APX for finding perfect matchings for $n = 100$ and $n = 200$. Then, we average results based on values of p_1 which is shown in Fig. 1. Accordingly, we see that Figs. 1(a) and 1(c) show that HAG runs faster than APX for $n = 100$ and $n = 200$ with every value of p_2. When p_2 increases from 0.0 to 0.9 for $n = 100$, HAG takes from $10^{-2.98}(s)$ to $10^{-2.68}(s)$, while APX takes from $10^{-2.85}(s)$ to $10^{-2.75}(s)$, but when $p_2 = 1.0$, the average execution time of HAG increases. When $n = 200$, the average execution time of HAG increases from about $10^{-2.58}(s)$ to $10^{-2.28}(s)$, while APX takes from $10^{-2.23}(s)$ to $10^{-2.16}(s)$. Figures 1(b) and 1(d) give the average number of iterations used by HAG and APX for $n = 100$ and $n = 200$ with every p_2. HAG needs fewer iterations than APX for p_2 from 0.0 to 0.9 for $n = 100$ and $n = 200$. This explains that HAG runs much faster than APX as shown in Figs. 1(a) and 1(c).

Next, we compare the percentage of perfect matchings and average number of unassigned students found by HAG and APX for every value of p_1 and p_2. Our experimental results show that when p_1 varies from 0.1 to 0.5 with step 0.1, both HAG and APX result in perfect matchings approximately, therefore we only show the results in Fig. 2 for values of p_1 from 0.6 to 0.8 and every value of p_2.

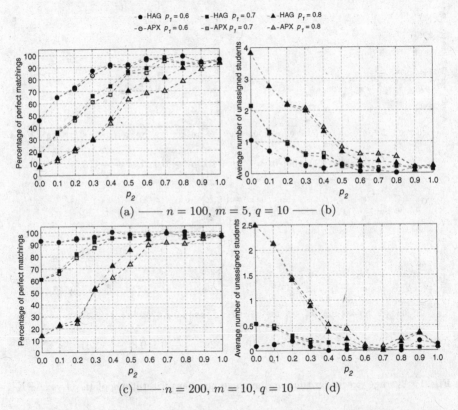

Fig. 2. Percentage of perfect matchings and number of unassigned students of HAG vs. APX

Figure 2 shows that when p_1 increases to 0.8, it is difficult for finding perfect matchings, but HAG finds better the percentage of perfect matchings than that found by APX. When p_2 varies from 0.1 to 1.0, HAG finds a much higher percentage of perfect matchings than APX as shown in Figs. 2(a) and 2(c).

Figures 2(b) and 2(d) show the average number of unassigned students found by HAG and APX. When p_2 varies from 0.1 to 1.0, HAG obtains a much smaller number of unassigned students than APX in non-perfect matchings. This means that HAG finds better maximum stable matchings than APX when p_1 increases in this experiment.

4.2 Evaluate the Variation of the Problem of Large Sizes

In this experiment, we set $n \in [1000, 10000]$ with step 1000, $m \in [0.02n, 0.04n]$ step 0.01, $q = 0.1n$, $p_1 = 0.9$, and $p_2 = 0.5$. The total capacity of projects and lecturers is set in two cases as follows:

Case 1: $C = 1.4n$ and $0.7C \leq D \leq 0.9C$, i.e., the capacity c_j of each project p_j is bound by $0.6C/q \leq c_j \leq 1.4C/q$ and the capacity d_k of each lecturer l_k is bound by $0.7 \sum_{k=1}^{|P_k|} c_j \leq d_k \leq 0.9 \sum_{k=1}^{|P_k|} c_j$, where P_k is a set of projects p_j offered by lecturer l_k.

Case 2: $C = 1.5n$, and $D = 1.2n$, i.e., we distribute C to the capacity c_j of each project p_j such that $0.6C/q \leq c_j \leq 1.4C/q$. Next, we set the capacity of each lecturer l_k to be $d_k = D/m$.

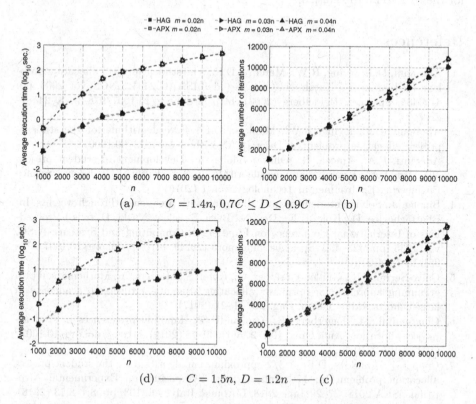

Fig. 3. Average execution time and average number of iterations HAG vs. APX

When $p_1 = 0.9$, 90% of projects or students do not rank in students' rank lists or lecturers' rank lists, respectively. However, because the number of projects and students is larger, both HAG and APX can find approximately 100% of perfect matchings. Figure 3 shows that HAG runs faster from 10 to 80 times and needs fewer iterations than APX in two cases of capacities.

5   Conclusions

This paper proposed an efficient heuristic algorithm HAG to solve the large size MAX-SPA-ST problem. Our algorithm starts from an empty matching and finds a solution for the problem based on two proposed *heuristic functions* to improve the performance of the searching process. If the algorithm reaches a *non-perfect* matching, we propose a heuristics strategy to increase the size of the stable matching by suggesting *unassigned* students to projects in the same ties with its partner in the current matching. Our experimental results show that our algorithm overcomes the APX algorithm [7] in terms of execution time and solution quality for the MAX-SPA-ST of large sizes. In the future, we will extend this proposed approach to find *strongly stable* or *super-stable* matchings for the SPA-ST problem [20].

References

1. Abraham, D.J., Irving, R.W., Manlove, D.F.: The student-project allocation problem. In: Ibaraki, T., Katoh, N., Ono, H. (eds.) ISAAC 2003. LNCS, vol. 2906, pp. 474–484. Springer, Heidelberg (2003). https://doi.org/10.1007/978-3-540-24587-2_49
2. Abraham, D.J., Irving, R.W., Manlove, D.F.: Two algorithms for the student-project allocation problem. J. Discrete Algorithms **5**(1), 73–90 (2007)
3. Aderanti, F.A., Amosa, R., Oluwatobiloba, A.: Development of student project allocation system using matching algorithm. In: International Conference Science Engineering Environmental Technology, vol. 1 (2016)
4. Binong, J.: Solving student project allocation with preference through weights. In: Bhattacharjee, D., Kole, D.K., Dey, N., Basu, S., Plewczynski, D. (eds.) Proceedings of International Conference on Frontiers in Computing and Systems. AISC, vol. 1255, pp. 423–430. Springer, Singapore (2021). https://doi.org/10.1007/978-981-15-7834-2_40
5. Calvo-Serrano, R., Guillén-Gosálbez, G., Kohn, S., Masters, A.: Mathematical programming approach for optimally allocating students' projects to academics in large cohorts. Educ. Chem. Eng. **20**, 11–21 (2017)
6. Chiarandini, M., Fagerberg, R., Gualandi, S.: Handling preferences in student-project allocation. Ann. Oper. Res. **275**(1), 39–78 (2019). https://doi.org/10.1007/s10479-017-2710-1
7. Cooper, F., Manlove, D.F.: A 3/2-approximation algorithm for the student-project allocation problem. In: 17th International Symposium on Experimental Algorithms, SEA 2018, 27–29 June 2018, L'Aquila, Italy, vol. 103, pp. 8:1–8:13 (2018). https://doi.org/10.4230/LIPIcs.SEA.2018.8

8. Gani, M.A., Hamid, R.A., et al.: Optimum allocation of graduation projects: Survey and proposed solution. J. Al-Qadisiyah Comput. Sci. Math. **13**(1), 58 (2021)
9. Gent, I.P., Prosser, P.: An empirical study of the stable marriage problem with ties and incomplete lists. In: Proceedings of the 15th European Conference on Artificial Intelligence, pp. 141–145. Lyon, France (2002)
10. Harper, P.R., de Senna, V., Vieira, I.T., Shahani, A.K.: A genetic algorithm for the project assignment problem. Comput. Oper. Res. **32**(5), 1255–1265 (2005)
11. Irving, R.W., Manlove, D.F., Scott, S.: The hospitals/residents problem with ties. In: SWAT 2000. LNCS, vol. 1851, pp. 259–271. Springer, Heidelberg (2000). https://doi.org/10.1007/3-540-44985-X_24
12. Iwama, K., Miyazaki, S., Yanagisawa, H.: Improved approximation bounds for the student-project allocation problem with preferences over projects. J. Discrete Algorithms **13**, 59–66 (2012)
13. Király, Z.: Linear time local approximation algorithm for maximum stable marriage. Algorithms **6**(1), 471–484 (2013)
14. Kwanashie, A., Irving, R.W., Manlove, D.F., Sng, C.T.S.: Profile-based optimal matchings in the student/project allocation problem. In: Kratochvíl, J., Miller, M., Froncek, D. (eds.) IWOCA 2014. LNCS, vol. 8986, pp. 213–225. Springer, Cham (2015). https://doi.org/10.1007/978-3-319-19315-1_19
15. Manlove, D., Milne, D., Olaosebikan, S.: An integer programming approach to the student-project allocation problem with preferences over projects. In: Proceedings of 5th International Symposium on Combinatorial Optimization, pp. 213–225. Morocco (2018)
16. Manlove, D., Milne, D., Olaosebikan, S.: An integer programming approach to the student-project allocation problem with preferences over projects. In: Lee, J., Rinaldi, G., Mahjoub, A.R. (eds.) ISCO 2018. LNCS, vol. 10856, pp. 313–325. Springer, Cham (2018). https://doi.org/10.1007/978-3-319-96151-4_27
17. Manlove, D.F.: Algorithmics of Matching Under Preferences, Series on Theoretical Computer Science, vol. 2 (2013). https://doi.org/10.1142/8591
18. Manlove, D.F., O'Malley, G.: Student-project allocation with preferences over projects. J. Discrete Algorithms **6**(4), 553–560 (2008)
19. Olaosebikan, S., Manlove, D.: An algorithm for strong stability in the student-project allocation problem with ties. In: Changat, M., Das, S. (eds.) CALDAM 2020. LNCS, vol. 12016, pp. 384–399. Springer, Cham (2020). https://doi.org/10.1007/978-3-030-39219-2_31
20. Olaosebikan, S., Manlove, D.: Super-stability in the student-project allocation problem with ties. J. Comb. Optim. **43**, 1–37 (2020). https://doi.org/10.1007/s10878-020-00632-x
21. Paunović, V., Tomić, S., Bosnić, I., Žagar, M.: Fuzzy approach to student-project allocation (spa) problem. IEEE Access **7**, 136046–136061 (2019)
22. Irving, R.W., Manlove, D.F.: Finding large stable matchings. J. Exp. Algorithmics **14**(1), 1–2 (2009)

Online File Caching on Multiple Caches in Latency-Sensitive Systems

Guopeng Li, Chi Zhang, Hongqiu Ni, and Haisheng Tan[✉]

LINKE Lab and the CAS Key Lab of Wireless-Optical Communications,
University of Science and Technology of China (USTC), Hefei, China
{guopengli,gzhnciha,nhq0806}@mail.ustc.edu.cn, hstan@ustc.edu.cn

Abstract. Motivated by the presence of multiple caches and the non-negligible fetching latency in practical scenarios, we study the online file caching problem on multiple caches in latency-sensitive systems, e.g., edge computing. Our goal is to minimize the total latency for all file requests, where a file request can be served by a hit locally, fetching from the cloud data center, a delayed hit, relaying to other caches, or bypassing to the cloud. We propose a file-weight-based algorithm, named OnMuLa, to support delayed hits, relaying and bypassing. We conduct extensive simulations on Google' trace and a benchmark YCSB. The results show that our algorithms significantly outperform the existing methods consistently in various experimental settings. Compared with the state-of-the-art scheme supporting multiple caches and bypassing, OnMuLa can reduce the latency by 14.77% in Google's trace and 49.69% in YCSB.

Keywords: Cache · Mobile edge computing · Delayed hits

1 Introduction

In computer architecture, *cache* is designed to address the gap between memory access latency and processor processing speed [9]. Today, the concept of cache has been extended to different areas and has played an important role in improving system performance. The structure located between different types of hardware and used to eliminate the impact caused by the disparities of access time can be called a *cache*, such as disk cache, CDNs cache, Web cache and DNS cache. Taking advantage of temporal locality, storing some files that are about to be accessed into cache is one way to improve performance by using cache [18]. In this context, in order to minimize the total cost, one of the first and most important problems is to select which files to store in the cache and which files to replace when the cache is full, *i.e.*, the *online file caching problem*. In the traditional online file caching problem, we are given a cache with a specified size k and a sequence of file requests, where each file has a specified size and a specified retrieval cost. The goal of the traditional online file caching is usually to minimize the cache misses or the total retrieval cost of file retrievals by maintaining

© The Author(s), under exclusive license to Springer Nature Switzerland AG 2023
T. N. Dinh and M. Li (Eds.): CSoNet 2022, LNCS 13831, pp. 292–304, 2023.
https://doi.org/10.1007/978-3-031-26303-3_26

files in the cache [21]. However, in practical scenarios such as Mobile Edge Computing(MEC) [16] and Content Delivery Networks (CDNs) [6], due to the long physical distance, the latency for fetching a file from the cloud data center can be up to 100 ms, with the increase in network bandwidth and system throughput, 1M file requests can be arrived in a second, *i.e.*, the average inter-time for two consecutive file requests could be as low as 1µs [2]. During the period when a missed file is retrieved from the cloud data center, the subsequent requests for the same file can not be served immediately, which is not a *hit*, and is also different from a *miss*, which is called a **delay hit** [23]. In MEC and CDNs, in addition to local hit and fetching files from cloud to local, file requests can also be served by *bypassing* and *relaying*: 1) In cloud-based scenarios, the request can be sent to and served at the remote cloud, which is called a *bypassing*. 2) Since there is more than one edge server or PoP (regarded as cache) in MEC and CDNs, respectively, when the requested file is not already stored in the cache that the request arrives, the request can be sent to a nearby cache that has the same file, which is called a *relaying*. An example of online file caching in MEC is demonstrated in Fig. 1.

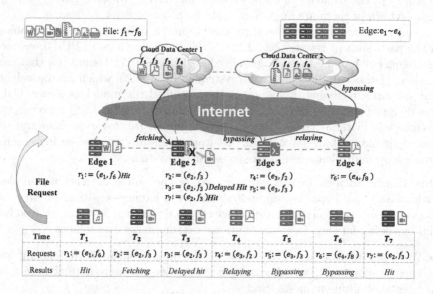

Fig. 1. An example in mobile edge computing system, where r_1 is served at e_1 locally, r_2 triggers the fetching and evict operation, r_3 is delayed served at e_2, r_4 is relayed to e_4, r_5 and r_6 are bypassed to the cloud data center, and r_7 is served at e_2 locally.

Motivating Example. As shown in Fig. 1, there are four edge servers and two cloud data centers connected through Internet, and eight different types of files with different sizes are stored in the cloud data center. Before T_1, each edge server has stored some files. $r := (e, f)$ represents the request arriving on an edge server e for file f. In order to demonstrate the ways (**hit, fetching, delayed hits, relaying and bypassing**) for serving requests on multiple caches in latency-sensitive systems, we have designed eight requests as an example. **1)** *hit*: If file f is already stored on e, the request can be served locally with no latency (as r_1 and r_7 illustrated in the figure). **2)** *fetching*: If file f has not been stored in e, edge server can fetch file from the cloud, due to the non-negligible fetching latency, file will not be stored in the edge server as soon as the fetching operation is triggered (as r_2 illustrated, and we assume the fetching operation will be finished at T_6). **3)** *delayed hits*: During the file fetching period, the subsequent requests can not be served until the fetching operation is finished and suffer delayed serve latency (as r_3 illustrated). **4)** *relaying*: If file f has not been stored on server e, but a nearby server e' has f, we can relay and serve the request r on e' with relaying latency (as r_4 illustrated). **5)** *bypassing*: When file f does not exist in the entire multiple caches system, in addition to fetching f from the cloud data center to the cache, we can bypass the request to and serve it at the cloud data center with bypassing latency (as r_5 and r_6 illustrated). Moreover, since the capacity of the edge server is limited, where some existing files might be replaced if the edge server is full. For r_2, when the fetching operation is triggered, we should check if there is enough empty space to store f_3. The capacity of cache e_2 does not allow it to store both f_3 and f_5 because both of them are large files, and we choose to evict f_5 from e_2.

Several algorithms have been proposed for the online file caching problem in the past, such as recency-based LRU [15], frequency-based LFU [4], recency and frequency based ARC [11], learning-based LeCaR [17], Camul [16] that uses marking methods in multiple caches system and CaLa [23] which handle weights and supports bypassing, but none of them studied the model as above. CaLa was designed for the one cache system, Camul focuses on fixed relaying and fetching cost without considering the non-uniform size of various files and the non-negligible fetching latency. In this work, we study the online file caching problem on multiple caches in latency-sensitive systems.

Our Contribution. In this work, we study the online file caching problem with relaying and bypassing on *multiple* caches in latency-sensitive systems, and propose an online algorithm to minimize the total relaying, fetching, delayed hits and bypassing latency. Our contributions are summarized as follows.

- We investigate a practical online file caching problem with relaying and bypassing on multiple caches in latency-sensitive system to minimize the total latency of all file requests (Sect. 3).
- We propose an online algorithm, called OnMuLa, to support delayed hits, relaying and bypassing. To the best of knowledge, OnMuLa is the first online algorithm for the online general file caching problem with delayed hits on multiple caches system (Sect. 4).
- We conduct extensive simulations on Google's trace and YCSB. Compared with Camul, the state-of-the-art algorithm that deals with multiple caches and bypassing, in default settings, OnMuLa can reduce the latency by 14.77% in Google's trace and 49.69% in YCSB (Sect. 5).

2   Related Work

Online file caching problem is often mentioned in computer and network systems. It can be described as follows, after being given a cache of k slots that each store one file, if the file has been cached in a slot when a request for a file arrives online, the request is served with no cost. If the file has not been cached, it has to be fetched into the cache with fetching cost. The online file caching algorithms maintain the contents of k slots to minimize the total fetching cost for all requests. LRU [15] is a classic algorithm widely used in practical scenarios. Considering the non-uniform fetching cost, Young et al. [21] proposed Landlord. In order to support bypassing, Landlord is extended to Landlord with Bypassing (LLB) [7]. Investigating general online caching problem on multiple caches with relaying and bypassing, Camul [16] has a high hit ratio with lower total cost than previous works.

Table 1. Related work of online caching problems

Algorithms	File size & Fetching cost	Delayed hits	Bypassing	Multiple caches
LRU [15]	Uniform	✘	✘	✘
Landlord [21]	Non-uniform	✘	✘	✘
LLB [7]	Non-uniform	✘	✔	✘
MAD [2]	Uniform	✔	✘	✘
CaLa [23]	Non-Uniform	✔	✔	✘
Camul [16]	Uniform	✘	✔	✔
OnMuLa [this work]	Non-Uniform	✔	✔	✔

In latency-sensitive systems, few works have paid attention to *delayed hits* so far. A representative work is the online paging problem studied in [2], Atre *et al.* creatively reveal the importance of delayed hits, and propose MAD, an online algorithm that combines file aggregation delays into existing caching algorithms (such as LRU and ARC). Zhang *et al.* [23] proposed CaLa, a general framework that imitates an existing file caching algorithm to get guaranteed performance in their work. Besides, caching problem has attracted a wide range of works from emerging application areas, such as mobile system [10,13], CDNs [3,20], container caching [8,12], and deep learning system [1,22]. For example, Beckmann *et al.* [3] proposed LHD to predict the hit density of each object to filter objects that have a small contribution to the cache hit rate. Yan *et al.* [20] proposed a timer-based mechanism that can optimize the mean caching latency. This work addresses the non-uniform file size and fetching latency, delayed hits, relaying and bypassing on multiple caches. We summarize some related results in Table 1.

3 Problem Formulation

Motivated by latency-sensitive scenarios such as MEC and CDNs, we consider the online general file caching model with multiple caches and cloud data centers. Let $\mathcal{F} = \{f_1, f_2, \ldots, f_N\}$ be the set of all kinds of files, and we assume all files are available in the cloud data center. Each file f_i has size s_{f_i} and fetching latency t_{f_i}. Without loss of generality, we assume all file sizes are integers. $\mathcal{E} = \{e_1, e_2, \ldots, e_M\}$ represents the set of the caches in the system, the size of each cache e_i is K_i. Naturally, the sum of sizes of files stored in each cache can not exceed the size of cache, i.e., $\sum_{f \text{ in cache } e_i} s_f \leq K_i$. Let $\mathcal{R} = (r_1, r_2, \ldots)$ be the sequence of file requests, a request r is a pair $(e, f) \in \mathcal{E} \times \mathcal{F}$, meaning a file f on cache e is requested. All requests arrive in an online manner, i.e., we can not get future information and no assumption is made on the arrival patterns. Time is divided into slots of unit size. Multiple different kinds of file requests might come within one time slot, while each file $f \in \mathcal{F}$ can be requested at most once in each slot. In the multiple caches system, when a request $r := (e, f)$ arrives at time T, the following 5 types of operations may be performed. The objective of this problem is to minimize the total relaying, bypassing, fetching and delayed hits latency to serve all requests.

- *Hit*: If the requested file f is already cached in cache e, then this request is served locally with no latency, i.e., called a *hit*.
- *Relaying*: If e does not have f stored but another cache e' does, the request may be relayed and processed at e' with relaying latency t_r.
- *Bypassing*: A request may be bypassed to the cloud with bypassing latency t_b. Note that in this case, the file is not necessarily fetched into the cache.
- *Fetching*: When file f does not exist in the entire system, f may be fetched to the cache with fetching latency t_f, i.e., request r can not be served until time $T + t_f$. Once f is cached, we need to decide which files should be replaced if the cache is already full.
- *Delayed Hits*: During the fetching period, i.e., from T to $T + t_f$, before file f is actually stored in the cache, all requests that require file f on cache e at time slot $t' \in \{T+1, T+2, \ldots, T+t_f-1\}$ can only be served at time $T+t_f$ and suffered a latency of $t_f - (t' - T)$, which are *delayed hits*.

4 Algorithm Design

In this section, we first propose a method to measure the importance of each file, called *file weight* (Sect. 4.1). Then, we present our algorithm OnMuLa and its version without bypassing OnMuLa⁻ in Algorithm 2.

4.1 File Weight

The central challenge of this problem is how to deal with delayed hits in multiple caches system. The method in [23] does not capture the relaying operation in the multiple caches scenario, in order to avoid the impact of the lack

---

Algorithm 1: Update Weight

---

1 **Input** *Parameter* γ, *cache* e, *file* f, *latency* l
2 **if** f_e.state = OUT **then**
3 | f_e.cumulativeDelay $\leftarrow$ f_e.cumulativeDelay $+ t_f$;
4 |_ f_e.numFetching $\leftarrow$ f_e.numFetching $+ 1$;
5 **if** f_e.state = FETCHING **then**
6 |_ f_e.cumulativeDelay $\leftarrow$ f_e.cumulativeDelay $+ l$;
7 f_e.aggregateDelay $\leftarrow \dfrac{f_e.\text{cumulativeDelay}}{f_e.\text{numFetching}}$;
8 f_e.weight $= (1 - \gamma) * f.aggregateDelay + \gamma * t_f^2$;

---

of consideration for relaying, as Eqn. 1 shown (assume start fetching f into e at time T) and AggDelay(e, f) = CumuDelay$(e, f)/$ # of fetching f into e, we refine the method to calculate CumuDelay to estimate the actual latency in the multiple caches system. Not only *delayed hits* and *fetching* of request on cache e contribute to CumuDelay(e, f), but the *relaying* from other caches also affect CumuDelay(e, f). In order to get a trade-off between CumuDelay and the upper bound of the total latency caused by the file's miss, t_f^2, we use $W_{f_e} = (1 - \gamma)$AggDelay$(e, f) + \gamma t_f^2$ to represent the weight of each file on each cache, where parameter γ is used to adjust between these two methods.

$$\text{CumuDelay}(e, f) = t_f + \sum_{1 \leq \tau \leq t_f - 1} (t_f - \tau)[f \text{ is requested at } T + \tau]$$
$$+ \sum (\max(t_f - \sigma, 0) + t_r)[\text{relay from } e' \text{to } e \text{ at } T + \sigma]. \tag{1}$$

4.2   Design of OnMuLa

The main algorithm is defined in Algorithm 2, and the weight update algorithm is defined in Algorithm 1. Initially, the caches in the system are initialized to empty. We use IN, FETCHING and OUT to represent the state of file f on cache e, where f_e.state = IN means f is already cached in e and f_e.state = FETCHING means f is already in a fetching period. When a new request $r := (e, f)$ that requests for file f on cache e arrives at T, OnMuLa decides how to serve r based on f_e.state. **1)** *hit*: If f_e.state = IN, then serve f on cache e with no latency (Line 13 to Line 14). **2)** *delayed hits*: If f_e.state = FETCHING, r will be served until the fetching operation of f is finished and sufferd the latency $t - T$ (Line 15 to Line 17). **3)** *relaying*: If there is a cache $e' \neq e$ has f in the multiple caches system, and the sum of waiting latency t_w and t_r is less than t_f, then relay r to e' with latency $t_w + t_r$. t_w represents the latency that r suffered on e' after be relayed to e' (Line 19 to Line 21). **4)** *fetching* and **5)** *bypassing*: We use Landlord [21] and Landlord with Bypassing(LLB) [7] as the replacement policy in OnMuLa$^-$and OnMuLa, respectively. It can be substituted with other algorithms which can handle weight and support bypassing. Landlord and LLB maintain a credit for each file to determine whether it should be evicted.

Specifically, in the implementation of `OnMuLa⁻`and `OnMuLa`, let f_e.weight be the credit in Landlord and LLB. When a request $r := (e, f)$ has not been served after the above process (Line 13 to Line 21), `OnMuLa⁻` checks the remaining size of e and uses Landlord to replace the files in e if e does not have enough to cache f, then fetch f to e. `OnMuLa` will first suppose f in e^α, the copy of cache e, and set the credit of f (Line 25). If f_{e^α}.weight > 0 in the end of the replacement, then fetch f to e, otherwise, bypass the request.

For Line 15 to Line 17 in Algorithm 2, we prove Theorem 1, *i.e.*, if there is more than one copy of file f in the multiple caches system, the request should be delayed served on e instead of being relayed.

Theorem 1. *If there is more than one copy of file f (the state of f is IN or FETCHING) in multiple cache systems, when f_e.state = FETCHING and $f_{e'}$.state = FETCHING or IN, request $r := (e, f)$ should be delayed served on e instead of being relayed to e'.*

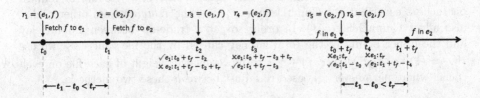

Fig. 2. An example of Theorem 1, after t_1, the request for f on e_2 should be served on e_2 instead of be relayed to e_1.

Proof. First, we prove Theorem 1 for `OnMuLa⁻`. In the multiple caches system, as shown in Fig. 2, suppose there is no request for f before t_0. At t_0, r_1 requests f on e_1, however, f_{e_1}.state = OUT, then start to fetch f into e_1, the fetching process will be finished at $t_0 + t_f$. At t_1, r_2 requests f on e_2, f_{e_2}.state = OUT, f_{e_1}.state = FETCHING, however, the condition in Line 19 of Algorithm 2 is not satisfied, $t_r + t_0 + t_f - t_1 > t_r$, i.e., $t_1 - t_0 < t_r$, start to fetch f into e_2, the process will be finished at $t_1 + t_f$. Now, there is more than one copy of file f in the multiple caches system. Next, we use the converse method to prove Theorem 1. During the time period t_1 to $t_1 + t_f$, requests r_4, r_5 and r_6 request f on cache e_2. For r_4, we will relay the request to cache e_1 if and only if $t_0 + t_f - t_3 + t_r < t_1 + t_f - t3$, i.e., $t_1 - t_0 > t_f$, however, this contradicts $t_1 - t_0 < t_r$. As a more direct example, for r_5, the condition is $t_r < t_1 - t_0$. For r_6, $t_r < t_1 + t_f - t_4$ should be satisfied, as Fig. 2 shown, it is impossible. For `OnMuLa`, since bypass is allowed, `OnMuLa` can fetch files into the multiple caches system or bypass the requests when the condition in Line 19 is not satisfied. If all requests are bypassed, file will not be cached in the system, the assumption of the theorem is not satisfied. If there is more than one copy of the file are cached in the system, as same as the above proof for `OnMuLa⁻`.   □

Algorithm 2: Main Algorithm

---

1 **Input** *Request* $r := (e, f)$, *the size of* f s_f, *Fetching Latency* t_f, *Bypass Latency* t_b, *Relay Latency* t_r, *bool* $\mathcal{B}$ *represents allow bypass or not* ;

2 $\mathcal{C} \leftarrow \varnothing$, $\mathcal{C}$ represents the set of files cached in e;

3 Fetching files $\mathcal{F}_f \leftarrow \varnothing$, $(e, f, t) \in \mathcal{F}_f$ means file f will arrive on e at time t;

4 Timer $T \leftarrow 0$;

5 **while** True **do**

6   **for** $(e, f, t) \in \mathcal{F}_{\text{fetching}}$ **do**

7   **if** $t <= T$ **then**

8   **if** f_e.state = FETCHING **then**

9   f_e.state $\leftarrow$ IN;

10                 $\mathcal{C} \leftarrow \mathcal{C} \cup \{f\}$;

11   Serve all the buffered and relayed requests for f on e;

12   **while** new request $r := (e, f)$ for file f on e arrive at T **do**

13   **if** f_e.state = IN **then** // hit

14   serve f on e with no latency;

15   **if** f_e.state = FETCHING **then** // delayed hit

16   delayed serve f on e at time t with latency $t - T$;

17   UpdateWeight($e, f, t - T$)

18   **if** f_e.state = OUT **then**

19   **if** there is a cache e' has f and $t_r + t_w$ ¡ t_f **then** // relay

20   relay $r := (e, f)$ to e' with latency $t_r + t_w$;

21   UpdateWeight($e', f, t_r + t_w$);

22   **else**

23   Let cache e^α be a copy of e ;

24   **if** $\mathcal{B}$ = Ture **then**

25   Let f in e^α, UpdateWeight($e^\alpha, f, 0$);

26   **if** remain size of $e^\alpha < s_f$ **then**

27                     $\mathcal{F}_{\text{evicts}} \leftarrow$ Replace(e^α, f);

28   **for** $f' \in \mathcal{F}_{\text{evicts}}$ **do**

29   Evict f' from e, $\mathcal{C} \setminus \{f'\}$;

30                         $f_{e'}$.state $\leftarrow$ OUT;

31   **if** f_{e^α}.weight > 0 or $\mathcal{B}$ = False **then**

32   f_e.state $\leftarrow$ FETCHING;

33                     $\mathcal{F}_{\text{fetching}} \leftarrow \mathcal{F}_{\text{fetching}} \cup \{(e, f, T + t_f)\}$;

34   UpdateWeight(e, f, t_f);

35   **else** // bypass

36   Bypass this request with latency t_b ;

37     $T \leftarrow T + 1$;

---

5   Evaluation

In this section, we evaluate the performance of OnMuLa and OnMuLa⁻ on two datasets: (1) the production trace from Google [14], (2) the Yahoo! Cloud Serving Benchmark (YCSB) [5]. We compare OnMuLa and OnMuLa⁻ with several caching algorithms *i.e.*, LRU [15], LRU-MAD [2], Landlord with Bypassing [7], Landlord [21], Camul [16], CaLa with Bypassing and CaLa [23]. The details of the results are shown in Sect. 5.2 and we highlight our key findings as follows.

- Compared with Camul, the state-of-the-art caching algorithm in multiple caches system. With default settings, in Google's trace, OnMuLa⁻ can reduce latency by 12.62%, this reduction will be increased to 14.77% if bypassing is allowed. In YCSB, OnMuLa reduces latency by 49.69% compared to Camul.
- The performance of the algorithm can vary significantly for different traces. For the one cache case, OnMuLa⁻ and CaLa get poor performance in Google's trace, and conversely, work well in YCSB.
- If the cache size is small (e.g., sum of 0.001% to 0.01% of the popular files), OnMuLa outperforms other algorithms by bypassing.

5.1   The Experiment Settings

By default, we set 400 caches for Google's trace and 200 caches for YCSB. The default cache size is the sum of the sizes of top 0.01% popular files. We let relay latency $t_r = 0.002 \times$ fetching latency t_f [16], bypassing latency $t_b = t_f$ [23]. We set the average inter-request time to $10^{-4}s$ [19], and the average default fetching latency of files is set to $0.1s$ [2]. For OnMuLa, OnMuLa⁻, CaLa and CaLa with Bypassing, the default value of γ is set to 0.1. The metrics used to evaluate the performance of algorithms is the total latency incurred of all requests. And we use the latency improvement relative to LRU to measure the performance of the algorithm when the parameters change, *i.e.*, Latency Improvement of A = (Latency(LRU) − Latency(A))/Latency(LRU), a higher latency improvement means better performance.

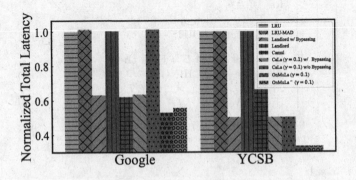

Fig. 3. Overall performance.

5.2 Experimental Results

Overall Performance. We first evaluate the overall performance of OnMuLa and OnMuLa⁻, and compare them with the baselines, where parameters are set as default values. The experimental results are shown in Fig. 3. In Google's trace, the latency improvement of OnMuLa⁻ to CaLa is 45.16%, and OnMuLa to Camul is 14.77%. For the results in YCSB, the latency improvement of OnMuLa⁻ to CaLa is 66.72%, and OnMuLa to Camul is 49.69%.

Sensitivity Analysis. In this part, we perform sensitivity analysis of the parameters in the experiment settings, including the number of caches, fetching latency, cache size and γ.

Impact of Number of Caches. Figure 4(a) and Fig. 4(b) illustrate the impact of the number of caches, which varies from 1 to 1000 for Google's trace and from 1 to 400 for YCSB. In Google's trace, OnMuLa⁻ performs poorly in the one cache system, with the number of caches increasing, the performance of OnMuLa⁻ becomes better. Due to the design for delayed hits and non-uniform file size, the performance of OnMuLa is better than Camul. In YCSB, for the one cache case, due to the design for non-uniform file size and delayed hits, OnMuLa⁻ and CaLa avoid bringing large and infrequent files into cache. For the multiple caches case, with the increasing number of caches, the performance of OnMuLa⁻, OnMuLa and Camul increase marginally. When the number of caches is large enough, 400 for Google's trace and 200 for YCSB, OnMuLa and OnMuLa⁻ achieve the best performance among the nine algorithms.

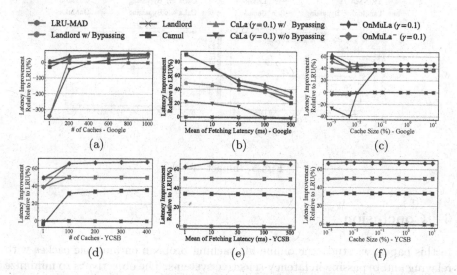

Fig. 4. Impact of number of caches, fetching latency and cache size.

Impact of Fetching Latency. We show the result of the impact of fetching latency in Fig. 4(b) and Fig. 4(e). The fluctuation of the curves reflects the different sensitivity of various algorithms to the fetching latency in different traces. In Google's trace, the performance of OnMuLa and OnMuLa⁻ gradually outperforms Camul when the fetching latency increases, since their awareness of delayed hits. In YCSB, the performance of other algorithms besides OnMuLa and OnMuLa⁻ is more likely the case without delayed hits. The reason is that the request locality of YCSB is too low, and there are few requests with delayed hits, especially when the requests are distributed to multiple caches.

Impact of Cache Size. Figure 4(c) and Fig. 4(f) show the impact of cache size for Google's trace and YCSB, respectively. In Google's trace, when the cache size is small, the performance of OnMuLa and Camul are far beyond other algorithms. This is because bypassing can avoid evicting some frequently requested files from the cache. As the cache size gradually increases, the performance of OnMuLa⁻, CaLa gradually catches up with OnMuLa. Due to the discreteness of files' size in the trace and the non-consecutiveness between different caches, the performance improvement brought by the additional cache size does not occur simultaneously as the cache size increases, which causes the fluctuations in performance curves.

Impact of γ. For Google's trace, as shown in Fig. 5(a), the best performance is achieved when $\gamma = 0.15$, which shows that it is better to use a value of γ closer to the aggregate delay for burst requests. For YCSB, as shown in Fig. 5(b), the performance of OnMuLa⁻ and OnMuLa remains stable as γ changes from 0 to 0.2.

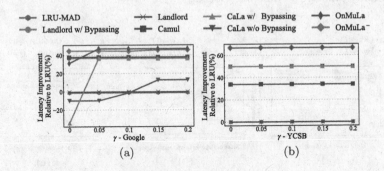

Fig. 5. Impact of γ.

6   Conclusion

In this paper, we study the online file caching problem on multiple caches with relaying and bypassing in latency-sensitive systems. The objective is to minimize the total latency to serve all requests. We first propose *file weight* to capture the potential impact of the fetching and relaying process. Then we propose an online algorithm OnMuLa to support delayed hits, relaying and bypassing, and its version without bypassing, OnMuLa⁻. We evaluate OnMuLa and OnMuLa⁻ on Google's trace and YCSB. The experiment results show that compare with Camul, OnMuLa can reduce the latency by 14.77% in Google's trace and 49.69% in YCSB.

Acknowledgements. The work is partially supported by NSFC under Grant 62132009, and the Fundamental Research Funds for the Central Universities at China.

References

1. Abdi, M., et al.: A community cache with complete information. In: USENIX FAST 2021, pp. 323–340 (2021)
2. Atre, N., Sherry, J., Wang, W., Berger, D.S.: Caching with delayed hits. In: ACM SIGCOMM (2020)
3. Beckmann, N., Chen, H., Cidon, A.: LHD: Improving cache hit rate by maximizing hit density. In: USENIX NSDI (2018)
4. Breslau, L., Cao, P., Fan, L., Phillips, G., Shenker, S.: Web caching and zipf-like distributions: evidence and implications. In: IEEE INFOCOM'99
5. Cooper, B.F., Silberstein, A., Tam, E., Ramakrishnan, R., Sears, R.: Benchmarking cloud serving systems with YCSB. In: ACM SoCC (2010)
6. Dilley, J., Maggs, B., Parikh, J., Prokop, H., Sitaraman, R., Weihl, B.: Globally distributed content delivery. IEEE Internet Comput. 6(5), 50–58 (2002)
7. Epstein, L., Imreh, C., Levin, A., Nagy-György, J.: Online file caching with rejection penalties. Algorithmica **71**(2), 279–306 (2015). https://doi.org/10.1007/s00453-013-9793-0
8. Fuerst, A., Sharma, P.: Faascache: keeping serverless computing alive with greedy-dual caching. In: ACM ASPLOS 2021, pp. 386–400 (2021)
9. Karlsson, M.: Cache memory design trade-offs for current and emerging workloads. Ph.D. thesis, Citeseer (2003)
10. Liang, Y., et al.: Cachesifter: sifting cache files for boosted mobile performance and lifetime. In: USENIX FAST 2022, pp. 445–459 (2022)
11. Megiddo, N., Modha, D.S.: Arc: A self-tuning, low overhead replacement cache. In: FAST 2003 (2003)
12. Pan, L., Wang, L., Chen, S., Liu, F.: Retention-aware container caching for serverless edge computing. In: IEEE INFOCOM 2022 (2022)
13. Ramanujam, M., Madhyastha, H.V., Netravali, R.: Marauder: synergized caching and prefetching for low-risk mobile app acceleration. In: MobiSys (2021)
14. Reiss, C., Wilkes, J., Hellerstein, J.: Google cluster-usage trace. In: Technical Report (2011)
15. Sleator, D.D., Tarjan, R.E.: Amortized efficiency of list update and paging rules. Commun. ACM **28**(2), 202–208 (1985)
16. Tan, H., Jiang, S.H.C., Han, Z., Liu, L., Han, K., Zhao, Q.: Camul: online caching on multiple caches with relaying and bypassing. In: IEEE INFOCOM (2019)
17. Vietri, G., et al.: Driving cache replacement with ML-based LeCaR. In: HotStorage 2018 (2018)
18. Wang, J., Hu, Y.: Wolf-a novel reordering write buffer to boost the performance of log-structured file system. In: FAST 2002 (2002)
19. Wendell, P., Freedman, M.J.: Going viral: flash crowds in an open CDN. In: ACM/USENIX IMC (2011)

20. Yan, G., Li, J.: Towards latency awareness for content delivery network caching. In: USENIX ATC 2022, pp. 789–804 (2022)

21. Young, N.E.: On-line file caching. Algorithmica **33**(3), 371–383 (2002). https://doi.org/10.1007/s00453-001-0124-5

22. Yuan, M., Zhang, L., He, F., Tong, X., Li, X.Y.: Infi: end-to-end learnable input filter for resource-efficient mobile-centric inference. In: MobiCom (2022)

23. Zhang, C., Tan, H., Li, G., Han, Z., Jiang, S.H.C., Li, X.Y.: Online file caching in latency-sensitive systems with delayed hits and bypassing. In: IEEE INFOCOM 2022, pp. 1059–1068. IEEE (2022)

Correction to: A Community Detection Algorithm Using Random Walk

Rajesh Vashishtha, Anurag Singh, and Hocine Cherifi

Correction to:
Chapter "A Community Detection Algorithm Using Random Walk" in: T. N. Dinh and M. Li (Eds.): *Computational Data and Social Networks*, LNCS 13831, https://doi.org/10.1007/978-3-031-26303-3_20

The original version of the book was inadvertently published without the acknowledgement in chapter. The acknowledgement has been added.

The updated original version of this chapter can be found at
https://doi.org/10.1007/978-3-031-26303-3_20

Author Index

© The Editor(s) (if applicable) and The Author(s), under exclusive license
to Springer Nature Switzerland AG 2023
T. N. Dinh and M. Li (Eds.): CSoNet 2022, LNCS 13831, pp. 305–306, 2023.
https://doi.org/10.1007/978-3-031-26303-3

Printed in the United States
by Baker & Taylor Publisher Services